Reviews in Modern Astronomy

2

Gerhard Klare (Ed.)

Reviews in Modern Astronomy 2

With 166 Figures

Springer-Verlag Berlin Heidelberg New York
London Paris Tokyo Hong Kong

Edited on behalf of the *Astronomische Gesellschaft* by

Dr. *Gerhard Klare*

Landessternwarte, Königstuhl,
D-6900 Heidelberg 1, Fed. Rep. of Germany

ISBN-13: 978-3-642-75185-1 e-ISBN-13: 978-3-642-75183-7
DOI: 10.1007/978-3-642-75183-7

Softcover reprint of the hardcover 1st edition 1989

2156/3150-543210 – Printed on acid-free paper

Preface

This second volume of *Reviews in Modern Astronomy* continues the new series of publications of the Astronomische Gesellschaft (AG).

In order to bring the scientific events of the meetings of the society to the attention of the worldwide astronomical community, it was decided to devote a new annual publication, *Reviews in Modern Astronomy*, exclusively to the invited reviews, the Karl Schwarzschild lectures, and the highlight contributions from leading scientists reporting on recent progress and scientific achievements at their institutes.

Volume 2 comprises the contributions presented during the spring meeting of the AG at Friedrichshafen in April 1989, which was dedicated to the topic "Astrophysics with Modern Technology – Space-based and Ground-based Systems", as well as those delivered at the fall meeting at Graz, Austria in September 1989.

The Karl Schwarzschild lectures constitute a special series by outstanding scientists honoured with the Karl Schwarzschild medal of the Astronomische Gesellschaft. Between 1959 and 1987 the following scientists presented Karl Schwarzschild lectures at meetings of the Astronomische Gesellschaft:

Martin Schwarzschild (1959)
Charles Fehrenbach (1963)
Maarten Schmidt (1968)
Bengt Strömgren (1969)
Antony Hewish (1971)
Jan H. Oort (1972)
Cornelis de Jager (1974)
Lyman Spitzer Jr. (1975)
Wilhelm Becker (1977)
George Field (1978)
Ludwig Biermann (1980)
Bohdan Paczynski (1981)
Jean Delhaye (1982)
Donald Lynden-Bell (1983)
Daniel M. Popper (1984)
Subrahmanyan Chandrasekhar (1986)
Lodewijk Woltjer (1987)

During the fall meeting of the Astronomische Gesellschaft at Graz in September 1989 the Karl Schwarzschild medal was awarded to Professor Martin Rees [1], Director of the Institute of Astronomy at the University of Cambridge. His lecture entitled "Is There a Massive Black Hole in Every Galaxy?" begins this volume.

Heidelberg, October 1989 *G. Klare*

[1] The complete text of the laudation will be published in *Mitteilungen der Astronomischen Gesellschaft* Vol. 73, 1990.

Preface

Contents

Is There a Massive Black Hole in Every Galaxy?

M.J. Rees

Institute of Astronomy, University of Cambridge,
Madingley Road, Cambridge, CB3 OHA, UK

The theoretical study of black holes in Einstein's theory dates back to Karl Schwarzschild's classic work in 1916. The realisation that such objects might actually exist is of more recent vintage. Recently, however, suggestive evidence has accumulated that black holes exist in galactic nuclei. In this lecture I shall discuss the dark objects, probably massive black holes, which lurk at the *centres* of such normal galaxies as M31, and perhaps even our own. This subject is approached 'historically' by considering the cosmic history of the quasar population. The evidence for massive black holes in the centres of some nearby galaxies is reviewed. Some interesting observational consequences (*e.g.* flares from tidally disrupted stars) are proposed and discussed.

The Quasar Population, and Expected Remnants

Investigations of quasars and active galaxies have established the luminosity function for such objects at different redshifts. Quasars were much more common at a redshift $z = 2$ than at the present epoch, in the sense that there were then, when the universe was 2 or 3 billion years old, almost 1000 times more luminous quasars per comoving volume than there are today. At still higher redshifts, corresponding to still earlier times, quasars seem to thin out, though the details are less clear beyond $z \simeq 4$. A current estimate of the evolution in the quasar density with redshift is shown on the left-hand side of Figure 1. This same data can, following Schmidt (1989), be presented much more dramatically in the manner shown on the right-hand side, where time and comoving density are plotted linearly rather than on logarithmic scales. Clearly quasar activity was sharply peaked at a particular cosmic epoch. It is an anti-anthropic irony that the most exciting time to have been an astronomer was when the universe was 2 billion years old, before the Earth had formed.

The reasons for the sharp rise and subsequent fall in quasar activity, presumably related in some way to galaxy formation and evolution, are not my subject today. However, one can, from quasar statistics, draw the important inference that about 10^7 solar rest masses of radiation were emitted by quasars, for every bright galaxy present in the universe today. Quasars have generated radiation amounting to about $3000 M_\odot c^2$ per cubic comoving Mpc (Soltan 1982, Phinney 1983). Many features of quasars remain enigmatic: active galactic nuclei display many phenomena on various scales and different wavebands, and it is hard to fit them into a single pattern. On the other hand, there is a much stronger consensus on what a *dead* quasar should be. Given that quasars derive their energy primarily via gravitation, there seems no way of evading the conclusion that a substantial fraction of the mass involved must eventually collapse to a massive black hole. Even if we optimistically assign an efficiency of 10% to the overall energy generation in quasars, we must then conclude that their black hole remnants have a total mass amounting to an *average of* $10^8 M_\odot$ per galaxy.

Even at the epoch of peak quasar activity, the comoving density of quasars is only 1 or 2 per cent of the present galactic density. One might therefore surmise that quasar remnants would be expected in only 1 or 2 per cent of galaxies, and that each would weigh around $10^{10} M_\odot$. However, we must remember that Figure 1 delineates the evolution of the quasar *population*, which decays on a timescale of $t_{Evo} \simeq 2$ billion years. While this may relate directly to the lifecycle of a typical quasar, there is the alternative possibility that individual quasars have much briefer lives, so that many generations flare and fade during the 2 billion year period of peak quasar activity.

Reviews in Modern Astronomy 2 Editor: G. Klare

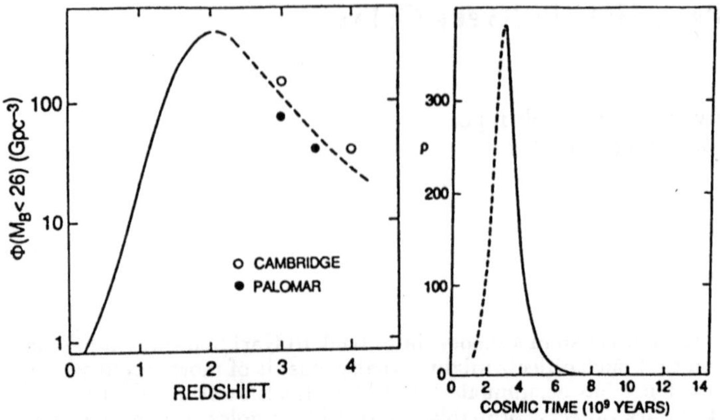

Figure 1 The comoving density of powerful quasars is given on the left as a function of redshift z. (Beyond z $\simeq$ 3 there is a fall–off, but the quantitative details are uncertain. The filled and open circles correspond to the results from two different surveys.) The same data are, following Schmidt (1989), replotted on the right on a linear scale, where the horizontal axis is cosmic time, assuming an Einstein–de Sitter cosmology with $H_0 = 50\,\mathrm{km\ s^{-1}\ Mpc^{-1}}$. This plot dramatises the relative brevity of the 'quasar era', when the Universe was 2–3 billion years old.

The contrasting implications of two different hypotheses about typical quasar lifetimes are displayed in Table 1. The masses scale with luminosity, given in units of $10^{47}\,\mathrm{erg\,s^{-1}}$, and inversely with the efficiency, given in units of 0.1. In practice, of course, it is likely to be an oversimplification to characterise quasars by a single typical lifetime – the highest luminosities may well correspond to shorter lifetimes, for instance. We also have to consider how lower level AGNs, such as Seyfert nuclei, fit into the scheme.

The table does, however, suffice to indicate that, if typical quasar lifetimes were a few times 10^7 years (a value which is supported by many theoretical models) one would expect dead remnants, massive black holes now starved of fuel, to lurk in the nuclei of most galaxies, including nearby ones.

Table 1

One generation of quasars	~ 50 generations of quasars
$t_Q \simeq t_{Evo}$	$t_Q \simeq 4 \times 10^7$ yr $\simeq 0.02\, t_{Evo}$
$M = 2.5 \times 10^{10} \varepsilon_{0.1}^{-1}\, L_{47}\, M_\odot$	$M = 5 \times 10^{8} \varepsilon_{0.1}^{-1}\, L_{47}\, M_\odot$
$L << L_{Ed}$	$L \simeq L_{Ed}\, \varepsilon_{0.1}$
Broad–line regions gravitationally bound	Broad–line region *not* gravitationally bound
Very massive remnants in ~2% of galaxies	~ 10^8 $M_\odot$ remnants in most bright galaxies

Effects of Central Mass on Surrounding Stars

A massive black hole will inevitably affect the orbits of stars passing close to it, and evidence for just such effects in the centres of several nearby galaxies has recently been reported by several observers. To appreciate the nature of this evidence, it is helpful to define a few characteristic length scales. These are illustrated and defined in Figure 2 and its caption. In numerical terms, r_h and r_c are approximately given by

$$r_h \simeq \frac{GM_h}{\sigma_c^2} \simeq 10^6 r_g \left(\frac{\sigma_c}{300\,\mathrm{km\ s^{-1}}}\right)^{-2} \tag{1}$$

$$r_c = \frac{GM_h}{v_*^2} = r_g \left(\frac{c}{v_*}\right)^2 \simeq 10^5 r_g \left(\frac{v_*}{1000\,\mathrm{km\ s^{-1}}}\right)^{-2} \tag{2}$$

The tidal radius r_T, defined as the radius within which a star gets tidally disrupted, obviously has a value depending on the type of star being considered. For solar–type stars it is approximately

$$r_T \simeq \left(\frac{M_h}{m_*}\right)^{\frac{1}{3}} r_* \simeq 3 \times 10^{13} \left(M_h/10^8\,M_\odot\right)^{\frac{1}{3}} (r_*/r_\odot)(m_*/r_\odot)^{-\frac{1}{3}} \tag{3}$$

Note that this is always much smaller than r_c, and is indeed inside the gravitational radius $r_g = 1.5 \times 10^{13}\,(M_h/10^8\,M_\odot)$ cm for black hole masses exceeding $10^8\,M_\odot$, for solar–type stars. If $M_h << 10^8\,M_\odot$ it is, however, sufficiently far outside r_g that the disruption can be approximated as a Newtonian process and it makes little difference whether the hole is described by a Schwarzschild or a Kerr metric. Tidal disruption is, of course, a complicated process and the details depend on the density profile within the star. Giants are subject to tidal effects even for hole masses $\gtrsim 10^{10}\,M_\odot$.

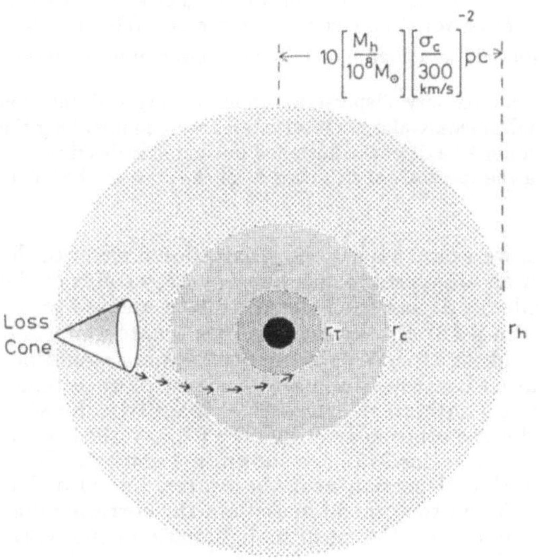

Figure 2 This diagram depicts, not to scale, various characteristic radii around a massive black hole in a stellar system. If the velocity dispersion in the core of the galaxy is σ_c the hole influences the stellar motions within a radius r_h (equation (1)). Within r_c (equation (2)) stars would be moving so fast that they would be more likely to experience (generally disruptive) physical collisions with each other rather than to undergo two–body encounters of the kind that can be treated by point–mass approximations. r_c is the radius where the escape velocity from the hole is comparable with the escape velocity v_* from the surface of a star. Tidal disruptions occur within a radius r_T (equation (3)). To be disrupted, a star must cross the sphere at $r \simeq r_h$ on a nearby radial 'loss cone' orbit.

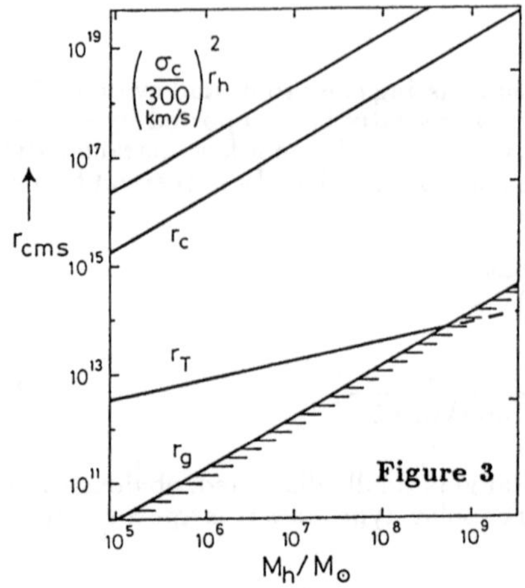

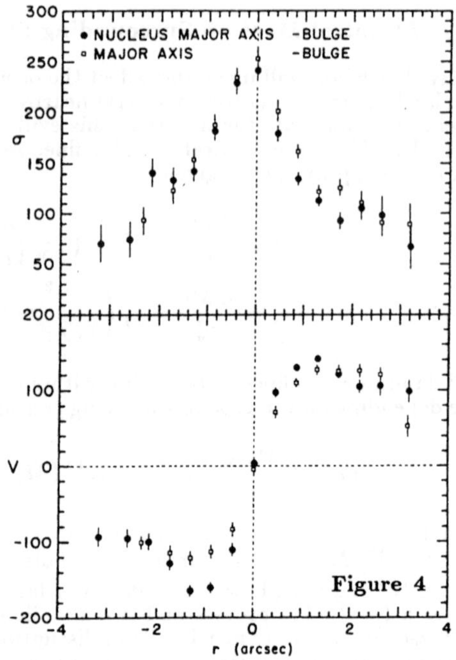

Figure 3 The various radii depicted in Fig. 2 are here plotted, on a logarithmic scale, as a function of the hole mass M_h. The radii r_c and r_T are plotted for a solar-type star with $v_* \simeq 1000\,\mathrm{km\ s^{-1}}$. Note that a hole of $\gg 10^8\,M_\odot$ can swallow solar-type stars (though not, of course, giants) without first disrupting them. When $r_T \simeq r_g$, tidal disruption effects would be restricted to the general-relativistic domain where the hole's tidal effects cannot be adequately modelled by a r^{-3} Newtonian approximation.

Figure 4 Data from Kormendy (1988) showing the velocity dispersion (upper panel) and rotation (lower panel) for stars in the nucleus of M31, the Andromeda Galaxy. Observations were made along the apparent major axis of the bulge and along the (misaligned) axis of the flattened nuclear star distribution; the plotted velocities were obtained after subtracting the spectral contribution from the stars in the bulge.

At the distances of even the nearest galaxies, we would expect the gravitational effect of the central black hole to be discernible only within an angular distance of a few arcseconds of the galactic nucleus. The first indications of such effects, dating from the late 1970s, related to the giant elliptical M87, where a mass of several times $10^9\,M_\odot$ was claimed (Sargent *et al.* 1977, 1978). Of course, at the M87 distance of around 20 Mpc, a more moderate mass would not have been discernible. This evidence remains rather controversial, partly because of uncertainty about the contribution to the central light from nonthermal processes. Recently, somewhat less ambiguous evidence has arisen for M31 and some other nearby galaxies (Tonry 1984, 1987; Kormendy 1988; Dressler and Richstone 1988). Data for M31 are shown in Figure 4. There is evidence not only for an increase in the velocity dispersion near the centre, but also that the rotation curve indicates a sharper concentration in the central mass than the corresponding central peak in the luminosity. One can directly infer that the mass-to-light ratio in the centre exceeds 35 solar units.

This limit does not self-evidently require a black hole. For instance, one might envisage a dense star cluster, consisting of stars with a different mass function from those in the body of the galaxy. The half mass relaxation time for such a cluster, with mass M, half-mass radius r, composed of stars each of mass m_*, is approximately

$$t_{rel} \simeq 10^9 \left(\frac{m_*}{m_\odot}\right)^{-1} \left(\frac{M}{10^8\,M_\odot}\right)^{-\frac{1}{2}} (r/1\,\mathrm{pc})^{\frac{3}{2}}\ \mathrm{yrs} \qquad (4)$$

This implies that a cluster of 10^8 neutron stars, within a radius of 1 parsec, could marginally survive for around 10^{10} years. If the dimensions could be shown (for instance, by HST observations) to be much less than 1 parsec, then this option could be ruled out. Moreover, for a cluster containing a range of different masses, evolution would take place in much less than 10 relaxation times, because the heavier stars would segregate towards the centre (Murphy and Cohn, 1988; Quinlan and Shapiro 1989). At first sight one might infer from this equation that a cluster of lower mass stars, correspondingly more numerous so as to produce the same total mass, would survive for longer. However, for very low mass stars, physical encounters would become more important than two-body 'coulomb' deflections in determining the evolution time for the cluster.

Apart from the data on Andromeda, there is evidence for a central mass of $5 \times 10^6\ M_\odot$ in M32 (Tonry 1984, 1987), and for $10^9\ M_\odot$ in NGC 4594 the Sombrero galaxy (Jarvis and Dubath 1988; Kormendy 1989). Dressler and Richstone (1988) conjecture that the masses may be proportional to the total mass in the bulge. Quasar remnants would then be found in ellipticals, the smaller holes in disc galaxies being relics of lower-level activity such as is manifested by Seyfert galaxies.

Although one cannot yet exclude alternative interpretations of the apparent dark central mass concentration – which could conceivably involve a dense cluster, or be an artefact of anisotropic velocities for the visible stars – massive black holes certainly seem the most natural inference from this body of recent evidence. (For a recent assessment, see Goodman and Lee (1989), or Binney and Petit (1989)).

One would like some independent corroboration of the black hole hypothesis, or, conversely, some way of ruling it out. *Stellar disruption* potentially offers such a test. The inner part of a galaxy could in principle be swept completely clean of gas. On the other hand, we directly observe a concentration of stars near the putative hole. As stellar orbits diffuse in phase space, it therefore seems inevitable that some may wander sufficiently close to the hole that they suffer tidal disruption. It is therefore of interest to estimate the rate for this process, and to explore the observational manifestations of a tidally-disrupted star.

Tidal Disruption of Stars

If the star density in the galactic nucleus, just outside r_h, is N^*, and the velocities are isotropic, then the rate of disruption is

$$\sim 10^{-3} \left(\frac{M_h}{10^7\ M_\odot}\right)^{\frac{4}{3}} \left(\frac{N^*}{10^5\ \mathrm{pc}^{-3}}\right) \left(\frac{\sigma_c}{300\,\mathrm{km\ s}^{-1}}\right)^{-1}\ \mathrm{yr}^{-1} \tag{5}$$

The actual rate could be lower than this approximate expression because radial loss cone orbits get depleted. Or it could be higher, because stars accumulate on orbits between r_h and r_c, and the density of stars in this cusp bound to the hole may exceed N^*. In fact, neither of these countervailing effects is likely to be of great importance for holes less massive than around $10^8\ M_\odot$ (Bahcall and Wolf, 1976; Frank and Rees, 1976; Lightman and Shapiro 1976). It may seem, however, that even the modest rate of stellar disruptions given above could have conspicuous consequences. If one supposed that the debris from a disrupted star were all swallowed by the hole, with efficiency 0.1 $\varepsilon_{0.1}$, and that the mean disruption rate led to steady accretion, then the resultant luminosity would be

$$6 \times 10^{42} \left[\frac{\text{disruption rate}}{10^{-3}\ \mathrm{yr}^{-1}}\right] \varepsilon_{0.1}\,\mathrm{erg\,s}^{-1} \tag{6}$$

This predicted luminosity does not seem to be observed in M31. Do we therefore have to abandon the black hole interpretation of stellar motions in this and other galactic nuclei? The answer is probably not, because of uncertainty about three things.

(i) The fraction of the debris which is swallowed, rather than expelled.
(ii) The radiative efficiency.
(iii) The timescale for accretion or expulsion of debris. Maybe we should expect bright flares with short duty cycles, rather than a steady luminosity?

The energy required to tear the star apart (*i.e.* the star's self-binding energy) is supplied at the expense of the orbital *kinetic* energy (which, at $r \simeq r_T$, is larger by $\sim (M_h/m_*)^{\frac{2}{3}}$). Unless there were some explosive energy input, which would be expected only if the star passed several times closer than r_T (Carter and Luminet 1982), the debris would be *on average* bound to the hole unless the star were initially on a hyperbolic orbit with asymptotic velocity $> v_*$, which is $\sim 1000\,\mathrm{km\ s^{-1}}$ for solar-type stars.

Several effects would, however, impart orbital energies *spread widely about* this mean to gas from different parts of the disrupting star; this spread crucially influences what we would actually observe. The dominant such effect is the following. While falling inwards towards the hole, the star would develop a quadrupole distortion which attains an amplitude of order unity by the time of disruption at $r \sim r_T$. The resultant gravitational torque would 'spin it up' to a good fraction of its corotation angular velocity by the time it gets disrupted: it would consequently, by that stage, be spinning at close to its break-up angular velocity. The parts on the 'outside track' *furthest* from the hole would therefore have an *extra* velocity, over and above the orbital velocity $v_{orb} \simeq (2GM/r_T)^{\frac{1}{2}} \simeq c(r_g/r_T)^{\frac{1}{2}}$, of order $v_* = (m_*/M_h)^{\frac{1}{3}} v_{orb}$; those *closest* to the hole would have a comparable velocity *deficit*. Moreover, the slower-moving gas on the 'inside track' is deeper in the potential well by an amount $\sim (GM_h/r_T)(r_*/r_T) \simeq (Gm_*/r_*)$. There would consequently be a spread of order $v_{orb}\Delta v$, where $\Delta v \simeq v_*$, in the energies of different bits of debris. (See Figure 5 and its caption.) Other processes during the flyby – for instance, impulses from shocks or nuclear energy released during the drastic compression and distortion of the stellar material (*e.g.* Luminet, 1987) – could further enhance Δv.

Even though the *mean* specific binding energy of the debris to the hole would be positive, and comparable with the self-binding energy (Gm_*/r_*) of the original star, the *spread about this mean* is larger by $(M_h/m_*)^{\frac{1}{3}}$ – a factor which is $\gtrsim 100$ for hole masses in the range relevant to galactic nuclei. (Lacy *et al.* 1982, Rees 1985, 1988.) Whenever a solar-type star passes within the tidal disruption radius r_T, some of the debris would be flung out on hyperbolic orbits with escape velocities up to $10^4\ \mathrm{km\ s^{-1}}$. When $M_h \gtrsim 10^6\,M_\odot$, the *bound* debris would be on

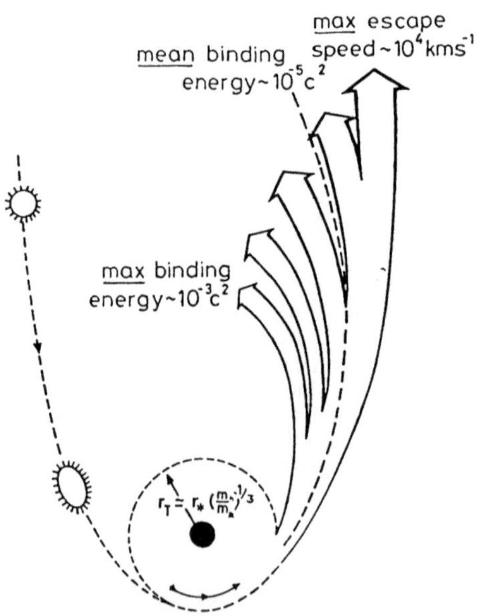

Figure 5 A solar-type star approaching a massive black hole on a parabolic orbit with pericentre distance r_T is distorted and spun up during infall, and then tidally disrupted. The average specific binding energy of the debris to the hole is $\sim 10^{-5}c^2$ (of order the self-binding energy of the original star). However, the *spread* in this energy [of order $v\Delta v$, where $v \simeq c(r_T/r_g)^{-\frac{1}{2}}$ and $\Delta v \simeq (Gm_*/r_*)^{\frac{1}{2}}$] is $\sim 10^{-3}c^2$ for hole masses $M_h \gtrsim 10^6\,M_\odot$. Almost half the debris would therefore escape on hyperbolic orbits with speeds up to $\sim 10^4\ \mathrm{km\ s^{-1}}$, the most tightly bound debris would traverse an elliptical orbit with major axis $\sim 10^3 r_g$ before returning to $r \simeq r_T$. Radiation from this debris, much of which may swirl down into the hole, creates a conspicuous 'flare'.

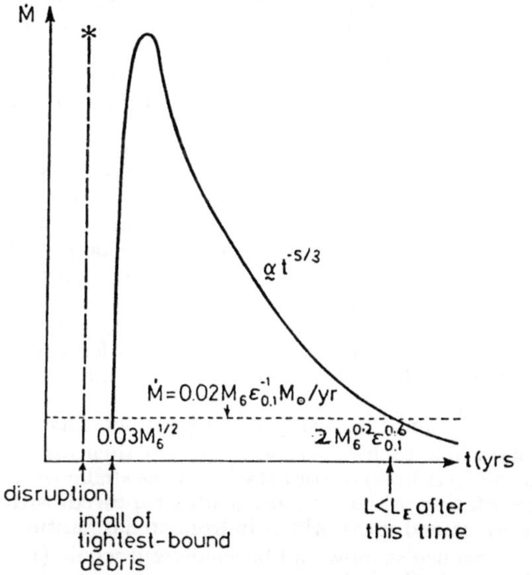

Figure 6 When a disruptive flyby occurs, the bound debris starts to rain down on to the hole after a timelag $\sim 0.03\left(M_h/10^6\,M_\odot\right)^{\frac{1}{2}}$ years. This diagram shows schematically the subsequent behaviour of $\dot{M}$. The peak infall rate is highly 'super-Eddington', but a thermal luminosity $\sim L_E$ can only be maintained for $\lesssim 1$ year. Thereafter the 'flare' would fade. At late times, the infall rate declines as $t^{-\frac{5}{3}}$ (the power law $-\frac{5}{2}$ given by Rees (1988) results from a stupid error). This is probably the dominant effect controlling the decline in luminosity after each flare.

orbits with characteristic specific binding energy $\gtrsim 10^{-3}c^2$ rather than $\sim 10^{-5}c^2$. The actual mass-fractions in a particular energy range, and the precise high and low energy cut-offs, would depend on the details. The bound orbits are very eccentric; for solar-type stars the orbital major axis of even the most tightly-bound debris is $\sim 10^3 M_6^{\frac{1}{3}} r_g$, and the period is only $0.03 M_6^{\frac{1}{2}}$ years. For $M_6 \lesssim 100$ this is certainly small compared with the interval between successive disruptions (equation (5)). Unless it takes many orbital periods to swallow the bound gas, the debris from each star would be digested separately – in contrast to Hills' (1975) 'debris cloud' model of quasars, where the disruptions were postulated to be frequent enough to generate (*c.f* equation (5)) a quasar-level luminosity, but the orbital periods of the debris $\left(\propto M_h^{\frac{1}{2}}\right)$ are longer. The fate of bound debris, where the quantitative details depend on viscosity, relativistic precession effects, etc., is discussed more fully elsewhere (Rees 1988).

When individual stars are being captured at the modest rate expected in relatively quiescent nuclei, the bulk of the debris from each would be swallowed or expelled *rapidly* compared with the interval between successive stellar captures – the only conspicuous luminosity being a flare (predominantly of thermal UV or X-ray emission) with $L = L_E$, fading within a few years (see Figure 6 and caption). The infalling debris forms a torus at $r \simeq r_T$, within which radiation pressure is dominant; its subsequent flow, controlled by viscosity, is likely to lead to accretion on a timescale which is a modest multiple of the rotation period at $r \simeq r_T$ (only a few hours). It is clear from the behaviour of $\dot{M}$ in Figure 6 that most debris would be 'fed' to the hole far more rapidly than it could be accepted if the radiative efficiency were high; much of the bound debris must then either escape in a radiatively-driven directed outflow or be swallowed inefficiently.

When a star passes close enough to be disrupted in a single flyby, almost half the debris escapes on unbound hyperbolic orbits. The kinetic energy of the ejecta is $\sim 10^{51}$ ergs. The material would be concentrated in a cone or 'fan' close to the orbital plane. Adiabatic cooling would severely reduce the internal radiative content before the debris became translucent.

The 'prompt' radiation from the unbound debris would therefore release less than the initial energy content of the star (just as a supernova would be optically inconspicuous were it not for continuing energy injection in the months after the explosion). There would therefore be no conspicuous flare until the *bound* debris fell back onto the hole. The main observable effects of the outflowing material would occur when it was decelerated by running into external diffuse matter.

The behaviour of stars passing *well within* the tidal radius r_T exhibits special features, (Carter and Luminet 1982, Luminet 1987, Luminet and Carter 1986, Luminet and Marck 1985). Such stars are not only elongated along the orbital direction but are even more severely compressed into a prolate shape (*i.e.* a 'pancake' aligned in the orbital plane). This compression is halted by a shock, raising the matter (which then rebounds perpendicular to the orbital plane) to a higher adiabat. Also, there is the possibility of explosive energy release. The resultant spread in the energy of the debris may then exceed $v_* v_{orb}$, but one needs a detailed model to quantify this because velocities perpendicular to the orbital plane yield only a second-order contribution, and are therefore less important than those in the plane. The orbits for which extreme compression occurs would enter regions where relativistic effects were more important than for those with $r_{\min} = r_T$.

For hole masses $M_h \gtrsim 10^8\, M_\odot$, solar-type stars cannot be disrupted without entering the strongly relativistic domain. The form of the black hole (Schwarzschild or Kerr?) then has an important quantitative effect, as does (for a rotating Kerr hole) the orientation of the stellar orbit relative to the hole's spin axis. Stars on counter-rotating orbits are more readily captured, with the result that Kerr holes would spin down if they gained mass primarily from stellar capture. When the hole mass is $>> 10^8\, M_\odot$ most main-sequence stars would be swallowed whole (*i.e.* $r_T \not> r_g$), and only giants would generate debris outside the hole.

Observability of Flares from Disrupted Stars

The most distinctive consequence of a $10^6 - 10^8\, M_\odot$ black hole's presence would be transient flares whenever bound debris from a star was swallowed, the luminosities being as high as $L_E \simeq 10^{44} M_6\,\mathrm{erg\,s^{-1}}$. In a given object, these flares would have a duty cycle of order 10^{-3} at peak luminosity. The rise time and the peak bolometric luminosity can be predicted with some confidence. However, the effective surface temperature (and thus also the fraction of the luminosity that emerges in the visible band) is harder to predict – this depends on the size of the effective photosphere that shrouds the hole, particularly when $\dot{M}$ is high. After each luminosity peak, there would be a decline as the 'dregs' were swallowed (see Figure 6), but the median luminosity would be far below that which would result from *steady efficient* accretion of the mass supply implied by equation (6). Therefore we would not yet expect to have detected such a flare. On the other hand, a sufficiently large sample of such galaxies should reveal some members of the ensemble in a flaring state. Such objects could be searched for out to large distances: they would last rather longer than supernovae and would differ from typical AGNs through the lack of any extended structure (emission line or radio components). The central location of the phenomenon, however, militates against its detection in supernova searches, which are notoriously incomplete in the inner high-surface-brightness regions of galaxies. If $10^6\, M_\odot$ holes were prevalent in small or even dwarf galaxies, the nearest such flares, in any given year, may be no further away than the Virgo Cluster.

The ejected matter, though less spectacular than the accretion-powered flares, could nevertheless have more sustained cumulative effects. There are 2 ways in which matter can be expelled:

(i) When a solar-type star is disrupted in a single flyby, the 'spray' of ejected debris on hyperbolic orbits (see Figure 5) moves outward at $\sim 10^4 M_6^{\frac{1}{3}}$ km s^{-1}. The kinetic energy output would be $\sim 10^{51} M_6^{\frac{2}{3}}\,\mathrm{erg\,s^{-1}}$ – for $M_6 \gtrsim 1$ (*i.e.* in all cases of interest here) this exceeds the energy of a supernova. Stars that penetrate well inside r_T would cause still more violent ejection.

(ii) The bound debris falls back to $r \simeq r_T$, where it can acquire centrifugal support. If energy is dissipated at a supercritical rate as this material swirls closer to the hole, some gas may then be ejected in a radiation-driven wind, or even in double jets aligned with the angular momentum of the debris. It is in principle possible for only a fraction $\varepsilon^{-1}\,(r_g/r_T)$ to be actually swallowed, ε being the efficiency, all the remainder being ejected. The characteristic speeds here – of order the escape velocity from r_T – would be even higher than those of the type (i) outflow discussed above. The debris from a star that does not approach close enough to be destroyed in a single flyby but is eventually disrupted by tidal dissipation could behave similarly.

Several aspects of the stellar disruption phenomenon call for detailed stellar-dynamical and hydrodynamic calculations (*c.f* Nolthenius and Katz, 1982; Evans and Kochanek 1989). Also, relativistic precession effects have an important effect on the flow at $r \lesssim r_T$. Obviously, precise modelling should allow for a realistic stellar population. The preliminary estimates presented here are nevertheless already relevant to the searches, with high sensitivity and spatial resolution, for mass concentrations in the centres of nearby galaxies, and to the properties of AGNs in general.

Mergers and Binary Black Holes

We have seen that there may be black holes in most galaxies. It is also implied by the data in Figure 1 that most of these had already formed by the time the universe was 2 or 3 billion years old. There have certainly been a substantial number of galactic mergers since that time. Indeed, according to some models for galaxy formation, the majority of large galaxies today will have experienced at least one merger since that time. This raises the question of what happens during a merger if *each* of the galaxies involved harbours a central black hole.

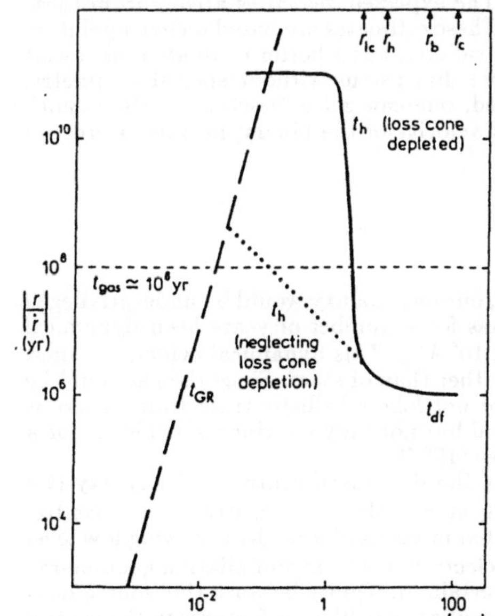

Figure 7 The timescales involved in the approach, and eventual coalescence, of a supermassive binary (after Begelman, Blandford and Rees, 1980). The core parameters chosen are those that might be appropriate to a giant elliptical: $\sigma_c \sim 300\,\mathrm{km\,s^{-1}}$, $N^* \sim 2\times10^3$ and $m_* \sim 1$. The members of the binary are taken to have masses $M_8 = 1$ and $m_8 = 0.3$. For this system the dynamical friction time is $t_{df} \sim 10^6$ yr. Within r_h the evolution timescale would be $(r_h/r)\,t_{df}$ if loss cone depletion could be ignored; however, unless collective effects permit replenishment of the loss cone on much less than the ordinary stellar relaxation time, the evolution within r_{lc} proceeds on a very much longer timescale. Influx of gas into the system at a rate $\dot{M} \sim 1\,M_\odot\,\mathrm{yr}^{-1}$ would however yield a timescale $t_{gas} \simeq 10^8$ yr. Gravitational radiation would take over as the dominant mechanism within $1.3\times10^{-2}\,(t_{gas}/10^8\,\mathrm{yr})^{\frac{1}{4}}$ pc, and the binary would then evolve towards coalescence. The recoil in the final burst may be enough to eject the newly-merged hole from the core of the galaxy.

Dynamical friction, operating on a timescale inversely proportional to the mass of the holes, quickly causes them to settle towards the centre of the merged galaxy. The holes eventually approach close enough to become a binary, where each is influenced more by the gravitational pull of the other than by the galaxies' overall gravitational field. The black hole binary continues to tighten, as it transfers energy to stars, until it gets sufficiently close for gravitational radiation, operating on a timescale proportional to the fourth power of separation, to bring about eventual coalescence. The shape of the curve in Figure 7 shows that a binary may spend a long time at a separation of order 10^4 or $10^5 r_g$, with orbital velocity in the range 1000 – 3000 km s^{-1}. For its orbit to shrink by a factor of 2 through this range, the binary must interact with $\sim (M_h/m_*)$ stars, each of which is imparted a velocity of order the orbital speed, and therefore probably ejected from the galaxy. The time taken for the binary to shrink can be straightforwardly calculated if the stellar orbits are isotropic, but in practice the binary may expel the stars on nearly radial orbits in the galaxy faster than these can be replenished, in which case loss cone effects will reduce the rate of orbital shrinkage. Perhaps interaction with gas, which could be associated with an epoch of nuclear activity, is needed to bring the binary close enough for eventual coalescence.

If black holes in galaxies are brought together as a result of galactic mergers, the interesting question arises of whether the resultant holes remain in the merged galaxies. There may be a recoil due to emission of net *linear* momentum by gravitational waves in the final coalescence (Redmount and Rees 1989). If the holes have unequal masses, a preferred longitude in the orbital plane is determined by the orbital phase at which the final plunge occurs. For spinning holes there may be a rocket effect perpendicular to the orbital plane, since the spins break the mirror symmetry with respect to the orbital plane. The expected velocities arising from these processes are several hundred kilometres per second. These estimates are based on extrapolating a weak field argument almost beyond the limits of its validity, but a better estimate must await full general relativistic numerical calculations in three dimensions without special symmetry. If a third hole drifts in before the binary has merged, one can get a Newtonian 'sling shot'. Ejection then occurs with speeds of order the orbital velocity of the binary, in other words up to 3000 km s^{-1}.

Our Own Galactic Centre

If there are indeed black holes in most nearby galaxies, our own Galaxy would be underprivileged if there was not also one in its own centre. There has for a number of years been dynamical evidence for a concentrated central mass of order $3 \times 10^6\, M_\odot$. This dynamical evidence comes, however, primarily from the motions of gas streams rather than of stars. Since the gas could be subject to nongravitational forces, and therefore need not follow ballistic trajectories, there is some ambiguity in this interpretation. The 'substantial but not fully convincing' evidence for a central mass has been reviewed by Genzel and Townes (1987).

A unique compact radio source appears to lie at the dynamical centre of the galaxy (Lo 1986). Its proper motion indicates that it is moving at < 40 km s^{-1} relative to the centre, and it can be naturally, though not uniquely, interpreted in terms of a model involving low level accretion onto a black hole (Rees 1982). The variable electron–positron annihilation gamma–ray flux may be related to this object. There remains a certain ambiguity because this source does not lie at the centre of the pattern delineated by the peculiar arm–like gas features in the central 2 parsecs. Nor is it located symmetrically with respect to the infrared sources which make up the components of IRS 16.

An energy input is required into the gas filling the volume in which the gas streams are embedded. The mean inferred input seems higher than the present input. One can speculate that this, and the gas streams, could be due to the last few tidally–disrupted stars. Equation (5) suggests that 1 star may be swallowed every $\sim 10^4$ years. Unless there has been such an event recently, we would expect that the current luminosity will be below the mean.

Let me mention finally two more speculative, but potentially more definitive, tests of whether there is indeed a massive black hole in our galactic centre.

First, one can ask whether a single star could be tidally captured into a circular orbit at $r = r_T$ without disruption. Such a star would have an orbital period of order 1 hour, but a remarkable velocity exceeding 0.1c. It is rather unlikely that a star could get into such an orbit without disruption, because it would have to radiate many times its own internal binding energy, and the time needed to do this would be many Kelvin times. Since the energy would be dissipated much more quickly than this, it seems almost unavoidable that the star would get disrupted. However, a related possibility, suggested by Hills (1988), is that the stars in the central parsec of our Galaxy may include a population of close binaries. If such a binary got sufficiently near to the hole, it would be disrupted, one component being left in an eccentric bound orbit with period less than a year (which may then circularise), its companion being expelled at $\sim 1000\,\mathrm{km\ s^{-1}}$. Such hypervelocity stars, if they were detectable, would be compelling testimony that there is indeed a massive central hole.

Conclusion

Discovery of a black hole candidate is of course just a preliminary step towards using it to test Einstein's theory in the strong–field regime – to check whether space–time around such objects obeys the metrics of Schwarzschild and of Kerr. There are, of course, *stellar*–mass black hole candidates in X–ray binaries within our own Galaxy. But in the long run these hugely massive 'dead quasars' in galactic nuclei (some as large as an entire Solar System), whose formation and environment can be studied without the complexities of high density physics or high opacities, offer the firmest hope of exploring the most crucial and remarkable consequence of Einstein's theory.

Much of this written text is adapted from a chapter in the book 'Baryonic Dark Matter', edited by G. Gilmore and D. Lynden–Bell (Kluwer, in press).

References

Bahcall, J.N. and Wolf, R.A. 1976. *Astrophys. J.* **209**, 214.
Begelman, M.C., Blandford, R.D. and Rees, M.J. 1980. *Nature* **287**, 307.
Binney J. and Petit, J.–M. 1989, in *'Dynamics of Dense Stellar Systems'*, ed. D. Merritt (CUP) p.43.
Carter, B. and Luminet J.P. 1982, *Nature* **296**, 211.
Dressler, A. and Richstone, D.O. 1988, *Astrophys. J.* **324**, 701.
Evans, C.R. and Kochanek, C.S. 1989, *Astrophys. J.Lett.* (in press).
Frank, J. and Rees, M.J. 1976, *Mon. Not. R. astr. Soc.* **176**, 633.
Genzel, R. and Townes, C.H. 1987, *Ann. Rev. Astr. Astrophys.* **25**, 377.
Goodman, J. and Lee, H.M. 1989, ApJ **337**, 84.
Hills, J.G. 1975, *Nature* **254**, 295.
Hills, J.G. 1988, *Nature* **331**, 687.
Jarvis, B.J. and Dubath, P. 1988, *Astr. Astrophys.* **201**, L 33.
Kormendy, J. 1988, *Astrophys. J.* **325**, 128.
Kormendy, J. 1989, *Astrophys. J.*(in press).
Lacy, J.H., Townes, C.H. and Hollenbach, D.J. 1982, *Astrophys. J.* **262**, 120.
Lightman, A.L. and Shapiro, S.L. 1976, *Astrophys. J.* **211**, 244.
Lo, K.Y. 1986, *Science* **233**, 1394.
Luminet, J.P. 1987 in *'Gravitation in Astrophysics'*, eds. B. Carter and J. Hartle, (Reidel Dordrecht), p.215.
Luminet, J.P. and Carter, B. 1986, *Astrophys. J.* **61**, 219.
Luminet, J.P. and Marck, J.A. 1985, *Mon. Not. R. astr. Soc.* **212**, 56.
Murphy, B.W. and Cohn, H.N. 1988, *Mon. Not. R. astr. Soc.* **232**, 835.
Nolthenius, R.A. and Katz, J.I. 1982, *Astrophys. J.* **263**, 377.
Phinney, E.S. 1983, Cambridge PhD Thesis
Quinlan, G.D. and Shapiro, S.L. 1989, *Astrophys. J.*(in press).
Redmount, I. and Rees, M.J. 1989, *Comm. Astrophys.* (in press).
Rees, M.J. 1982 in 'The Galactic Center', eds. G. Riegler and R.D. Blandford, (AIP, New York) p.166.

Rees, M.J. 1985 in *'The Milky Way Galaxy'*, eds. H. van Woerden *et al.*, (Reidel, Dordrecht) p.379.
Rees, M.J. 1988, *Nature* **333**, 523.
Sargent, W.L.W., Schechter, P.L., Boksenberg, A. and Shortridge, K. 1977, *Astrophys. J.* **212**, 326.
Sargent, W.L.W., Young, P.J., Boksenberg, A., Shortridge, K., Lynds, C.R. and Hartwick, F.D.A. 1978, *Astrophys. J.* **221**, 731.
Schmidt, M. 1989 *'Highlights of Astronomy* **8**, 31.
Soltan, A. 1982, *Mon. Not. R. astr. Soc.* **200**, 115.
Tonry, J.L. 1984, *Astrophys. J.Lett.* **283**, L 27.
Tonry, J.L. 1987, *Astrophys. J.* **322**, 632.

European and Other International Cooperation in Large-Scale Astronomical Projects

C. Patermann

BMFT, Heinemannstraße 2, D-5300 Bonn 2, Fed.Rep.of Germany

Thank you very much for kindly inviting me to present a paper before such a distinguished audience as the members of the Astronomical Society. I feel honoured also because, as someone who is not an astronomer, I have been given the opportunity to speak to you on a technological and scientific subject with which you are far more familiar than I am - namely, European and other international cooperation in large-scale astronomical projects.
I consider this subject particularly challenging because there is a great deal of public debate today about large-scale projects for European cooperation, for example, in the sectors of communications, information technology, space activities and high-energy physics. I need only mention the cooperation with the European Space Agency ESA, with CERN and in EC research projects and, more recently, also to an increasing extent, the cooperation carried out within the framework of EUREKA. It would appear that astronomy has been somewhat neglected. This is regrettable because I consider it to be an excellent field - owing to its transboundary, indeed, literally infinite dimension - for publicizing European and international cooperation in science, research and high technology.

In the 1970s I was Scientific Attaché at the German Embassy in Washington. On one occasion, the former head of the American Academy of Sciences, P. Handler, said to me while giving a general picture of scientific activities in the USA for the benefit of an important German visitor: "After all, if I talk about astronomy in the US, it is the same like everywhere: Astronomy is everybody's second love."

I have often had occasion in recent times to remember this statement. Everyone is fascinated by astronomy. We all consider this field necessary, interesting - but, particularly from the point

Reviews in Modern Astronomy 2 Editor: G. Klare

of view of astronomers, it frequently ranks only as "second option" when financial or research-policy priorities are at stake. But I see no reason why this should remain so.

Now let me turn to my subject. I consider that large-scale astronomical projects carried out in European and other international cooperation fall under three major categories:

a) Projects carried out in bilateral cooperation, for example with NASA or SERC in Great Britain, or ROSAT or GRO (Gamma Ray Observatory) and also those carried out via bilateral cooperation in IRAM in Grenoble, in bilateral cooperation between the Max Planck Institute of Astronomy in Heidelberg with the Spanish Commission of Astronomy at the Calar Alto observatory in Spain. Then there are the Gallex Neutrino Experiment and the solar telescope on Teneriffe, both of which are projects being carried out in bilateral cooperation.

b) The participation of the Federal Republic of Germany in astronomical space projects within the framework of ESA. Here I am referring to the Hubble Space Telescope, Hipparcos, the European astrometry satellite ISO (infra-red observatory) as well as to planned projects such as XMM (the X-ray spectroscopy telescope) and FIRST (Far Infra-Red Space Telescope), which, though they can only be implemented at some time in the future, nonetheless constitute major pillars of the ambitious ESA science programme Horizon 2000.

c) German participation in large-scale astronomical projects carried out within the framework of ESO, the European Southern Observatory, e.g. NTT, VLT.

I consider this division of large-scale astronomical projects into groups important because it is the best way of showing international interdependence and the cooperation potential regardless of what branch of astronomy is concerned - whether terrestrial astronomy, space-related astronomy and optical astronomy or radio astronomy. Finally, for the sake of completeness, there are preparatory studies, for example, with regard to LEST, the Large European Solar Telescope - a project which has likewise not yet been definitely agreed, as well as a number of important bilateral projects with NASA which are currently the subject of intensive deliberations:

- Orbiting Solar Laboratory (OSL)
- SOFIA (Submm/infra-red range)
- SPEKTROSAT (follow-up project to ROSAT).

However, with regard to all these projects which may be carried out in cooperation with NASA, no decision will be forthcoming until next year at the earliest - and the USA must express its desire to become involved in such projects before they can be agreed. Finally, a brief word on the USSR. We have already received offers of participation with regard to the X-ray satellite SPEKTRUM and the "SPEKTRUM GAMMA" gamma astronomical satellite, but so far the Federal Republic of Germany has not yet received adequate information regarding the details.

But all these projects furnish impressive evidence of the current state-of-the-art as regards the programme objectives of astronomical research throughout the world.

Up to the time when space research opened up new possibilities, astronomical research had been confined to those optical and radio spectral areas which are accessible from the Earth.

As a result of the possibilities opened up by **space research**, i.e. in particular the **observations** carried out outside the **Earth's atmosphere** on account of its absorption and atmospherics, observation projects are now being tackled continuously in all the spectral ranges. These activities are shared as far as possible in **international cooperation** - among other reasons on account of the heavy specific outlay required for the implementation of the projects. A crucial factor in this connection is that **modern astrophysics** depends on the results of observations in all the spectral areas in order to solve outstanding issues (star formation and final phases, the structure of the universe, the development of galaxies, cosmology).

This is why the **exploration of all the spectral ranges** (gamma, X-ray, UV, infra-red) constitutes an important goal of current astronomical activities.

The **exploration of the individual spectral ranges** is carried out in several phases including mapping the celestial sphere, observing single sources and the application of high-resolution spectroscopy to these sources.

Now it would be "carrying coals to Newcastle" if I were to describe the projects I have mentioned in detail. This would also exceed considerably the time allowed me for presenting this paper. I should therefore like to devote my attention particularly to one organization - namely the European Southern Observatory, ESO, in Munich - and give you some more information about this organization's projects, particularly the VLT. By selecting this organization I shall be able to give you a graphic picture of the projects being carried out within the framework of European cooperation. I also consider the VLT particularly appropriate as the central subject of my paper because I consider that the elaboration of this project and its implementation best show how the pooling of national resources at the European level can enable top-quality European research to remain in the lead throughout the world. In addition, this project clearly reveals that large-scale astronomical projects today represent a highly interesting combination of different functions, involving those responsible for research, for science and for technology, and that they also constitute a factor for cooperation and thus also for political integration. Let me now speak in greater detail about Europe's ambitious endeavour to reach for the stars with the aid of VLT.

When discussing Europe's recent success in research and technology the following keywords usually come to mind: CERN, the European centre for high-energy physics in Geneva with the construction of the biggest particle accelerator - the LEP - which is to be commissioned at the end of this year, and the European launcher ARIANE. I think that very soon another flagship of European research and technology will be added to this list - namely the VLT - Very Large Telescope - which I have already mentioned. The VLT will be the world's largest ground-based optical telescope. The Council of the European Southern Observatory ESO with its headquarters in Munich - which is known to only a small section of the public - agreed unanimously through its Council, the supreme body of this organization, on December 8, 1987 that a very large telescope of this kind is to be set up in the mountains of Chile at a cost of approximately DM 350 million on the basis of the prices prevailing in 1987. ESO now numbers eight European states among its members: the Federal Republic of Germany, Belgium, Denmark, France, Italy, the Netherlands, Sweden and Switzerland. Contacts have also been established with Austria, Portugal and Turkey. The observator - or rather the group of individual telescopes maintained by ESO - are situated 600 km north of Santiago de Chile at the southern end of the Atacama Desert in the proximity of La Silla. 13 telescopes are in operation there and of these the 2.2 m telescope on loan from the Max Planck Society is regarded as one of the most modern telescopes in the world today. At the moment, ESO's annual budget consists of contributions amounting to DM 70 million. Of this sum, the Federal Republic of Germany and France each contribute 26.5 %, while the remaining contributions are raised by the other member states in proportion to the gross national product of each country. European astronomers are required to submit to a rigorous selection procedure carried out by the appropriate ESO boards in order to be assigned observation time on the basis of relevant scientific proposals. As a rule almost all the telescopes are heavily overbooked.

I need hardly tell you astronomers just how successful ESO has been in recent years. Whether - what is generally less well-known - the exact information concerning the position of Halley's comet over Chile which is sent from Munich daily to the European Space Operations Centre maintained by ESA in Darmstadt (ESOC) in connection with the spectacular approach of the European comet probe GIOTTO to within approximately 600 km of the core of Halley's comet, particularly during the last 14 days of the encounter, whether we call to mind ESO's fantastically rapid organizational and scientific recording activities approximately 1 1/2 years ago in connection with the Supernova, which are unparallelled, whether we recall the activities carried out within the framework of ESA's European coordination agency in order to coordinate the astronomical activities of European participants in the Hubble Space Telescope, the ECF (European Coordinating Facility), which will hopefully commence work before the end of this year. Finally, at the end of this year, ESO will commission what is probably the world's most modern technological telescope, the NTT, a telescope 3.5 m in diameter, which, it is interesting to know, has been exclusively financed by the "entry fees" paid by Italy and Switzerland when they joined ESO a few years ago. The NTT will be equipped with a light-weight computer control system based on a construction method which will drastically reduce construction costs. This telescope is therefore considered to be the prototype of a new generation of high-tech telescopes. Its costs only amount to roughly one third of the costs of the largest ESO telescope operated so far in Chile - a telescope with a mirror 3.6 m in diameter, which was erected 13 years ago. The first images were produced with unexpected precision on March 23, 1989: a major success!

For approximately the past eight years, astronomers in Europe, the USA, the Soviet Union and also in Japan have been discussing the next step to be taken with large-scale optical telescopes. The manufacture of mirrors with a maximum diameter of 5 - 6 m has meant that the development of ground-based optical telescopes has been at a stand-still as regards their two major characteristics, namely, their collecting capacity and resolution. Most of the observers of the technological scene who are not astronomers have so far been unaware of this fact. In order to investigate fainter sources of light situated at far greater distances from the Earth, and thus also older parts of the cosmos which enable important conclusions to be drawn concerning the origin of matter, mirrors with a greater collecting capacity and higher resolution are required - i.e. mirrors of a far greater diameter than hitherto. For this reason discussion has been going on for the past decade concerning the construction of large-scale telescopes equipped with mirrors of between 10 m and 25 m in diameter. It is, however, not necessary to actually construct mirrors of this diameter. With the aid of interferometry methods, several smaller mirrors of 8 m or 10 m in diameter can be set up in a group, thereby attaining the same collecting capacity. After intensive preparation in the European astronomical community, ESO decided in 1987 to submit to its competent bodies a concept envisaging the linear grouping of four telescopes, each 8 m in diameter, which will have a collecting capacity equal to that of a mirror 16 m in diameter. At that time it had still not been decided what material should be used for manufacturing the mirrors - whether they should be made from glass ceramics (Zerodur), which has so far proved to be the best material for constructing mirrors for optical telescopy because, among other characteristics, it has an extremely low coefficient of expansion, or whether metal, pyrite or other materials should be used. The Federal Ministry for Research and Technology promoted the development of Zerodur more than ten years ago and this has been successful.

At the time of the VLT development, the site had not been selected for erecting the world's largest optical telescope. Since the measurement of infra-red rays is becoming increasingly important also in the field of ground-based astronomy, the content of water vapour in the atmosphere at future sites will have to be low. Above and beyond this, an increasingly important role is played by so-called "seeing" - the absence of so-called atmospheric distortion of images. For this reason, measurements are currently being carried out in the north of Chile in the Atacama Desert, as well as at other sites, and the result may be that, in the long term, the site selected at La Silla will be abandoned. However, the ESO member states have clearly stated that a new site should be selected elsewhere only if this could mean a dramatic improvement in the levels recorded, i.e. regarding the number of cloudless nights, the "seeing factor" and also the presence of an extremely low amount of vapour in the atmosphere.

Above and beyond this, the new VLT telescope will also incorporate other technological innovations. For the first time, it will be possible to introduce so-called adaptive optics for industrial application and, with the aid of computers, the telescope will correct very slight distorting movements of the mirror caused by atmospheric influences. In addition, the so-called "no-astronomer telescope" will operate for the first time on an industrial scale. The data measured will be computerized and transmitted via satellite or cable to Munich, together with the observation results, for compilation, evaluation and storage. The astronomer will be able to control and conduct observations from a great distance from Munich or from his particular institute. This is why Munich will acquire more importance for ground-based astronomy in Europe than has hitherto been the case.

The new Very Large Telescope will also set new requirements as regards the manufacture of the mirrors and also with regard to the mirror-polishing facility. Since no mirrors of this dimension have been manufactured before, nothing is known about their durable life, the way in which they can be assembled and transported, and how they are to be used. The same applies to the so-called mirror grinding and mirror polishing operations, which have to be carried out with utmost precision and which will therefore also provide impulses for new computerized production techniques.

The new Very Large Telescope is to be commissioned with the use of all four mirrors in the year 2000. The first mirror is to be mounted in Chile in 1995. It is expected that the operation of the NTT at the end of next year will also provide important operational data which can be incorporated into the new overall concept. Funding of the annual ESO budgets is ensured. Only one issue is as yet uncertain: whether Denmark will share in the financing of the ESO Very Large Telescope.

In the meantime, the first major contract, namely, for the manufacture of the four large 8 m diameter mirrors, has been awarded. A German firm will manufacture the mirrors from the glass ceramics material Zerodur, which neither contracts nor expands, even in the event of changes in temperature. During the past ten to fifteen years, most large telescope mirrors have been manufactured along these lines throughout the world. According to ESO, a fifth mirror, which might even be the first mirror to go into operation as early as 1993, could conceivably be manufactured on the basis of aluminium. But this is currently only being discussed on an informal basis by a few experts. The next major contract, that of grinding the mirrors, will be awarded very soon. The German firm of Zeiss and the French firm of REOSC in Paris are competing for this contract. It must be submitted by April 10, 1989. The value of the contract will probably be between DM 30 million and DM 40 million.

The VLT will enable optical astronomy in Europe to gain the lead over the rest of the world. ESO has so far been successful in considerably reducing the major lead of US-American astronomy over European astronomy during the period following World War II. Now the VLT will enable Europe to assume the international lead also in the field of astronomy, in addition to high-energy physics, selected fields in molecular biology, X-ray synchrotron radiation and other sectors of basic research. This should be remembered by all those who like to make glib comments about Eurosclerosis.

Ladies and gentlemen, I should now like to make a few comments about the German ASTRO network and also to point out that the international manned space station, whose European part is called COLUMBUS, will certainly offer interesting perspectives for astronomy within the framework of space flight activities.

As you know, the Federal Ministry for Research and Technology cannot promote the entire area covered by astronomy. Astronomy is a discipline of basic research which is promoted chiefly by the German Research Society within the framework of science's autonomy in competition with the remaining disciplines of basic research. The Länder are responsible for providing the basic equipment of the institutes of astronomy in so far as these are university institutes, while the Federal and Länder Governments together support the Max Planck society and the institutes on the Blue List. In addition, the Federal Ministry for Research and Technology also supports astronomy via its funding of space equipment and space missions at the national level as well as within the framework of ESA, in accordance with the delimitation agreement concluded with the German Research Society in 1964. Furthermore, the Federal Ministry for Research and Technology also supports the European Southern Observatory ESO on a proportionate basis together with the other member states, as well as via the Max Planck institute for millimetre wave astronomy, IRAM, which I have already mentioned. Recently, however, a new course has been adopted. As in the case of the funding of groups of universities which use large-scale equipment for work in the field of high-energy physics, the government intends to promote selected groups of astronomers who work with large-scale telescopes. Within the framework of its competence for space astronomy projects, the Federal Ministry for Research and Technology has launched a measure which will improve the infrastructure for astronomical research by promoting, together with the Länder and the German Research Society, a computer network for German astronomers (German Astro network), at the same time using the German research network, which is intended in particular to facilitate the use of future large-scale equipment by university institutions. This promotion was to be subject to the readiness of the Länder to provide the institutes concerned with the necessary funds for purchasing computers. To a considerable extent, this has now been done. I consider this modest, but nonetheless important, measure an interesting step towards the systematic development of the activities of the Federal Ministry for Research and Technology as regards research and promotion.

Now let me conclude with a few general remarks.
Astronomy is a fascinating science which has a quite special importance, particularly for young people, because it helpsus see our planet in perspective and because young people learn to regard the universe with awe. It is not without good reason that many people regard astronomy as the world's oldest classical science. I think that astronomers therefore also have an important duty towards society. I think that in taking the example of the Very Large Telescope I have shown you that astronomy has aided the development of outstanding high technologies, as it were, unintentionally and without attracting attention. The Very Large Telescope is not merely an instrument for expanding knowledge, it is not merely the fascination of that which is new, the search for our origins in a boundless dimension - astronomy is also a vehicle for technology and provides the impetus for innovations. This should lighten astronomers' duty towards society. In any case, I consider that, from the ideological aspect, astronomy has an easier time than, say, particle physics. After all, we can hardly confuse astronomy with nuclear physics. In parliament it is frequently the case that, when CERN is the subject of discussion, there are always some members who think that nuclear power stations are developed there. These problems do not arise for astronomy. Moreover, astronomy holds out no prospect of military application, as does space flight to some extent. I consider that this is an opportunity which astronomers should use to the full in the public debate on acceptance. Remember that the approximately 400 German astronomers at universities, institutes and

international organizations will probably be able to use scientific equipment of between DM 7 thousand million and 9 thousand million in value in the year 2000 - the fruit of international cooperation and interdependence, for which politicians alone are responsible. This is quite a considerable sum and means that we must be fully aware of our responsibility. The results of a survey carried out in 1982 revealed that, at that time, approximately 100 astronomical associations in the Federal Republic of Germany numbering 10,000 members were operating approximately 80 public observatories. But this great display of interest in the public observatories - I cannot give you the newest figures - is not enough. The same applies to the descriptions for laymen of the night sky which are published each month particularly by those newspapers which have a reputation to uphold. Astronomers should rather press ahead with scientific publications written for the general public which describe their tasks and findings. They must develop a strategy for the earlier publication in a more popular form of the fruits of long years of research and deliberately set out to attract the attention of the public. Why are there no special achievement courses in astronomy - at least at the grammar schools in German cities numbering more than 500,000 residents? The further education programme for teachers, which in your Society traditionally follows on your conferences, is an excellent institution, but it is actually no more than a start in the right direction. Have you established contact with private television companies? Do you maintain intensive contacts with scientific programme broadcasts by public television companies? When big competitions are held, why are not prizes awarded in the form of trips to the important observatories, which are often situated in remote places? This could be done in cooperation with firms. Why are there no excursions, say, to La Silla, to Calar Alto, to Teneriffe, which could be organized by travel agencies in order to arouse the interest of the large number of German tourists who holiday in these parts of the world? Despite the fact that the event took place at midnight, the "night of the comet" in March 1986 none the less drew an audience of 4 million to their television screens. The interest is there. Use it to advantage: there is competition from other branches of science and research. For this reason, regard Supernova and Halley's comet also from this aspect as "heaven-sent gifts".

A Decade of Stellar Research with IUE

H.J.G.L.M. Lamers

SRON Laboratory for Space Research and Astronomical Institute,
Princetonplein 5, NL-3584 CC Utrecht, The Netherlands

1. INTRODUCTION

The International Ultraviolet Explorer is one of the most successful astronomical satellites. It has been operating now since more than 10 years. IUE was launched on 26th January 1978. The satellite consists of a spacecraft of 312 kg with 122 kg of instruments. It is an elliptical orbit with an altitude varying between 26000 and 46000 km. The telescope, a 45 cm Cassegrain has a 16 arc.min field of view. There are two spectrographs: a short wavelength spectrograph, operating between 1150 and 1950 Å with a high solution mode of 0.15 Å resolution, and a low resolution mode of 6 Å resolution. The long wavelength spectrograph operates between 1900 Å and 3200 Å, with a high resolution mode of 0.25 Å, and a low resolution mode of 7 Å. A detailed description of IUE, its history and its instruments can be found in "Exploring the Universe with IUE " (Kondo,1987).

This instrument has been oparating now for 11 years. During the first 10 years about 70000 images have been taken. It is easy to calculate that this comes down to a rate of about one image per 1.2 hours.

The observations of IUE have resulted in 1500 papers in refereed journals already, and the number is still increasing rapidly. The brightest object observed with IUE was Venus at V = - 4 mag, and the faintest object observed up to now was a 20-th magnitude central star of a Planetary Nebula. It is impossible for me to mention or discuss in this review all the results of IUE, and the impact IUE has had on almost every field of astronomy: planetary astronomy, interstellar medium, stellar evolution, galactic and extragalactic astronomy. I will limit myself to those things which have been learned from IUE in the field of stellar astronomy. I will use the evolution of the stars as a guideline to discuss the results of IUE. The evolution of a low mass star and a massive star are shown schematically in Fig.1.

The low mass star burns hydrogen on the main sequence. Then developes a degenerate helium core with a hydrogen burning shell when the star is a red giant. After the ignition of helium in the center of the star, the star developes a carbon/ oxygen core, surrounded by two shells, a He- and a H-burning shell. In this phase the star is on the Asymptotic Giant Branch, AGB, and ejects an enormous amount of mass. When the star has ejected most of its H outer envelope due to pulsation, the core contracts and the star moves to the left in the Hertzsprung-Russell diagram. When the star is hotter than about 30000 K, its ultraviolet radiation will ionize the material which was ejected previously and the star evolves into a hot central star of a planetary nebula. At this stage, the two burning shells loose their efficiency and finally the star cools down and becomes a degenerate white dwarf.

For the most massive stars, say $M > 40\ M_\odot$, the evolution is quite different. During the main sequence phase, when the star is burning H in its core, the star looses about 10 to 20 percent of its mass

Reviews in Modern Astronomy 2 Editor: G. Klare

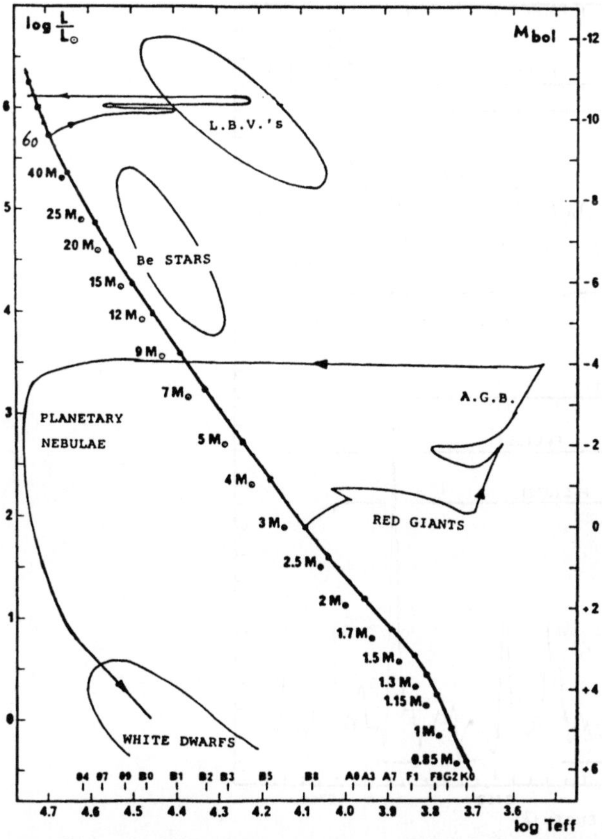

Fig. 1: A schematic picture of the evolutionary tracks of a low mass star and a massive star. The results of IUE will be discussed in the context of these two evolutionary tracks

in a stellar wind of about 10^{-6} $M_\odot$/yr. After the H-core burning when the star starts to expand, the mass loss increases drastically to 10^{-4} to 10^{-5} $M_\odot$/yr. For some reason which is not understood the mass loss becomes very chaotic. The star is ejecting matter in eruptions as a Luminous Blue Variable. During this phase, and the previous main sequence phase, the star can get rid of as much as 30% to 40% of its initial mass. What is left over then is a helium-rich star with very little hydrogen remaining in the envelope. The star is then a hot Wolf-Rayet Star.

We will follow the evolution of these two stars through the ultraviolet eyes of IUE, starting with the low mass stars.

2. CHROMOSPHERIC ACTIVITY IN COOL STARS

Low mass stars are cool stars throughout most of their life, and it is surprising to see that IUE has made such a significant contribution to our knowledge about these cool stars. One might have expected that IUE would be mainly used to observe hot objects which emit most of their radiation in the UV. The reason for the success of IUE for the cool stars can be demonstrated most easily by looking at the UV

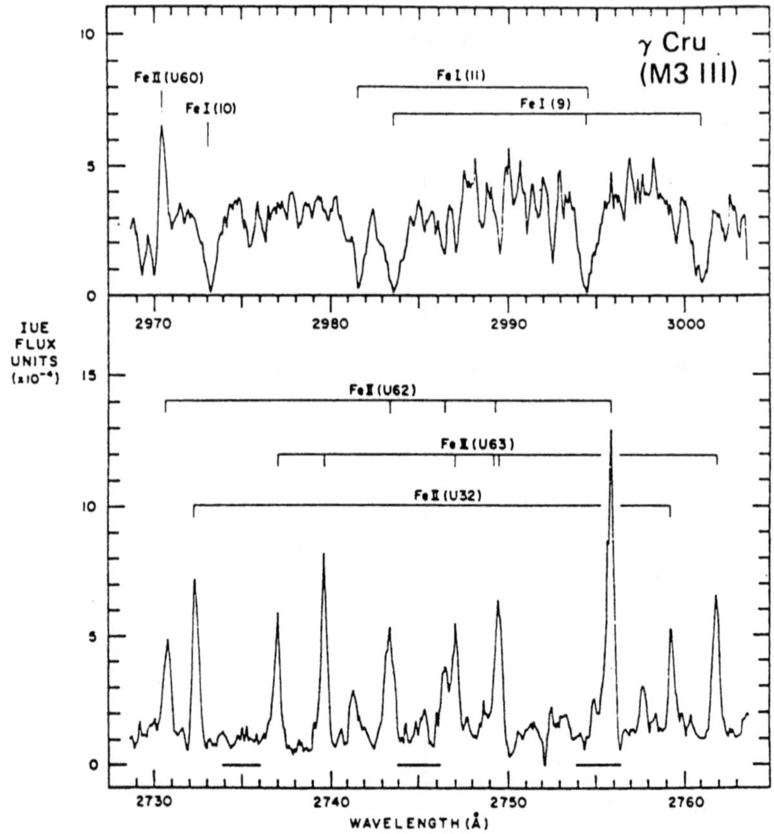

Fig. 2. IUE high resolution spectra of γ Cru illustrating in this giant the change around 2900 Å from a photospheric absorption line spectrum at $\lambda > 2970$ Å to a chromospheric emission line spectrum at $\lambda > 2730$ Å (Böhm-Vitense and Querci, 1987)

spectrum of a typical M giant which, after the burning of hydrogen in its core, has developed into a hydrogen shell burning star.

Fig. 2 shows the spectrum of the star γ Cru (M 3 III), observed with IUE in a wavelength range between 3000 Å and 2700 Å. The top part of the figure shows the spectrum between 2900 Å and 3000 Å. In this wavelength range the spectrum is typically a photospheric spectrum, consisting of a continuum with absorption lines, which in this case are due to Fe I. The lower part of the Fig.- shows the spectrum, below 2800 Å. It is completely different in signature. There is a very faint continuum, and the spectrum is dominated by strong emission lines mainly due to Fe II. At these shorter wavelengths, the photospheric spectrum has decreased in intensity, we do not recognize it anymore, but the chromospheric emissions become clear now. So by going to shorter wavelength from the optical spectrum into the ultraviolet spectrum of late type stars, we see the change of the spectrum from a photospheric to a chromospheric spectrum! This means that IUE is a perfect instrument for providing information about the outer layers of cool stars, i.e., their chromospheres.

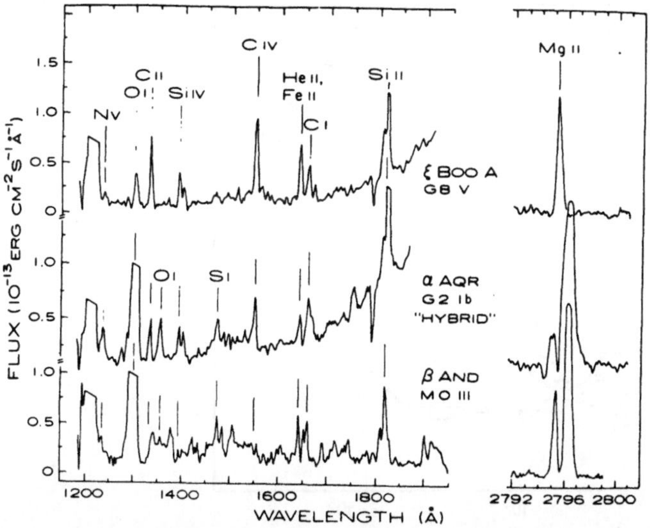

Fig. 3. IUE spectra of three cool stars showing different chromospheric emission features. ξ Boo is an active dwarf with high ionization emission lines (C IV, N V) and symmetric Mg II lines, indicating no mass loss; β And is an M0 giant with low ionization emission lines (O I, etc.) and strongly asymmetric Mg II lines, indicating high mass loss; α Aqr is a G2 supergiant with hybrid chromospheric emission lines (C IV, N V, O I, etc.) and asymmetric Mg II lines (Dupree, 1986)

Before IUE, the only source of information about the chromospheres, or I should say chromospheric activity in cool stars, were the visual lines of Ca II, Na I and the Balmer lines of H I. With IUE, it is possible to study many more ions in the chromospheres of stars: ions with very different ionization and excitation potentials, This in turn makes it possible to determine the structure of the chromospheres by studying the amount of material in the different temperature regimes through the atmospheres of cool stars.

Fig. 3 shows an example of the large differences that are found in the chromospheric activity of the cool stars. All three stars show emission lines in the short λ spectrum observed with IUE and the spectra of the two G stars ξ Boo and α Aqr also show an increase of the flux at $\lambda > 1700$ Å due to the photospheric flux. The M star β And shows mainly emission lines of low ionization stages only, such as O I, Si II and Fe II. The G8 V star ξ Boo has lines from higher ionization stages such as C II, C IV and N V. The G2 Ib star α Aqr has emission lines of both high and low ionization.

Since the emission lines are due to recombination or collisional excitation followed by photo-deexcitation, their strength can be expressed in the product of the emission measure and a known function of the temperature. It is thus possible to model the chromosphere of a star by studying the strength of the lines which arise in different temperature regimes.

Fig. 4 shows a result of such a modelling, a model of the chromosphere of the G8 V star ξ Boo . It shows the distribution of the emission measure as a function of temperature from about 8000 K up to 3 10^5 K in the chromosphere of this star.

We know from detailed studies of the sun that the chromospheric emission and chromospheric activity of late type stars is closely related to the magnetic activity which in turn depends on the internal

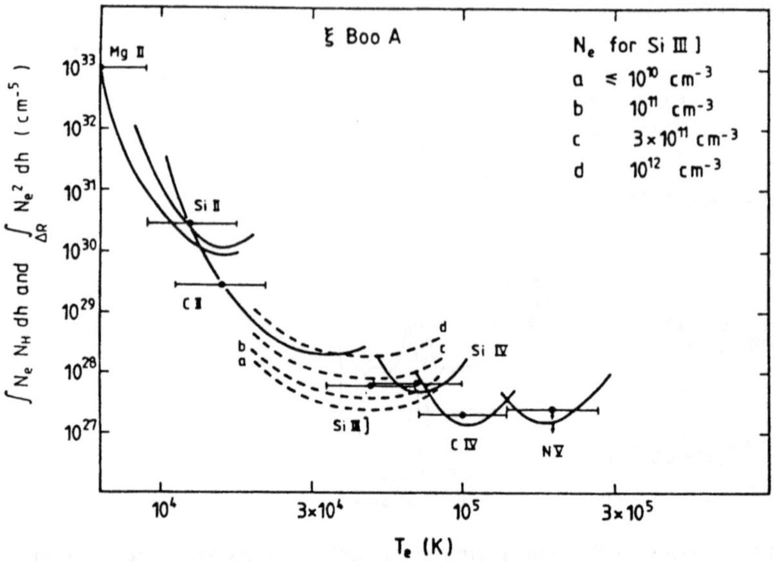

Fig. 4. The distribution of the emission measures of different chromospheric lines as a function of temperature derived from IUE observations of ξ Boo (G8 V). From such a distribution the temperature and density structure of stellar chromospheres can be derived (Jordan and Linsky, 1987)

structure of the star via the convection and the dynamo mechanism. So by studying the chromospheric structure of cool stars it is possible to obtain information about the effects which are occuring inside the cool stars.

3. CHROMOSPHERIC ACTIVITY AND ROTATION

Some of the cool stars have very active chromospheres because these stars have large surface magnetic fields. In most cases this is due to the fact that the star is a rapid rotator.

An example of such an chromospheric active star is V471 Tau. This K2 V star has strong chromospheric emission lines. The star is a rapid rotator which is due to the fact that it is in a close binary system with a white dwarf companion. The companion has been used to trace the chromospheric structure of the K star by means of IUE observations in a unique way (Fig. 5a and b). The ultraviolet light of the hot white dwarf is used as a light source which can be monitored when it passes behind the K dwarf. If the K dwarf is surrounded by an active chromosphere, then before the white dwarf disappears behind the photosphere of the star, the spectrum should show the absorptions which are formed in the extended chromosphere of the K star. The observed spectrum does indeed exhibit strong absorption lines due to hot material in the chromosphere of the K star which absorbs the light from the white dwarf. These chromospheric lines are due to ions such as Lyman-α, Si II, Si III, O I, Si IV, C IV, etc. By studying the change of these absorptions as the white dwarf passes behind the K star, on can monitor the distribution of the material around the K star. In this case, the observations show that the K star is surrounded by a chromosphere which probably consists of large prominences. These prominences seem to be similar to those found in the sun, but they are

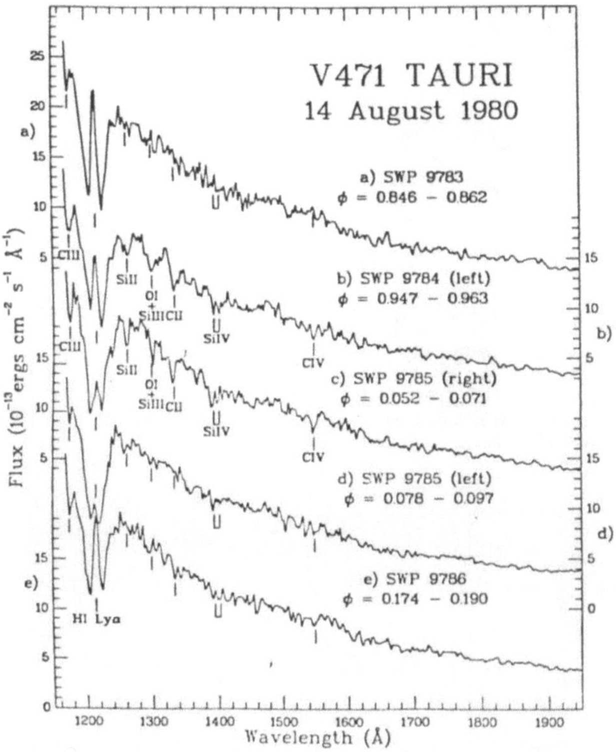

Fig. 5a. Carefully timed short wavelength IUE spectra of the eclipsing system V471 Tauri showing the increasing strength of absorption features of Si II, O I, Si III, Si IV, and C IV as the white dwarf passes behind the active K-dwarf star

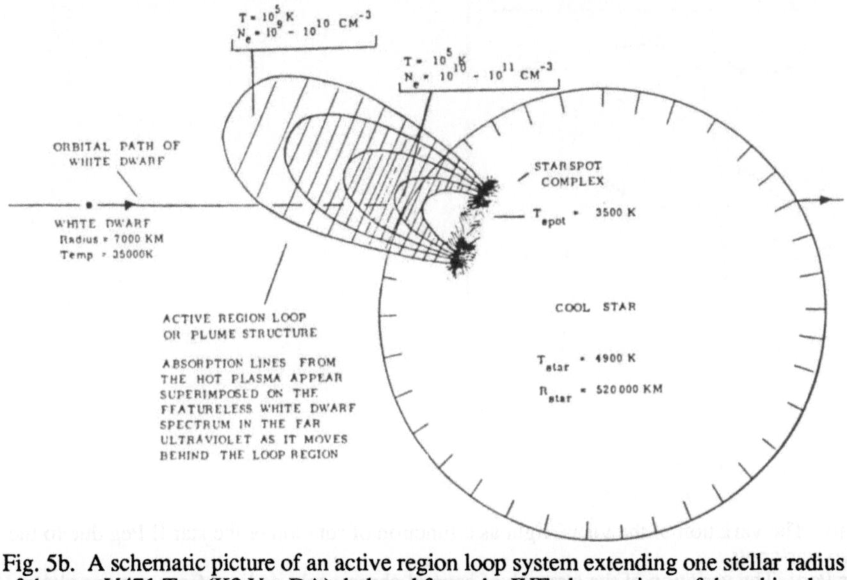

Fig. 5b. A schematic picture of an active region loop system extending one stellar radius above the limb of the star V471 Tau (K2 V + DA) deduced from the IUE observations of the white dwarf as it passed behind the K2 V star on the indicated trajectory (Guinan et al. 1986)

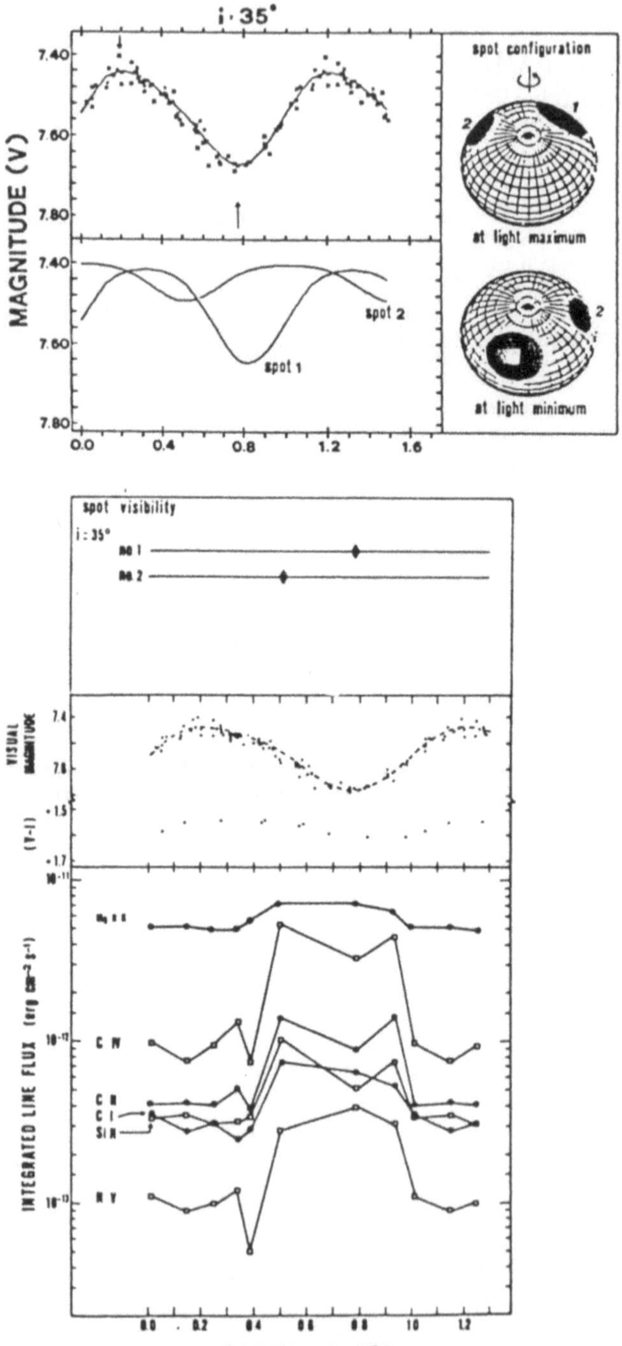

Fig. 6. Top Part: The variation of the visual light as a function of rotation of the star II Peg due to the presence of two star-spots

Lower Part: The variation of the strength of several chromospheric lines from regions above the star-spots as the star rotates (Rodono, 1988)

considerably more extended since as they reach a height of about one stellar radius above the photosphere. The structure of these prominences, at least in the one dimension which is scanned by the path of the white dwarf, can be observed and inferred directly from the IUE observations (see Fig. 5b).

For the most active cool stars the activity seems to be concentrated in star-spots, similar to those in the solar photosphere, but much more intense. These star-spots can be modelled by means of Doppler-mapping, a technique that has been successfully applied to IUE observations of a number of cool stars. Results are shown in Fig. 6 for the star II Peg, where the activity is found to be concentrated in two star-spots on the surface of the star, and the chromospheric emissions arise from regions just above these star spots. The location of the spots on the surface and the rotation of the star determine the wavelength and the Doppler shifts of the emissions formed above the spots. As a result one will observe chromospheric emission lines which are not symmetric in wavelength, but consist of several components. Each component is formed by one star spot, and is observed at the wavelength corresponding to the Doppler velocity of the spot at the rotating surface of the star. The figure shows the spectrum of II Peg in which two star-spots have been recognized. These-star spots can be followed when the star rotates by following the shift in the wavelength of their components. In this way the location, the number, the magnetic strength and the chromospheric activity of these star-spots has been determined for this star as well as for several other cool stars.

Up to now, information about star-spots and their location was only available for the sun. With IUE it has been possible to extend these studies to more active late type stars which provide information about the effects below the stellar surface, which are responsible for the magnetic activity. IUE observations have shown that stars with similar spectral type can show very large differences in their chromospheric activity. We now know that this is mainly due to the rotation of the stars. In fact, it has been shown that the strength of the chromospheric emission lines which can be used as an indication of the strength of their magnetic activity, is correlated with the Rossby number, i.e., the ratio between the rotational period and the characteristic rise time for convective elements in a convection zone below the

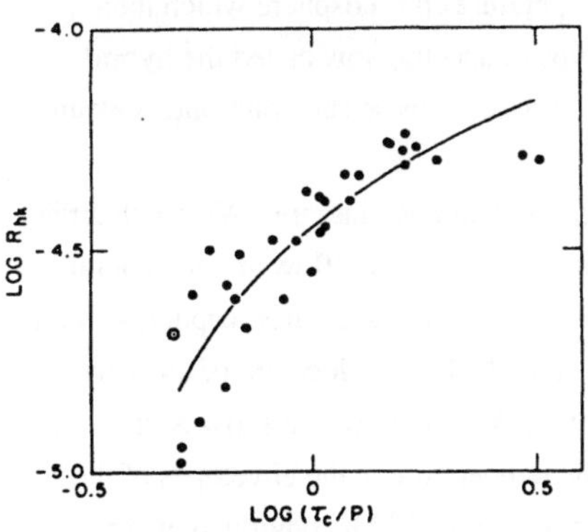

Fig. 7. The strength of the chromospheric emission lines of Mg II for cool stars as a function of the inverse Rossby number. The Rossby number $R_o = P_{rot}/\tau_c$ is the ratio between the rotational period P of the star and the rise-time τ_c of the cells of the convection layer. Large values of τ_c/P imply twisting of the magnetic field in the convection zone due to differential rotation. The observed relation shown in this figure between the chromospheric lines and τ_c/P supports the theory that the chromospheric activity is due to magnetic fields from the convection zones (Hartmann et al. 1984)

photosphere. When the ratio between the rotation time and the convective rise time is small, then the Coriolis forces can produce a severe twisting of the rising cells in the convection zone, and this, as we know, is one of the ingredients of the dynamo mechanism. So when the rotation is large, the dynamo mechanism can be more efficient, which means that the magnetic flux will be large at the stellar surface, and as a result of this, the star will show a stronger chromospheric activity. Fig. 7 shows that there is indeed a strong correlation between the chromospheric activity of lines such as C IV, Si IV and N V and the Rossby number. This relation provides very strong evidence for a dynamo mechanism which is a result of the interplay between the convective rising cells and the rotation of the star.

4. CHROMOSPHERES AND WINDS OF COOL GIANTS AND SUPERGIANTS

Up to now we have looked at late type stars on or just off the main sequence. Let us now look at slightly advanced stages of evolution. When we consider the low mass stars at different stages of evolution and look at their chromospheric spectra, it turns out that we can distinguish three types of stellar chromospheres. The first type is mainly found along the main sequence, or in the slightly evolved low luminosity stars. They show chromospheres which are characterized by low excitation lines such as O I and C II, and high excitation lines, such as C IV, Si IV, usually called transition region lines indicating $T \simeq 10^5$ K. There is no indication of a stellar wind. Even the stronger chromospheric lines, for instance those of Mg II, are not shifted and strictly symmetric.

A completely different type of chromospheric spectrum is found for the more luminous and more evolved stars, for instance the M0 supergiants. In this case the chromospheric spectrum shows only the low excitation lines with no sign of the high excitation lines. This indicates that the temperature in the chromosphere does not rise above 10000 or 20000 K. For these stars, however, the Mg II lines are usually shifted or split by a central absorption which is shifted to shorter wavelengths. This clearly indicates that these stars are loosing mass at a considerable rate. So the difference between these two major types of spectra are that the one class has high and low excitation lines with no mass loss, i.e., a stable hot chromosphere, whereas the other type has a low temperature chromosphere which then develops into a stellar wind. A number of stars have a third type of spectra, now called the hybrid spectra, which fall in between the previously mentioned two classes. In these stars one finds low and high excitation lines and also an indication for a stellar wind.

Fig. 8 shows the distribution of these different kinds of spectra in an HR-diagram. We see that the cooler, more luminous stars have chromospheres which are characterized by outflow and in general rather low temperatures, whereas the less luminous stars or the hotter stars have chromospheres which reach temperatures of a few 10^5 K. If in the same diagram we then look at the location of the stars which show X-rays from a corona, it is not surprising to see, that those stars which show X-rays are the same ones as those which show hot chromospheres. So the temperature in the envelopes of these stars, which increase outward from the photosphere to about a few times 10^5 K, obviously increases to even higher temperatures in the layers above the chromosphere and then form the typical coronal layers of a few 10^6 degrees which are responsible for the X-rays.

If we now combine all this information which has been derived from IUE and X-ray observations about the chromospheres of cool stars, we arrive at the following picture: in cool stars the convection

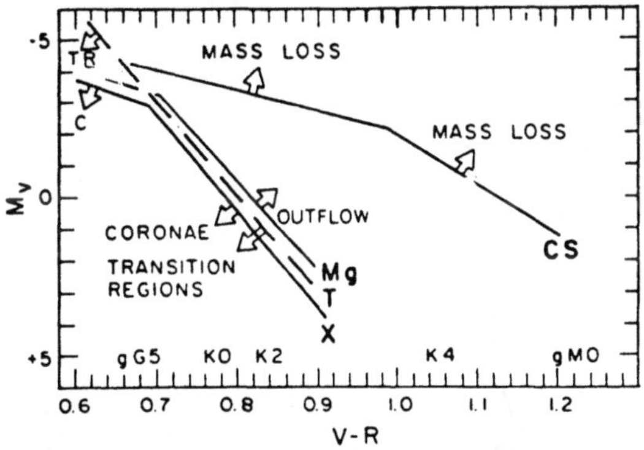

Fig. 8. The location of stars with high mass loss, and with chromospheric lines from the transition region (T ~ 10^5 K) are shown in an HR diagram. High mass loss rates are observed in the upper right hand side = cool giants and supergiants. Transition regions, detected from IUE C IV lines and coronae, detected by X-rays occur in the lower left hand side = G and early K dwarfs and giants (Mullan and Stencel, 1982)

zone below the photosphere can be twisted by the stellar rotation and then forms a magnetic field by the dymano mechanism which is stronger for rapidly rotating stars. This magnetic field produces star spots and prominences with hot plasma which are responsible for the chromospheric emission lines. In those stars which have a high gravity, e.g. in the main sequence stars, this hot plasma can be retained and so the chromosphere is more or less stable. However, in the more evolved low mass stars, which have expanded to a much lower surface gravity, the hot plasma cannot be retained by the gravity and results in an outflow in the form of a stellar wind.

This stellar wind is most obvious in the late type supergiant stars, and in fact IUE has increased our knowledge of the mass loss from these late supergiant stars very drastically. Fig. 9 shows an example of how IUE observations have provided new information about the winds of late type supergiants. The K supergiant 32 Cyg is a binary with a B star companion. This B star has been used as a light-probe for measuring the distribution of matter around the K supergiant. When the spectrum of the B star is studied in various phases of the orbital motion around the K star, the wind of the K-star produces absorption lines in the spectrum of the B star due to the material along the line of sight to the B star. From the strength of the lines and the wavelenght shifts of the absorption lines, one can determine the amount of material in the line of sight to the B-star and its velocity in the direction to the observer. By following the B star in its orbit around the K-star it is possible to probe the densitiy and velocity distribution of the wind of the K-star at various distances from the star. From this study, a model in terms of the distribution of density as a function of the distance for the wind of the K-star has been derived. This is shown in Fig. 9, lower part.

Let us go to even later stages of the evolution of low mass stars, the Mira stars, which have a Helium and a Hydrogen burning shell. These stars at the top of the AGB are pulsationally unstable with periods in the order of 100 to 1000 days. They have very high mass loss rates on the order of 10^{-5} $M_{\odot}$/yr. It had been proposed that the high mass loss of the Mira stars is driven by their pulsation, and there are models which predict that a pulsating star produces shocks in the photosphere which

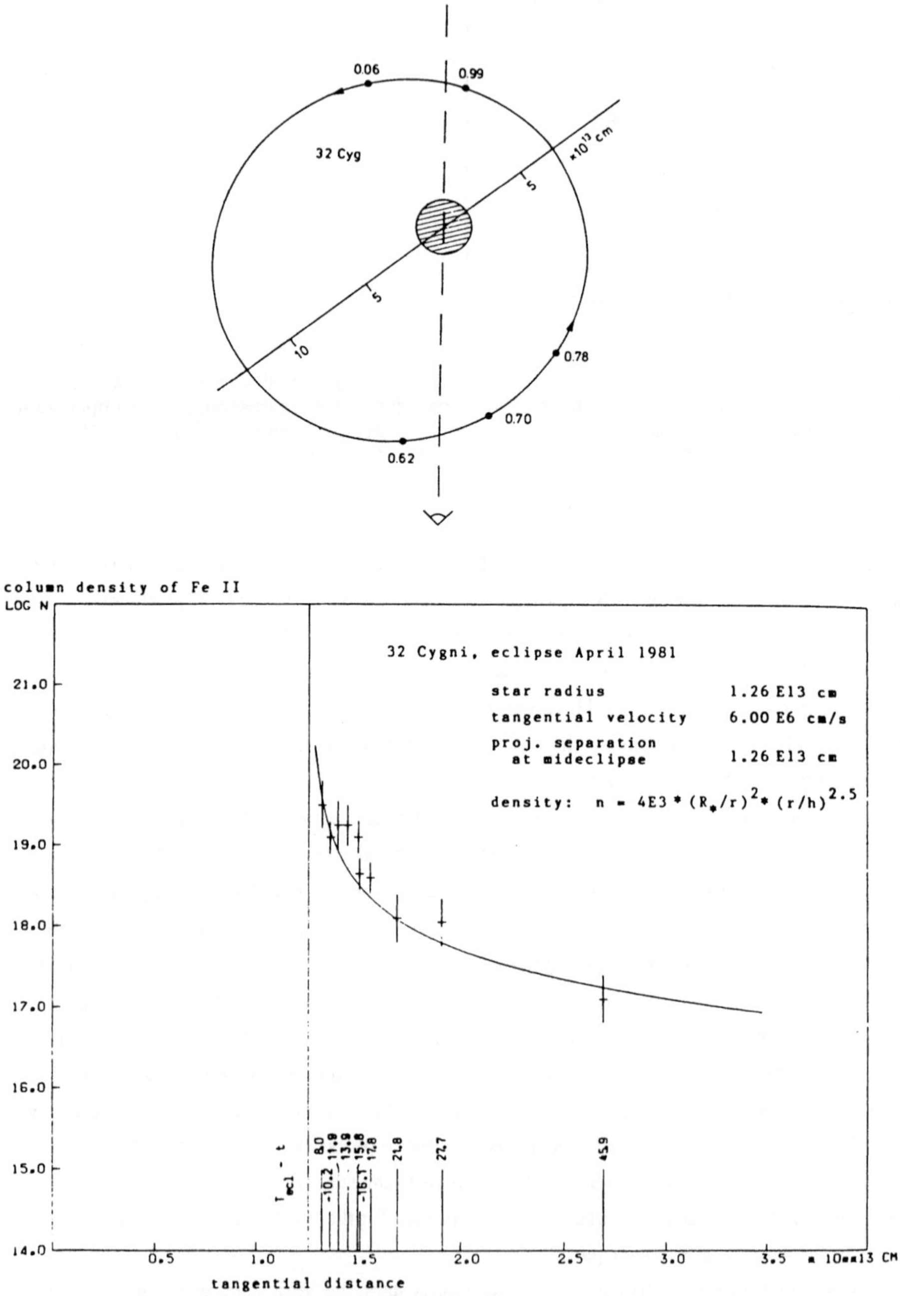

Fig. 9. Top Part: The orbital phases of the IUE observations of the B star relative to the K supergiant (roughly on scale) of the system 32 Cyg

Lower Part: The density distribution of the wind around the K5 Iab star 32 Cyg, derived from the absorption of Fe II lines in the IUE spectrum of the B-type companion during its orbit around the K star. (Dupree and Reimers, 1987)

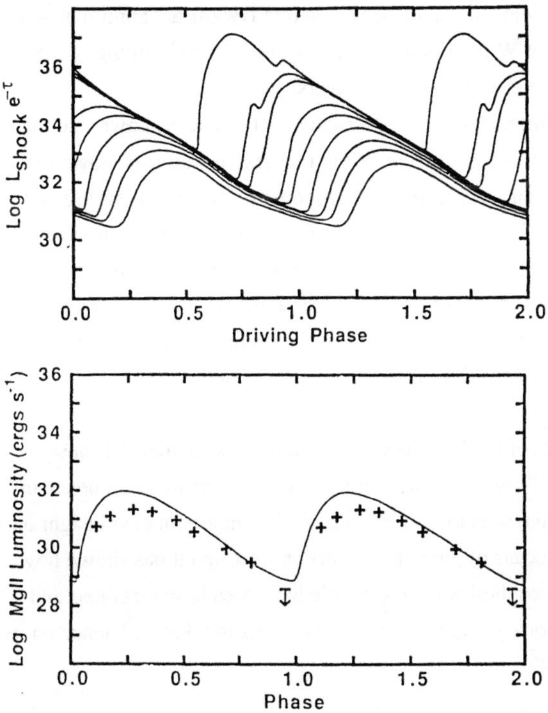

Figure 10. Top: Predicted emergent flux at 2800 Å assuming that most of the shock flux L_S emerges in the MG II h and k lines, for κ (2800 Å)/κ_R = 1, 50, 100, 500, 1000, 2000, 4000, 6000, and 10,000 Bottom: Fit to observed Mg II flux for κ/κ_R = 1000 (Brugel et al. 1987)

temporarily heat the gas and provide energy for the outflow. The IUE observations shown in Fig.10 show a beautiful confirmation of this theory. They show the strength of the Mg II emission lines which are a measure of the amount of the heated material as a function of the phase of pulsation. It is clear from this figure that the Mg II line strength shows a variation very close to what was predicted from the pulsation driven mass loss model. So the IUE observations of the Mg II emission lines now provide strong evidence that the mass loss of the Mira stars is indeed due to pulsations which produce shocks, which produce hot gas, which produce the Mg II lines, and then finally these shocks develop into an outflow in the higher layers.

5. WHITE DWARFS

Now we follow the evolution of the low mass stars to even later stages, when most of the Hydrogen layers have been ejected by means of the Mira pulsation mechanism. The remaining star contracts, becomes very hot and finally develops into the central star of a planetary nebula and later cooles down to become a White Dwarf.

The IUE observations have shown very interesting and unexpected results about White Dwarfs. One of the most surprising results of IUE observations of the White Dwarfs is the very large range of chemical abundances. It was known that the chemical abundances in White Dwarfs are abnormal and that this is due to the fact that the high gravity of White Dwarfs produces gravitational settling of the heavier elements in deeper layers and the light elements in the top layers.

The IUE observations have shown a much more complicated structure. It has been shown that the range of chemical peculiarities is much larger than predicted in White Dwarfs. It was also shown that some White Dwarfs suffer mass loss despite their large gravity. Consequently, when measuring and interpreting the chemical abundances on the surface of White Dwarfs, one now has to take into account three competing processes: gravitational settling, radiation pressure on selective ions which can produce outward diffusion some specific ions, and mass loss. These three processes together are in some way, which is not clear at all at the moment, responsible for the surprising variety and the characteristics of White Dwarfs that have been derived from spectroscopic studies.

To conclude the summary of the IUE results of the low mass stars: we have seen that IUE has increased our knowledge about the evolution of low mass stars all the way from the main sequence up to the end products, the White Dwarfs. We have seen that IUE has provided important new insight into the internal structure of the late type stars by means of the dynamo mechanism, and it has shown how the outer layers of the late type stars are chromospherically active. We have seen how mass loss and chromospheric activity depends on the evolutionary stage of the star and on its rotation and hence on its evolution.

6. VARIABLE MASS LOSS FROM O AND B STARS

We now turn our attention to the massive stars. I will discuss how IUE observations have provided us information about the physical processes in the outer layers of the massive stars and how our knowledge about these effects has given new insight into the evolution of the massive stars.

Long before IUE previous ultraviolet instruments in rockets, balloons, the TD1 satellite and the Copernicus satelite had already shown that massive stars suffer mass loss, i.e., they eject their outer layers in a stellar wind. The terminal velocities of the winds are typically between 1000 and 3000 km/sec when the stars are hot (T > 20.000 K). This mass loss continues throughout most of the evolution of the massive stars. These previous observations had shown that the mass loss depends on the luminosity of the star as a rather steep function, $\dot{M} \sim L^{1.6}$, and that the mass loss is typical on the order of 10^{-6} to 10^{-5} $M_\odot$/yr. Such a high mass loss rate over a main sequence lifetime on the order of several million years, immediately implies that the mass loss will effect their evolution because the stars can lose 10 to 30 percent of their mass. When the star has lost a considerable fraction of its mass the nuclear products which are formed in the stellar interior and brought to higher layers by convection may reach the photosphere. Hence, in various phases of massive stars we find chemical abundances which differ from those of typical population I stars. All this was well known before IUE was launched. The main contibution of IUE to our knowledge of massive stars is the insight that it has provided into the physical mechanism of the mass loss phenomenon and its origin during the various evolutionary stages.

One of the most important results of the mass loss studies with IUE is the fact that the IUE observations have shown that mass loss in luminous stars is highly variable.

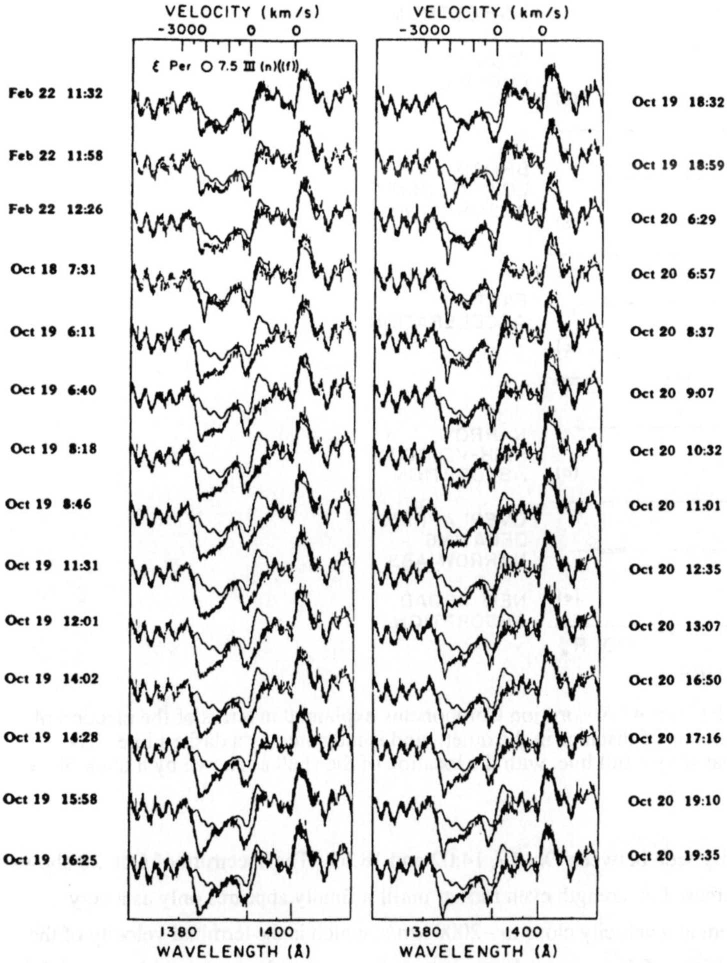

Fig. 11. Variations in the Narrow Absorption Components of the Si IV resonance lines of ξ Per 07.5 IIII ((f)). The long tick marks at the horizontal axis indicate the laboratory wavelengths of the two components. The two short tick marks indicate the wavelengths of 1380 and 1400 Å. The shape of the underlying P Cygni profile is shown for comparison in each observation. Notice the appearance of a strong absorption at v $\simeq$ - 2100 km s^{-1} on Oct 19 6:11 and its gradual shift to shorter wavelength (v $\simeq$ 2100 km s^{-1}) and narrowing within one day. On Oct. 20 8:37, a new component appears at v $\simeq$ -1000 km s^{-1} (Prinja et al. 1987)

Let me show you an example of the variations of the UV resonance lines of Si IV in the typical O-type star ξ Per. Figure 11 shows the profiles of these lines over a period of about half a year, but concentrated in a period of only 3 days from Oct 18 to Oct 20, 1985. We see in these profiles, if you look at them carefully, the appearence and disappearence of narrow absorption components which change in character and in velocity as time progresses. For instance the spectrum of Oct 19, 6:11 shows a deep and rather wide absorption component appearing at a velocity of about - 1000 km/s. As time progresses, this wide absorption component shifts to more negative velocities and becomes

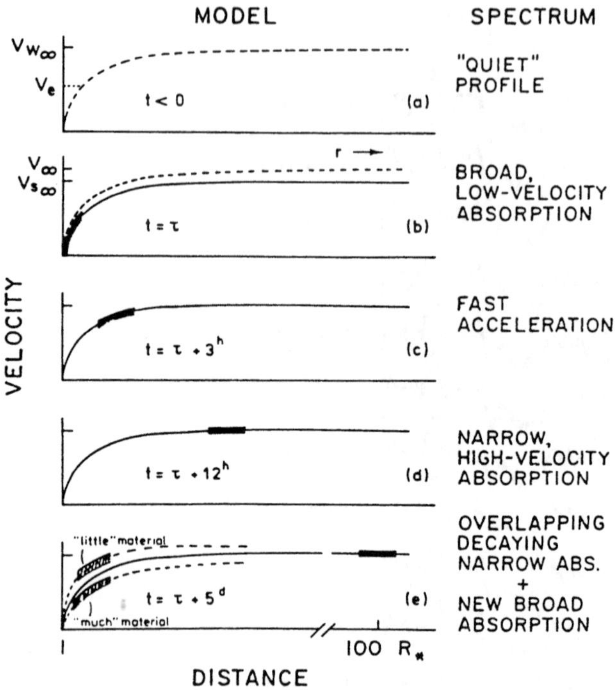

Fig. 12. The variations of the Narrow Absorption Components explained in terms of the ejection of shells. The velocity as a function of distance in the quiet wind is indicated by a dashed line. The velocity of the shell is indicated by a full line, with the location of the shell as shown by a thick block (Henrichs et al. 1983)

narrower. This is most easily seen between Oct 19, 14:02 and 18:59. The spectrum of Oct 20 shows that this component has decreased in strength even further untill it finally appeares only as a very narrow absorption component at a velocity close to - 2000 km/s, which is the terminal velocity of the wind of this star. This behaviour of the narrow absorption component can be explained in terms of the ejection of a shell from the star (see Fig. 12). When the shell is just ejected and is still at low altitude above the photosphere, its outer parts are rapidly accellerated by radiation pressure. So the shell covers a wide range of velocities from almost 0 km/s up to about the velocity which is reached by the outer edge of the shell. This produces a wide, deep absorption component, observed on Oct 19, 6:11 in the UV resonance lines.

As time progresses and the shell is accellerated further, reaching larger distances and higher velocities, the velocity dispersion within the shell decreases because the outer layers of the shell and the inside of the shell finally will reach the same terminal velocity. In the spectrum of the star the UV resonance lines show that the shell-components are moving to higher negative velocities and becoming narrower.

When the shell finally reaches its terminal velocity and keeps moving outward with an almost constant velocity, the column density decreases and so the absorption components, which are then narrow, become weaker untill they finally disappear. We can see this in Fig. 11 where this occurs

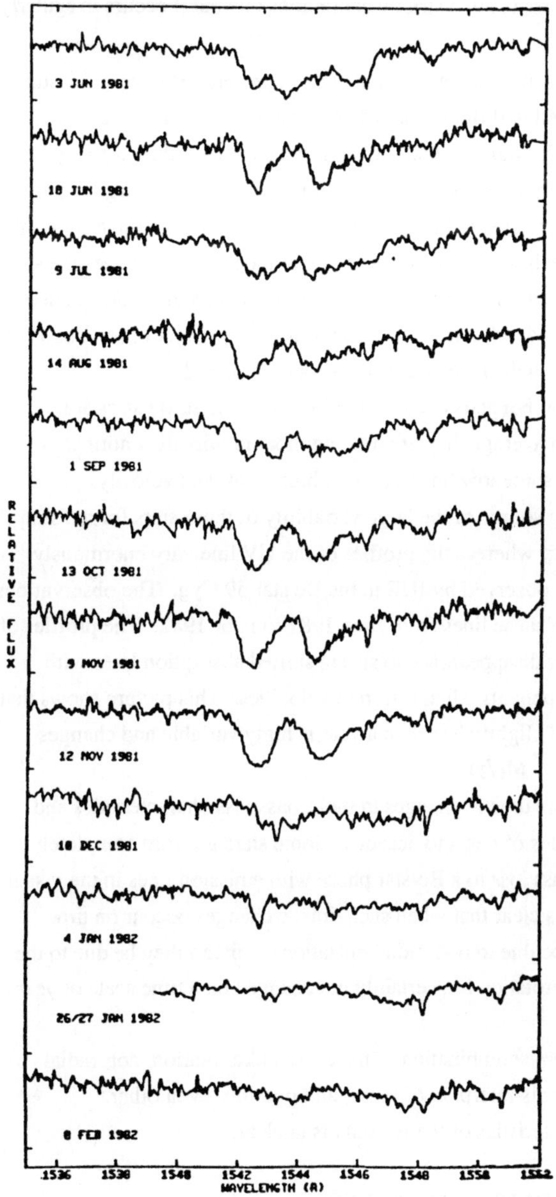

Fig. 13. Variations in the C IV profile in 59 Cyg. This time sequence, covering a nine-month period, illustrates many of the modes of variation seen in Be star ultraviolet line profiles (Doazan et al. 1985)

around Oct 20, 6 to 11 hours. However, at that time a new shell has already been been ejected which again appears as a wide, rather broad absorption component at lower velocities. So we see from these IUE observations of ξ Per. that the star ejects shells which move away from the star and can be followed for a period of a few days. After a few days a new shell is ejected. These observations clearly demonstrate that the process of mass loss is not a strictly continuous one but it occurs irregularly by the ejection of many shells.

The reason for this shell ejection, the physical cause of it, is not clear at present. But from the study of the time scales at which the shells are ejected in different stars there is an indication that the ejection process is related to non-radial pulsations of the stars. If that turns out to be the case then we may conclude that the mechanism of mass loss from hot stars is due to a combined effect of a large radiation pressure and a variability of the underlying star. It should be mentioned however that the nature of this relation between mass loss and a non-radial pulsation is not clear at the moment. In some stars the shell ejections may be in phase with non-radial pulsations themselves while in other stars the shell ejection appears to be related to the changing of the mode of non-radial pulsations. Very recent observations of a few stars suggest that the appearance of the shells is related to their rotation period.

The most dramatic changes in the spectra of hot stars occur in the Be stars. These Be stars are usually rather close to the main sequence. On average they are rotating more rapidly than normal B stars and it is assumed that their variability is somehow linked to their high rotational velocity. However, rotation alone cannot be the driving source of the large variability of these stars for the simple reason that the rotation of a stars does not vary whereas the profiles of the UV line vary enormously. Fig. 13 shows the example of the C IV lines, observed by IUE in the Be star 59 Cyg. The observations show an enormous variation in the profiles of these lines from June 1981 to Febr 1982. The profiles of the C IV lines change from almost completely disappearence to strong shifted absorption lines with wide components, or sometimes narrow components, all at different velocities. This picture shows that the mass loss of this Be star on a time scale of slightly less than a year is higly variable and changes from the detection limit to about 10^{-9} to 10^{-7} $M_{\odot}$/yr.

Be stars are known to go through even more drastic changes in their mass loss characteristics and their photospheric characteristics on time scales of years to decades. Some stars go from completely normal B type characteristics without any mass loss to a Be-star phase with emission lines in the visual spectrum indicating high mass loss rates. It is clear that when such drastic changes occour on time scales of years to decades, that they can not be due to non-radial pulsations, nor can they be due to the fact that the star is a rapid rotator because the rotation has certainly not changed on a time scale of years to decades.

It is possible that the changes are due to some combination of the effects like rotation, non-radial pulsations, and radiation pressure. But how this interplay of these, and possibly even other, mechanisms can produce the changing characteristics of the Be-stars is unclear.

7. MASS LOSS FROM LUMINOUS BLUE VARIABLES

We now follow the massive stars to slightly later stages of evolution. The stars have evolved away from the main sequence, and they are now burning Hydrogen in a shell around a Helium core.

The distribution of the luminous stars ($L > 10^4$ $L_{\odot}$) in our Galaxy and also in other galaxies show that stars with initial masses higher then about 50 or 60 $M_{\odot}$ do not evolve to the upper right hand portion of the HR-diagram. The stars are expected to increase their radius when the core contracts during the H- shell burning phase, so all original evolutionary calculations predicted that these stars would evolve into Red Supergiants. However, the most luminous stars never seem to become a red supergiants, because the part of the HR-diagram above an absolute bolometric magnitude of M_{bol} = - 9.5 and temperatures cooler than 8000 K is empty. So obviously the stars do not reach a stage where they develop extended convective H-layers as red supergiants. The only way for stars to avoid becoming supergiants after the H-burning phase in the core is to get rid of almost all of their remaining outer H-layers. Evolutionary calculations have shown that the stars have to get rid of about one third of their mass if they want to avoid becoming red supergiants. Obviously they succeed in it!

During the main sequence phase the stars lose only 10 to 20 percent of their mass. So after the main sequence phase there must be a time when the stars are able to eject the other 10 or 20 percent, which is typically 5 to 10 $M_{\odot}$. They must get rid of this in a short time (10^4 - 10^5 years) before they develop into a red supergiant.

This phase has been identified with the Luminous Blue Variables, LBV's. The LBV's are luminous stars with $L > 10^5$ $L_{\odot}$ and $M_{bol} < -9.5$. They are variable on all time scales from weeks, to years, to decennia, and possibly even centuries.

Fig. 14 shows the variation of a typical LBV, the star AG Car from 1969 to 1985. It shows a large photometric variation of this star on many time scales whith an amplitude up to about 2.5 magnitude.

Spectroscopic observations of other luminous variables have shown that the stars change their spectral type during the photometric variations. The spectral types can change from a hot O-star to an A or F supergiant. When the LBV's are in the hot phase they are visually faintest and their mass loss rates are typically 10^{-5} $M_{\odot}$/yr. When they are A of F stars they are visually at maximum and their mass loss rate is at least an order of magnitude larger. We do not know which mechanism is able to produce such a large variation in photometric magnitude, in spectral type, and in mass loss rate.

The IUE observations, however, have shown a very interesting property which is an important clue to the variability mechanism of LBV's. IUE has shown that during these drastic variations in photometric magnitude and spectral type the total bolometric luminosity remains the same.

Fig 15 shows how the UV spectrum of AG Car is changing when the visual magnitude of the star is changing. The star is very faint in the UV when the star is visually at its brightest. When the star is visually at minimum, the UV fluxes are much larger. A combination of the visual and the UV energy

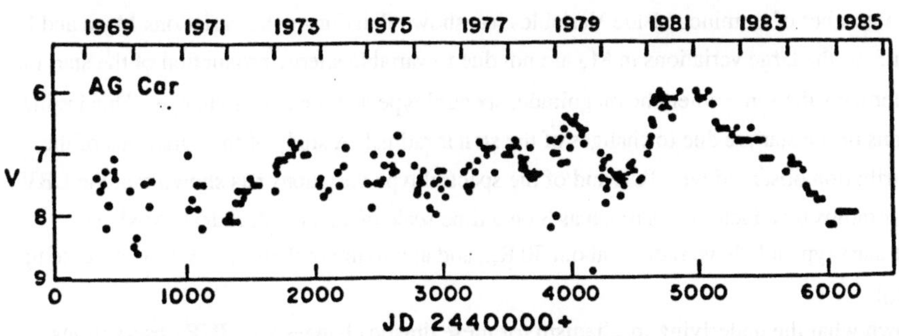

Fig. 14. The lightcurve of a typical Luminous Blue Variable, AG Car. The visual magnitude varies by 2.5 magnitudes in an irregular way. The IUE observations have shown that LBV's are bright in the UV when they are faint in the visual, in such a way that their bolometric magnitude is constant (Lamers, 1987)

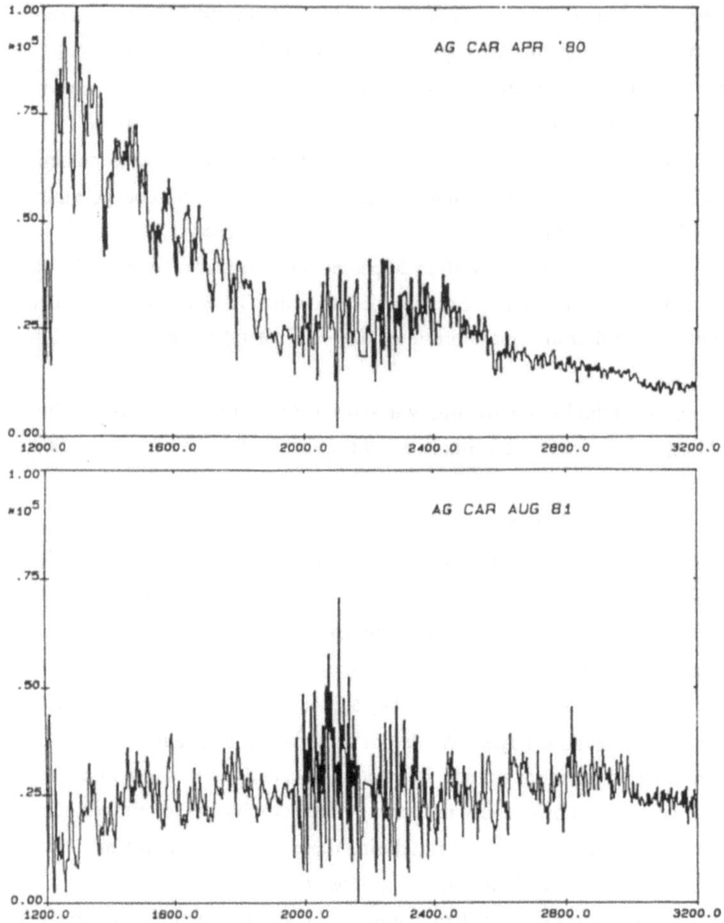

Fig. 15. The drastic changes in the UV energy distribution of the LBV AG Car, observed by IUE. In April 1980 when the star was visually faint (V = 7.14) the UV flux at $\lambda < 2000$ Å was considerably larger than in August 1981 when the star was visually bright (V = 6.40). The bolometric magnitude remained constant (Lamers et al. 1989)

distribution of a number of Luminous Blue Variables has shown that during the variations M_{bol} and L remain constant. So the large variations in M_V are not due to variable energy production of the star: the energy output remains the same when the magnitude, spectral type and the colors change. This implies that the variations of the star are due to changes of the stellar radius! A study of the variations of the UV energy distribution observed with IUE and of the spectral type variations has shown that the LBV's can change their radius by a factor of as much as 8 on a time scale of years to decades! At visual minimum these stars typically have radii of about 30 $R_\odot$, and at maximum their radius may increase up to about 200 $R_\odot$!

It is not known what the underlying mechanism for these drastic changes are. IUE observations have shown that the radius variations are correlated with large variations in the mass loss rate: when the radius of the star is large, the mass loss rate of the star is large.

What is the causal relation between the radius and massloss changes? Is there some mechanism in the star which produces periods of high mass loss rate, and when the rate of mass loss is very high, the photosphere moves outward and the star appears larger and of later spectral type? Or is the situation just the reverse? Is there some mechanism in the star which produces changes in the radius of the star, making the star larger or smaller on time scales of years to decades, and then, when the star is larger and the gravity is smaller, the mass loss rate is larger? We do not know the solution of this "chicken and egg"-problem. But we do know from IUE observations that all these variations must occur at constant luminosity.

We are still waiting for a good explanation for these variations on time scales of years to decades, but some of these stars also vary on time scales of weeks. They seem to eject shells on time scales of weeks to months. Some or possibly all LBV's suffer large outbursts where they eject as much as one solar mass of material during a big eruption. Such eruptions have been observed during the last centuries in the stars Eta Carinae and P Cygni. Other LBV's such as AG Car are surrounded by ring nebulae which suggest that they also had similar large eruptions in the past. The cause of these variations is unknown.

We do know, however, that during this highly unstable LBV-phase in their life the massive stars can get rid of as much as a few tens of solar masses. It is this LBV mass loss which makes that the massive stars cannot develop extended convective H-layers and evolve into red supergiants after their main sequence, because they lost most of the remaining H-layers. When the H-layers are lost the evolutionary track is reversed. Instead of moving away from the main sequence to the region of the red supergiants, the stars will contract into the hotter regions of the HR-diagram where the He-rich Wolf-Rayet stars are located.

8. SUPERNOVA 1987A

I have followed the evolution of the massive stars from the main sequence phase to the latest stages. Any talk about IUE would not be complete if we would not mention the IUE observations of Supernova 1987a.

Figure 16 shows the light curve of Supernova 1987a in the ultraviolet at wavelengths around 3100 and 2700 Å. The picture clearly shows the very rapid decrease of the ultraviolet flux on a very short time scale.

This very rapid decrease of the UV flux is due to the rapid cooling by the expansion of the ejected shell. The UV flux decreased by about a factor 1000 in the first 3 days. During these first 3 days a few highly resolution IUE observations were made in which the supernova was used as a bright background source to study the interstellar medium in our Galaxy along the line of sight to the Magellanic Clouds.

Prior to the first IUE observations of the supernova, the star must have gone through an even more drastic phase of the UV flash. This phase was not observed, but we can see the consequences of this UV flash by its echo produced in the circumstellar matter around the Supernova. Figure 17 shows the IUE spectrum of the Supernova, observed on Oct 8, 1987. The spectrum of March 1987 is subtracted. This differential spectrum shows the appearence of many emsission lines in the spectrum of the supernova between October and March 1987. The emission lines from hot ions, such as N V, Si IV,

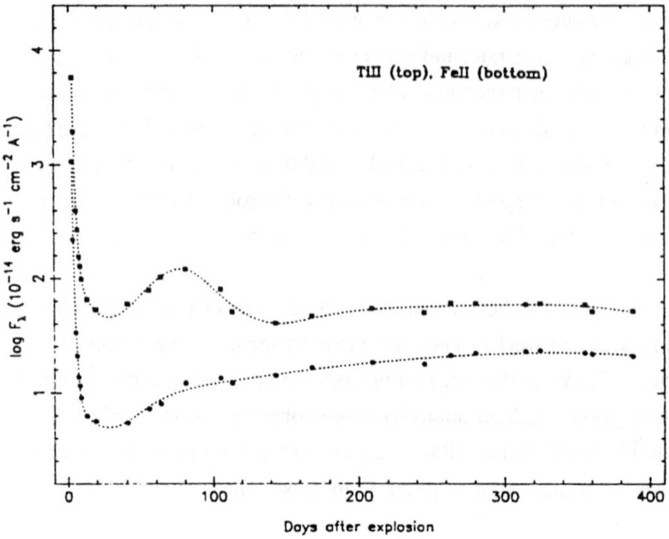

Fig. 16. The IUE light curve of the Supernova 1987A in two wavelength regions of the Ti II lines around 3100 A (top) and of the Fe II lines around 2700 A (bottom) (Sanz et al. 1988)

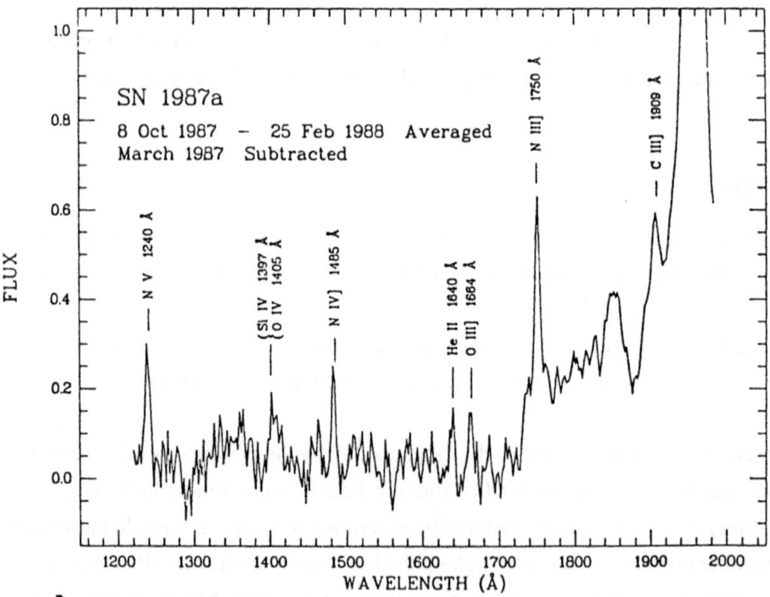

Fig. 17. The "echo" of the UV-flash of SN 1987A observed in the emission lines from excited circumstellar material about half a lightyear away from the supernova. These emission lines were found by subtracting the circumstellar spectrum of March 1987 from the circumstellar spectrum of Oct 1987 to Feb 1988 (Kirshner, 1988)

O IV, etc., are formed in the circumstellar medium around the Supernova. They are due to the fact that the radiation from the first UV flash ionizes the circumstellar matter which then produces these emission lines. So although IUE has missed the UV flash which is assumed to have occured immediately in the first stages of the supernova explosion we can see the echo of this flash. When this circumstellar spectrum is studied in detail, one can see that the abundance ratio is rather abnormal: the circumstellar matter has a high N over C ratio which is about 20 times as large as in the normal IS matter. This means that the matter which we see now is not interstellar matter of the Magellanic Cloud, but that it is material which must have been ejected from the star in the previous evolution when the star ejected N-rich layers. So by following the emission from the echo of the UV-flash one can derive the composition of the circumstellar material around the SN and thus trace back the previous mass loss history!

But there is more to come! The first echo occurred approximately half a year after the real explosion, so the matter which now produces the emission lines must be about half a light year away from the Supernova. The material which was ejected by the supernova explosion is travelling with a speed of about 10 percent of the speed of light. So one can expect that the material which was ejected by the supernova will reach this circumstellar matter in about five to ten years after 1987. When this happens, the circumstellar matter will be shock-heated to high temperatures and will produce many interesting highly ionized emission lines in the UV and also X-rays from the material around the supernova.

I am sure that by that time somebody else will be here speaking for the Deutsche Astronomische Gesellschaft to report on "A Decade of Research of the Supernova 1987a".

Acknowledgements: The text of this paper is a transcript from a tape with the review presented in Friedrichshafen on April 12, 1989. I am grateful to Michael Grewing and Renate Straumann in Tübingen who took care of the transcription, and to Sheri Pittman in Madison for preparing the final manuscript.

REFERENCES

Most of the results reported here are from the reviews of the various ESA and NASA IUE conferences and from the book:

Exploring the Universe with the IUE Satellite, 1987, Y.Kondo (Reidel; Dordrecht).

Boggess, A. and Willson, R.: The History of IUE, p. 3.

Bohm-Vitense, E. and Querci, M.: Intrinsically Variable Stars, p. 223.

Cassinelli, J.P. and Lamers, H.J.G.L.M.: Winds from Young Hot Stars, p. 139.

Dupree, A.K. and Reimers, D.: Mass loss from Cool Stars, p. 321.

Jordan, C. and Linsky, J.L.: Chromospheres and Transition Regions, p. 259.

Snow, T.P. and Stalio, R.: Be-Star Phenomena, p. 183.

Vauclair, G. and Liebert, J.: The End Stages of Stellar Evolution: White Dwarfs, Hot Subdwarfs and Nuclei of Planetary Nebulae, p. 355.

Other References:

Brugel, E.W., Beach, T.E., Wilson, L.A. and Bowen, G.H.: "1987 IAU Colloquia: The Symbiotic Phenomenon".

Dupree, A.K. 1986, Ann. Rev. Astron. Astrophys., 24, 377.

Guinan, E.F. et al.: 1986 in "New Insights in Astrophysics: Eight Years of UV Astronomy With IUE", ESA SP-263, p. 197.

Hartmann, L., Balinnas, L.S., Duncan, D.K. and Noyes, R.W.: 1984, Astrophys. J., 279, 778.
Henrichs, H.F., Hammerschlag-Hensberge, G., Howarth, I.D. and Barr, P.: 1983, Astrophys. J., 268, 807.
Kirschner, R.P.: 1988 in "A Decade of UV Astronomy With the IUE Satellite", ESA SP-281, Vol. 1, p. 3.
Lamers, H.J.G.L.M.: 1987 in "Instabilities in Luminous Early Type Stars", eds. Lamers and de Loore, (Reidel, Dordrecht), p. 99.
Mullan, D.J. and Stencel, R.E.: 1982, in "Advances in UV Astronomy: Four Years of IUE Research", NASA CP-2238, p. 235.
Prinja, R.K., Howarth, I.D. and Henrichs, H.F.: 1987, Astrophys. J., 317, 389.
Rodono, M.: 1988, in "A Decade of UV Astronomy With the IUE Satellite", ESA SP-281, Vol. 2, p. 45.
Sanz Ferandes de Cordoba, L. et al.: 1988, in "A Decade of UV Astronomy With the IUE Satellite", ESA SP-281, Vol. 1, p. 115.

Astrophysics with GRO

V. Schönfelder

Max-Planck-Institut für extraterrestrische Physik,
D-8046 Garching b. München, Fed. Rep. of Germany

1. Introduction

The Gamma-Ray Observatory GRO is one of "The Great Observatories" which presently are in the phase of development and planning in the USA. The Hubble Space Telescope and GRO, both are in the final phase of development and hopefully will have been launched by 1990 by the Space Shuttle. The other two observatories (The Advanced X-Ray Astrophysics Facility: AXAF, and the Space Infrared Telescope Facility: SIRTF) both are still in the planning phase. The four Great Observatories will cover the electromagnetic spectrum from infrared to gamma-ray energies and will lead to an enormous progress in space astronomy.

The Gamma-Ray Observatory GRO is the first satellite mission that covers the entire gamma-ray range from about 100 keV to 30 GeV, more than five orders of magnitude in photon energy. Therefore, simultaneous observations over the full dynamic range will be possible. The coverage of such a broad spectral range cannot be achieved by one single instrument. Instead, GRO will contain 4 different instruments with complementary properties.

Experience has shown that the gamma-ray fluxes from celestial objects are in general very small. This is easily understood, because a 100 MeV gamma-ray combines the same energy in one single photon as about 10^{10} infrared photons. Therefore, gamma-ray telescopes must in general be large in size, and long observation times are required. GRO will be one of the largest and heaviest astronomy space projects ever built.

A schematic view of GRO is shown in Fig. 1. You see the platform carrying the 3 major instruments OSSE, COMPTEL, and EGRET (from left to right) and the fourth instrument BATSE, which actually consists of 8 discrete detectors - two at each corner of the spacecraft. The observatory is 9 m long, and the total weight is about 15 tons.

GRO will be launched by the Shuttle. The Shuttle will carry GRO into a circular orbit of 450 km. GRO will stay there for at least 2 years. GRO has a self-contained propulsive system to enable the spacecraft to maintain the 450 km orbit. In addition, the system will allow the spacecraft to undergo a controlled re-entry at the end of the mission.

Though GRO is a US-space project, Germany is involved to quite some extent. Germany has the main responsibility for COMPTEL, which actually was built by a collaboration between US-, Dutch-, ESA-, and German Research Institutes; in additon Germany is also involved in the EGRET-project. The other two instruments are pure American.

The GRO mission will start with an all-sky survey during the first year of the mission. This will be the first complete gamma-ray sky survey. For this survey at least 23 pointings of 2 weeks duration, each, will be needed; 9 pointings are needed to map the galactic plane. From the second year onwards, the detailed study of selected celestial objects is foreseen with observation times, which may well exceed the 2 weeks interval. In addition, the gamma-ray observations will be complemented by ground based observations at other spectral ranges.

Reviews in Modern Astronomy 2 Editor: G. Klare

Fig.1: Schematic View of GRO

2. Astrophysics with GRO

Due to the 10 to 100 times higher sensitivity of the GRO-Instruments in comparison to previously flown instruments, a significant progress in the exploration and understanding of the gamma-ray sky can be expected.

The main targets of interest are listed below. These are:

- Discrete objects in the Galaxy that show steady gamma-ray emission.
- The diffuse gamma-ray continuum emission from interstellar space.
- Gamma-ray line emission from discrete or extended sources.
- External galaxies, especially Seyferts and Quasars.
- The diffuse cosmic gamma-ray background.
- Cosmic gamma-ray bursts (their localization and the study of their energy spectra and time histories).
- The Sun (solar flare gamma-ray and neutron emission).

Each of these topics are briefly addressed below:

Galactic Gamma-Ray Sources

The key question in this field is: *"What kind of objects do we see as gamma-ray sources in the Milky Way?"*

At present radio pulsars and X-ray binaries are the most promising candidates. Both kinds of objects contain stars in the end stage of their evolution (neutron stars, and in case of binaries perhaps black holes). We know that the Crab and Vela pulsars, both, radiate 5 orders of magnitude more power at gamma-ray energies than in the radioband. If this is true in general for pulsars, then the key for an understanding of the pulsar radiation mechanism may be found at gamma-ray energies. GRO should be able to see more than these two pulsars at gamma-ray energies. The estimates range from a few to a few dozens depending on the pulsar model which one uses.

X-ray binaries are known to be very common X-ray sources which are powered by accretion. A few of them have also been observed at very hard X-ray energies around 100 keV. Among these are Cyg X-1 - the best black hole candidate, Cyg X-3 - the source which has attracted so much attention by its possible TeV and PeV gamma-ray emission, and other sources like GX 5-1, GX 1+4, GX 339+4. Except for Cyg X-1 and possibly Cyg X-3 none of these sources so far could be detected at gamma-ray energies. GRO- and especially OSSE should see many of them, and therefore, should add important information to an understanding of binaries.

Another object of high interest is the Galactic Center - one of the other most promising black hole candidates. First positive gamma-ray observations from the galactic center have already been made during the last ten years. Further observations will be needed to derive firm conclusions about the nature of this source.

Finally, of course, the puzzle of the unidentified COS-B sources has to be solved. Though, about half of the originally 22 objects contained in the COS-B source catalogue are now known to be simply regions of enhanced interstellar matter, about half a dozen of these sources remain unidentified till now. Nobody at present knows the nature of these sources! Do they represent a new class of celestial objects which mainly radiate at gamma-ray energies? We expect that GRO will be able to answer this question.

The prospects of GRO for studying galactic gamma-ray sources can be judged from the sensitivity diagram in Fig. 2. Here the sensitivities of the 3 instruments OSSE, COMPTEL, and EGRET over their energy ranges are compared with the fluxes of known gamma-ray sources, e.g. the Crab, Cyg X-1 in its low and high intensity state, and the COS-B source Geminga. OSSE has the highest sensitivity up to about 1 MeV. Between 1 and 30 MeV COMPTEL will be able to detect sources that are 20 to 50 times weaker than the Crab. The sensitivity of EGRET to some extent depends on the position of the source within the Galaxy. On average EGRET will be able to see sources which are a factor of 10 below the sensitivity limit of COS-B.

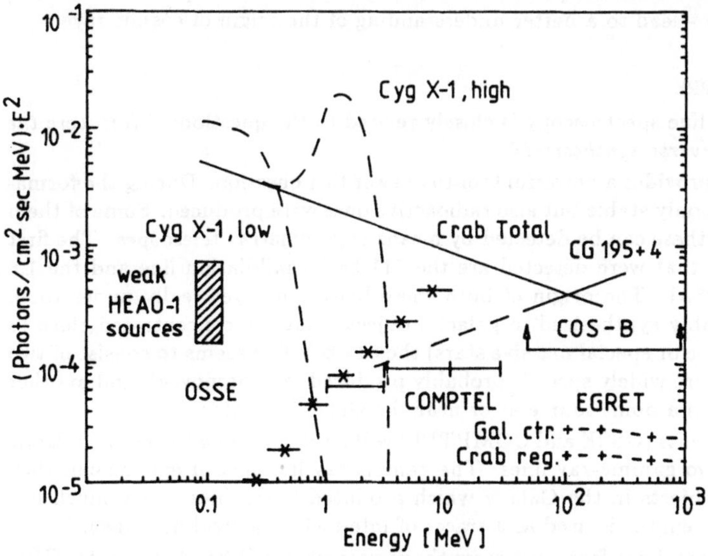

Fig.2: Sensitivity of GRO to detect galactic objects within a 2 weeks observation period.

Gamma-Ray Bursts

Though the gamma-ray bursters are the strongest gamma-ray sources in the sky during their short outburst, their nature is not yet known. Most people believe that - again - a neutron star is somehow involved. This thinking is mainly based on lines found in some of the burst spectra - though the physical trigger for the outburst is not yet known. Under discussion are at present starquakes, impact of an asteroid or comet onto the NS, magnetic instabilities or accretion of matter onto the NS.

The main burst experiment BATSE will:

- locate the burst within a few degrees, so that an identification may become possible
- measure the celestial distribution of bursts
- detect weaker bursts than has been possible before (down to fluences of 10^{-7} erg/cm^2)
- perform gamma-ray line spectroscopy of burst spectra, measure short time fluctuations and spectral variations.

Also the other three GRO instruments have burst detection capabilities, which in many cases provide complementary information on the bursts. Of special importance is the capability of COMPTEL to locate those cosmic gamma-ray bursts to an accuracy of about 1 degree which happen to be within the field-of-view of COMPTEL.

Diffuse Galactic Gamma-Ray Emission

The diffuse galactic gamma-ray emission so far has been studied at high gamma-ray energies (around 100 MeV), only. From the interpretation of the COS-B sky map we know that the cosmic ray density is not constant throughout the Galaxy, but higher in the inner part and lower in the outer parts.

EGRET will repeat this survey of the Milky Way at high gamma-ray energies with much higher sensitivity and better angular resolution. COMPTEL will extend the survey to lower energies down to about 1 MeV and therefore allow to study the electron-induced gamma-ray component. The better angular resolution of both instruments will allow to separate the diffuse interstellar gamma-ray emission from the point source distribution. The interpretation of these measurement will - hopefully - lead to a better understanding of the origin of cosmic rays.

Gamma-Ray Line Spectroscopy

The field of gamma-ray line spectroscopy is closely related to the question: *"How were the chemical elements in the Universe synthesized?"*

Gamma-ray astronomy provides a powerful tool to answer this question. During the formation of chemical elements not only stable but also radioactive ones were produced. Some of them are gamma-ray emitters and these can be detected by means of gamma-ray telescopes. The first two cosmic gamma-ray lines that were detected are the 511 keV annihilation line and the 1.8 MeV line from radioactive ^{26}Al. The origin of both these lines is not yet really understood. Whereas the ^{26}Al was probably synthesized in galactic objects that are concentrated close to the Galactic Center (like novae or special massive stars) the 511 keV line seems to consist of two components - one which is more widely spread (probably produced in supernovae), and another one which seems to come from a point source at or near the Galactic Center.

Two of the GRO instruments (OSSE and COMPTEL) will be able to map the entire galactic plane in the light of these two gamma-ray lines. The gamma-ray line measurements can then be used as tracers of those objects in the Galaxy which produced these lines - very much like the 21 cm line at radio-wave lengths is used as a tracer of interstellar neutral hydrogen.

Probably other gamma-ray lines from nucleosynthesis processes will be observed by GRO in addition: the recent supernova in the Large Magellanic Cloud is of course a target of prime interest. From the recent positive detection of the two ^{56}Co-lines we now definitely know that nucleosynthesis processes took place during the explosion. GRO has a good chance to see also lines from other radioactive isotopes which were produced in this event (e.g. ^{57}Co).

In addition, there is quite some chance that GRO will observe gamma-ray line emission from some of those supernovae bursts which can be expected within the mission life-time of GRO in external galaxies within the Virgo cluster.

External Galaxies

In the extragalactic sky there is a good chance that a few nearby normal galaxies (namely Andromeda, LMC and SMC) will be seen by at least some of the GRO-telescopes.

The most interesting objects in the extragalactic gamma-ray sky are - however, the nuclei of active galaxies. At least some of the known active galaxies and quasars do have their maxima of luminosity at gamma-ray energies. The situation is illustrated in Fig. 3, where the sensitivities of the three GRO instruments OSSE, COMPTEL, and EGRET for detecting AGN's are compared with the observed spectra of the radio galaxy Cen A, the quasar 3C273 - which both peak at gamma-ray energies - and the X-ray spectra of 12 AGN's - mostly Seyferts - that were observed by HEAO-A1. From the diagram we can estimate that GRO will be able to study - say a dozen or even more AGN's - if all AGN's have spectra similar to Cen A and 3C273. We may get the first measurement of the luminosity function of AGN's at gamma-ray energies. This function - together with the measured properties of individual galaxies - may lead to a better understanding of the engine that powers the objects. The question of the nature of the central source in AGN's is one of the most fascinating one in modern astronomy. Many theoreticians believe that the energy source is a mass accreting black hole. The GRO observations may be crucial for an understanding of the central source.

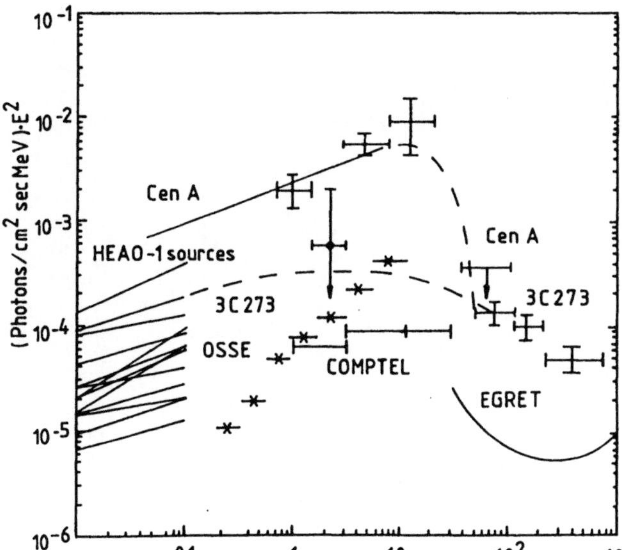

Fig.3: Sensitivity of GRO to detect Active Galactic Nuclei within a 2 weeks observation period.

The Diffuse Cosmic Gamma-Ray Background

A cosmic background radiation exists at practically all wavelengthes. Best studied is the microwave background at 2.7 K. The origin of the cosmic gamma-ray background is not yet really understood.

GRO will not only provide an accurate measurement of the background spectrum, it will also allow to address the question of its origin and to decide between the two classes of models which presently are under discussion: an unresolved source origin (e.g. from unresolved AGN's) or a really diffuse origin (e.g. from matter-antimatter annihilation in a baryon symmetric universe).

The Sun

The Sun is not a primary objective of GRO, all four GRO instruments do have, however, certain capabilities to measure gamma-rays and neutrons from the Sun during solar flares. Such observations in conjunction with those of other instruments (interplanetary particle detectors, ground based neutron monitors) provide the best possible observation of acceleration, production, and propagation aspects during solar flares.

3. Conclusions

The previous discussion has illustrated, what kind of question will be addressed by GRO. In addition, it can be expected that also new, unexpected phenomena will be discovered - as was always the case, when a new field was explored. After GRO gamma-ray astronomy will probably no longer be restricted to the rather small group of gamma-ray astronomers, but rather be attractive to the whole community of astronomers and astrophysicists.

The Infrared Space Observatory ISO

D. Lemke[1] *and M. Kessler*[2]

[1]Max-Planck-Institut für Astronomie, D-6900 Heidelberg, Fed. Rep. of Germany
[2]European Space Agency, ESTEC, NL-2200 AG Noordwijk, The Netherlands

The ISO satellite and its four scientific instruments are currently in an advanced hardware-building stage. ISO will offer unprecedented sensitivity, spectral and spatial resolution from near- to the far-infrared wavelengths. About two-thirds of the observing time will be available to the community when ISO is launched in 1993. The first call for observing proposals will be issued in 1991.

1. From IRAS to ISO

The first cooled infrared satellite IRAS, launched in 1983, opened a new window to the cold universe. In addition to many famous discoveries, the four-color far-infrared catalogue of the sky resulting from its survey mission is now a treasure of astronomy. Even today, six years after the mission, the number of scientific publications triggered by IRAS results does not seem to be declining.
Early in 1983, at the same time as IRAS was launched, the European Space Agency ESA decided, after years of careful preparations, to adopt a pointed cooled infrared observatory its next scientific mission. The selection of ISO's scientific payload, consisting of four instruments, was completed in 1985. Shortly after this date an industrial consortium, eventually to include 35 companies started the development of the largest and most complex scientific satellite ever built in Europe.
Since two-thirds of ISO's time is open to the scientific community and the launch in 1993 is fairly close, this paper is designed to present an overview of ISO's status; its capabilities; and a brief description of how it will be made accessible to a wider public of potential observers.

2. The ISO-satellite

ISO is essentially a large cryostat with a toroidal tank containing 2300 l of superfluid helium at a temperature of 1.8 K (fig. 1 and 2). The evaporating gas cools the scientific instruments, the telescope and a set of radiation shields surrounding the helium tank, therefore allowing for a cryogenic lifetime of more than 18 months in orbit. A cooled baffle surrounding the 60 cm telescope ensures that the thermal background and straylight are $<10^8$ photons cm^{-2} s^{-1} on the detectors. The telescope beam is divided by a pyramidal mirror following the primary into four sectors each containing one scientific instrument (fig. 3). Therefore all the instruments view the sky simultaneously, but their unvignetted fields of view of 3 arc min each are ~10 arc min apart from each other in the sky. The experiments form a coherent set of observatory instrumentation (tab. 1). Their combined capabilities allow an almost complete study of celestial infrared sources.

3. The Four Scientific Instruments

The photopolarimeter (ISOPHOT) has arranged its detector groups around three filter wheels (fig. 4). Appropriate setting of these wheels allows selection of the observing mode (photometry, polarimetry, imaging), the wavelength band and the aperture. Of particular interest are two subsystems: (i) a 200 μm-camera, covering a region unexplored by IRAS, and (ii) a spectrophotometer for simultaneous observations of two bands in the middle ir rich in interstellar dust features (fig. 5). ISOPHOT is the only instrument containing a chopper for differential measurements. In combination with internal calibration sources and a beam blocker it will be responsible for cancelling instrumental drifts and offsets.

The camera (ISOCAM) houses two imaging arrays covering the near- and mid IR wavelengths (fig. 6). Appropriate setting of selection wheels enables the user to choose wavelength bands (various filters) and imaging scales (lenses of different magnifications) (fig. 7). Spectrophotometry by a circular variable filter and polarimetry are also possible.

The short wavelength spectrometer (SWS) and the long wavelength spectrometer (LWS) cover the region 2.4 to 180 μm with an overlap

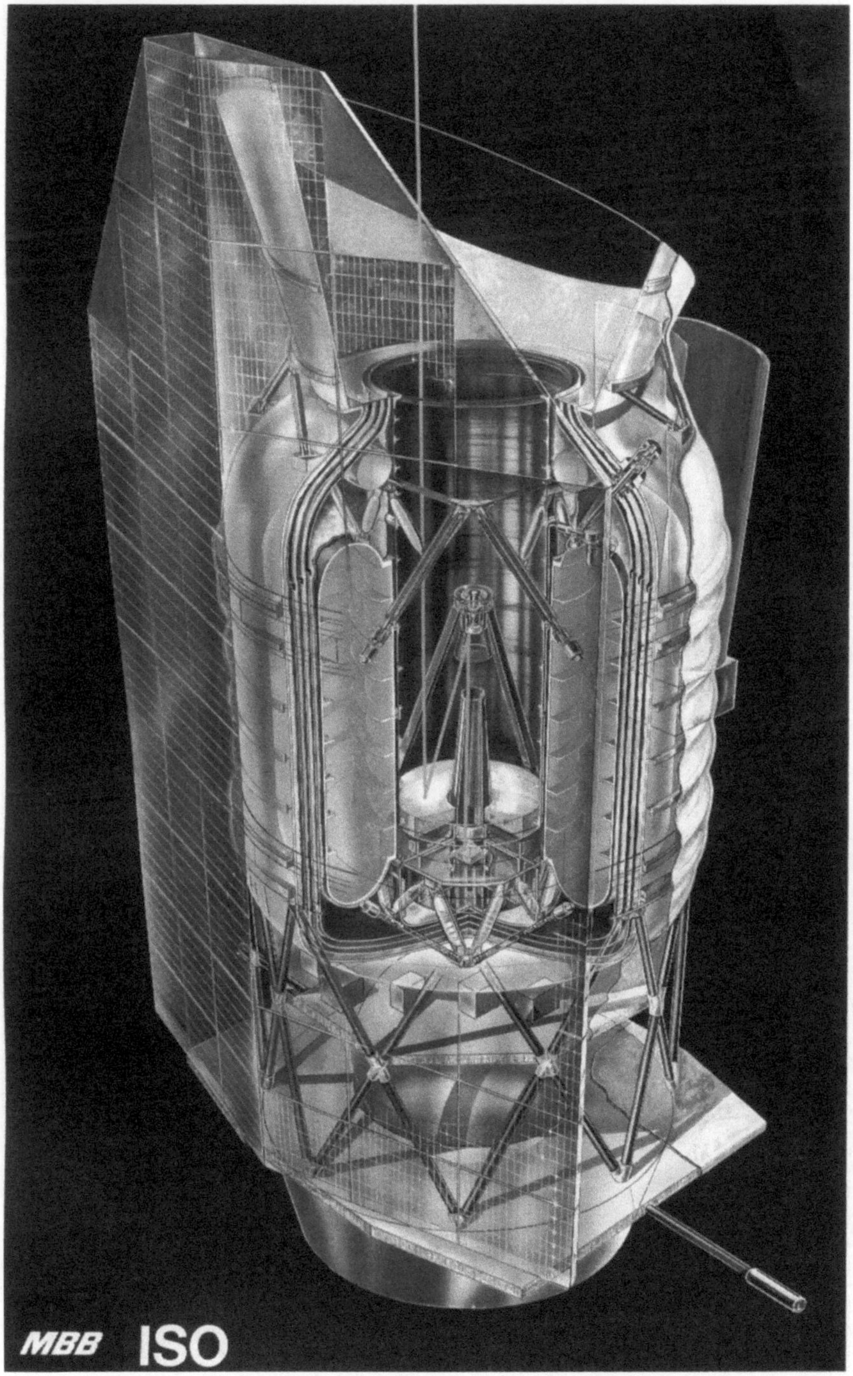

Fig.1. The ISO satellite is about 5 m high and has a mass of 2.4 tons. (MBB)

Fig.2. The 2300 l liquid helium tank of ISO (here under assembly) has successfully been tested meanwhile.

Table 1. The four ISO experiments

Instrument and Principal Investigator	Main Function	Wavelength (Microns)	Spectral Resolution	Spatial Resolution	Outline Description
ISOCAM (C. Cesarsky, CEN-Saclay, F)	Camera and Polarimetry	2.5 - 17	Broad-band, Narrow-band, and Circular Variable Filters	Pixel f.o.v.'s of 1.5, 3, 6 and 12 arc seconds	Two channels each with a 32x32 element detector array
ISOPHOT (D. Lemke, MPI für Astronomie, Heidelberg, D)	Imaging Photo-polarimeter	2.5 - 200	Broad-band and Narrow-band Filters. Near IR Grating Spectrometer with R = 90	Variable from diffraction - limited to wide beam	Four sub-systems: i) Multi-band, Multi-aperture photo-polarimeter (3-110 μm) ii) Far-Infrared Camera (30-200 μm) iii) Spectrophotometer (2.5-12 μm) iv) Mapping Array (18-28 μm)
SWS (Th. de Graauw, Lab. for Space Research, Groningen, NL	Short-wavelength Spectrometer	2.5 - 45	1000 across wavelength range and 2×10^4 from 15 - 30 μm	10×20 and 20×30 arc seconds	Two gratings and two Fabry-Pérot Interferometers
LWS (P. Clegg, Queen Mary College, London, GB)	Long-wavelength Spectrometer	45-180	200 and 10^4 across wave-length range	1.65 arc minutes	Grating and two Fabry-Pérot Interferometers

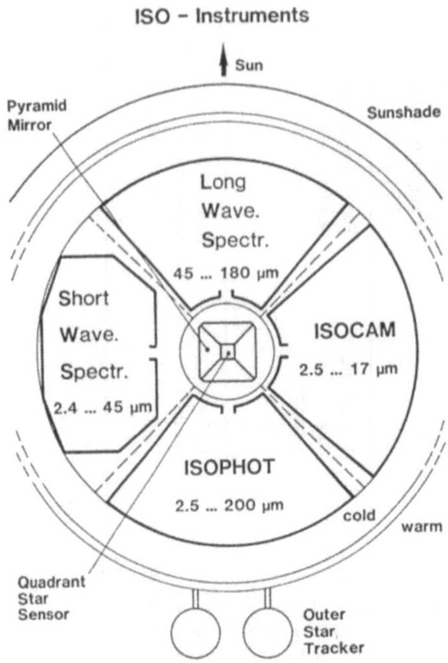

Fig.3. The 4 ISO-experiments in the 3 K focal plane chamber underneath the primary mirror. The pyramidal mirror devides the beam into 4 adjacent fields on the sky.

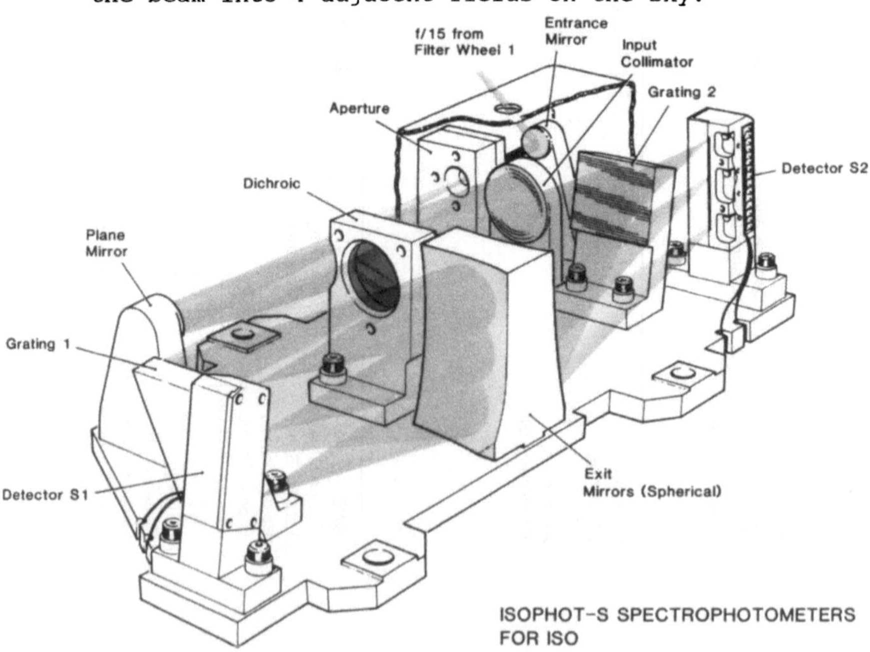

Fig.4. The ISOPHOT experiment.

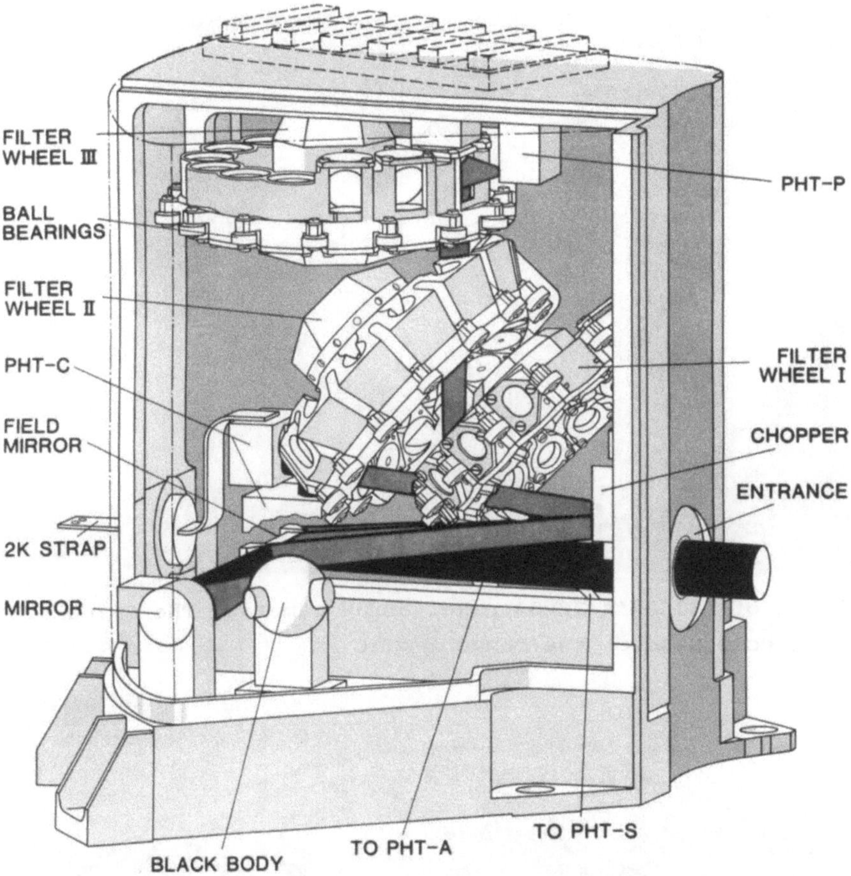

Fig.5. ISOPHOT-S houses two fixed gratings and two linear 64 pixel Ge:Si-arrays for simultaneous measurements of 2.5...5.0 and 6.0...12.0 μm spectra. The device is mostly a Spanish contribution.

point near 45 μm (fig. 8 and 9). Both instruments are grating spectrometers allowing medium spectral resolution and both are capable of introducing a Fabry-Perót-Etalon for high resolution ($\approx 10^4$) into the beam (fig. 10). Scanning mechanisms are used in both instruments in both modes to move the interesting wavelength region across the detectors.

Fig.6. One of ISOCAM's 32x32 pixel array cameras. The Si:Ga detectors cover the range $\lambda < 16\ \mu m$.

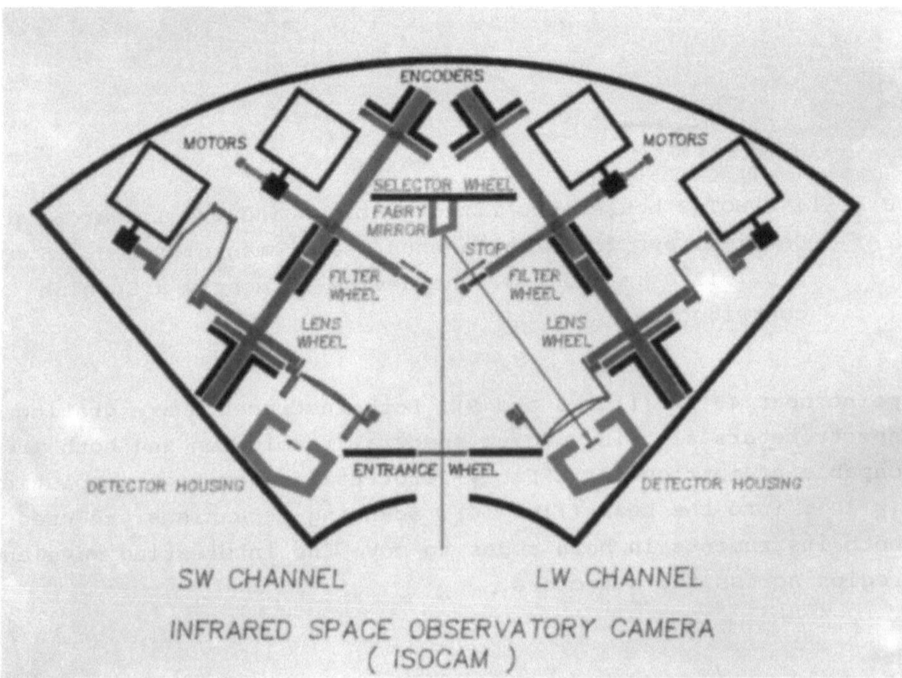

Fig.7. Schematics of ISOCAM.

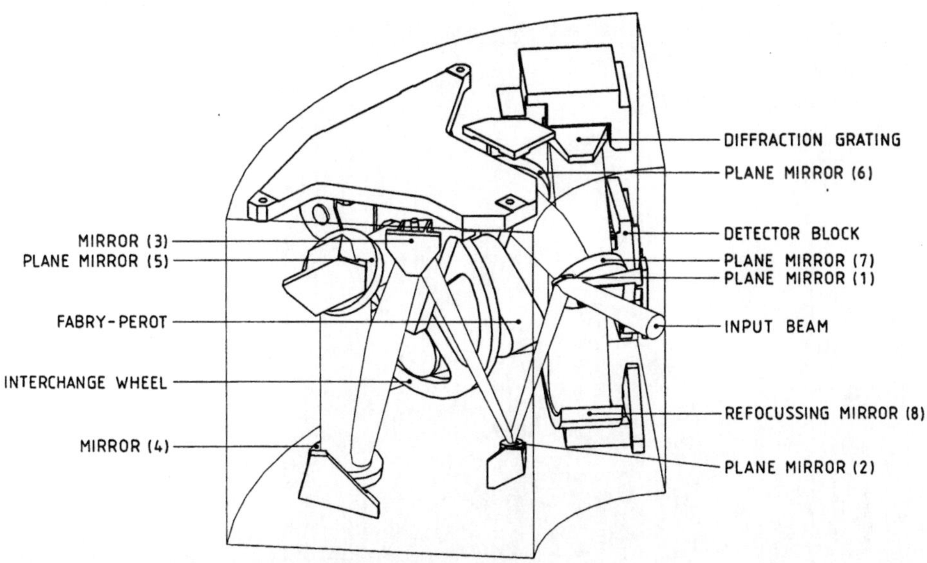

Fig.8.The Long Wavelength Spectrometer LWS.

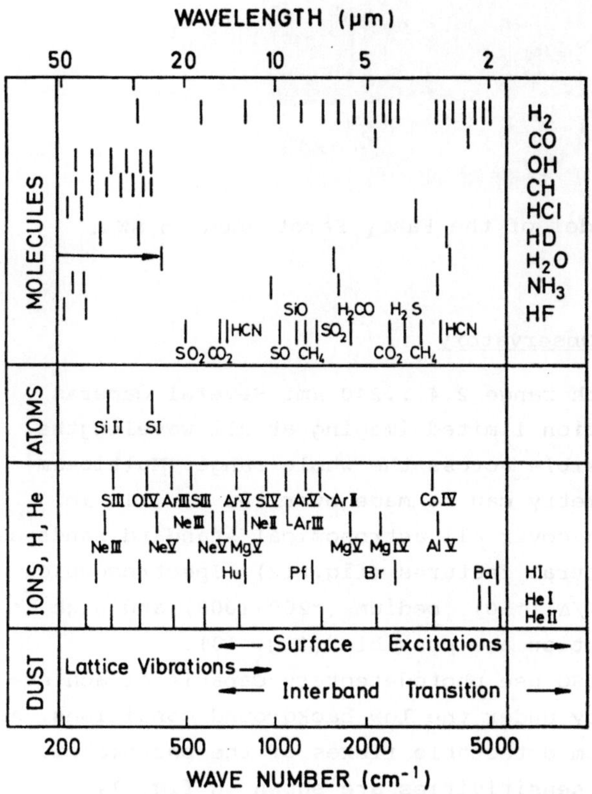

Fig.9.Lines and spectral features observable by SWS.

Fig.10. The prototype model of the Fabry Perót used in SWS.

4. Capabilities of the Observatory

ISO covers the wavelength range 2.4...240 μm. Several cameras (fig. 11) allow diffraction limited imaging at all wavelengths. Also polarimetry is possible across the whole range. Multibeam and multiaperture photometry can be made over the 3...110 μm region. Numerous filters cover all astronomical standard bands as well as interesting spectral features (fig. 12). Spectroscopic measurements with low ($\lambda/\Delta\lambda$ ~50), medium (~200-1000) and high (~10000) spectral resolution are possible (fig. 13).
All instruments aboard ISO use photodetectors capable of achieving very high sensitivity under the low background conditions. As an example of the minimum detectable fluxes of the photometric instruments the ISOPHOT sensitivities are shown in fig. 14.
In order to cover extended objects ISO can raster scan as explai-

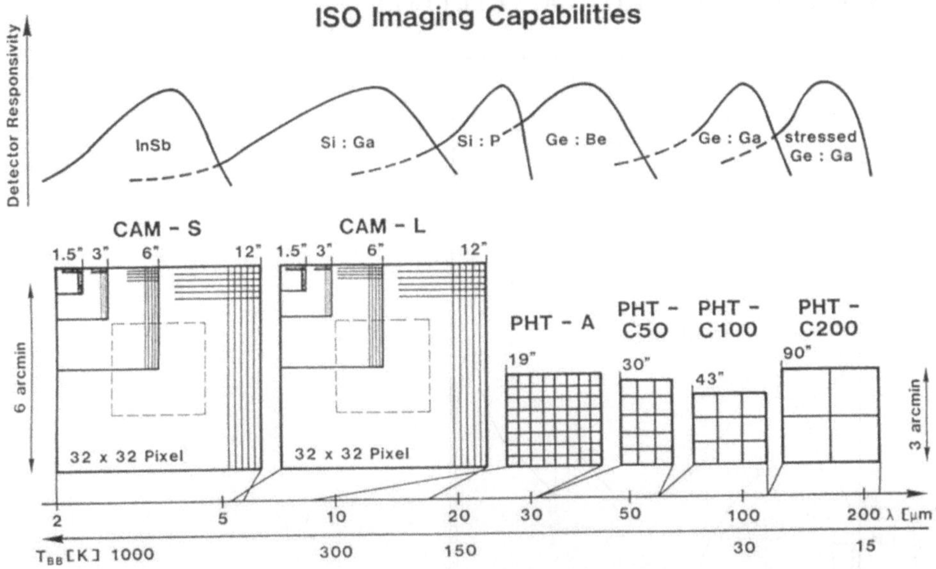

Fig.11.The 6 imaging cameras are housed in ISOCAM and ISOPHOT. The plate scale of the two leftmost cameras can be changed by a lens wheel, while the unvignetted field is always limited to 3 arc min. The increasing pixel size reflects the increasing diffraction image.

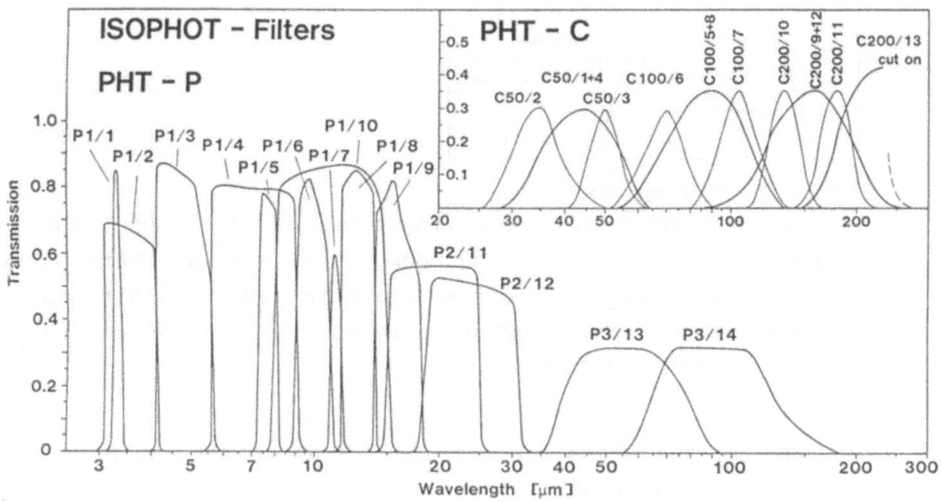

Fig.12.ISOPHOT and ISOCAM contain all standard filters of ground-based IR astronomy as well as the IRAS bands and cover the cosmological windows. The FIR filters are based on resonant meshes and are developed at Max-Planck-Institut für Radioastronomie, Bonn.

ISO - SPECTROSCOPIC CAPABILITIES

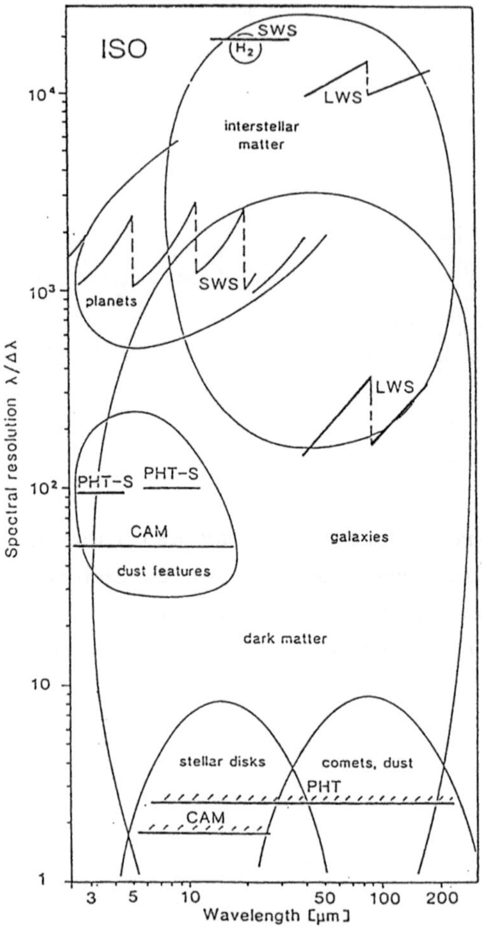

Fig.13. Spectroscopy with a resolving power of $\lambda/\Delta\lambda \sim 10^2 \ldots 10^3$ is possible with the grating mode of SWS and LWS. Both instruments can increase the resolution to $\sim 10^4$ by additional Fabry-Peróts. Spectrophotometry with $\lambda/\Delta\lambda < 90$ can be made with ISOCAM and ISOPHOT.

ned in fig. 15; the scans are carried out as an automatic sequence of fine pointings.

During ISO's lifetime of 18 months all parts of the sky will be accessible; however, there are ±60° avoidance angles around the sun and the antisolar direction and ISO must not point too close to the Earth.

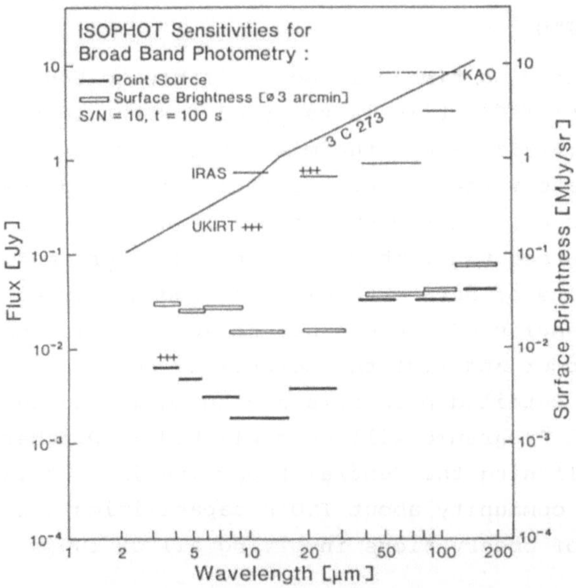

Fig. 14. Photometric sensitivities for 100 s observing time and a signal to noise ratio of 10. For comparison the sensitivity of the IRAS sky survey, of UKIRT and the Kuiper Airborne Observatory are also plotted.

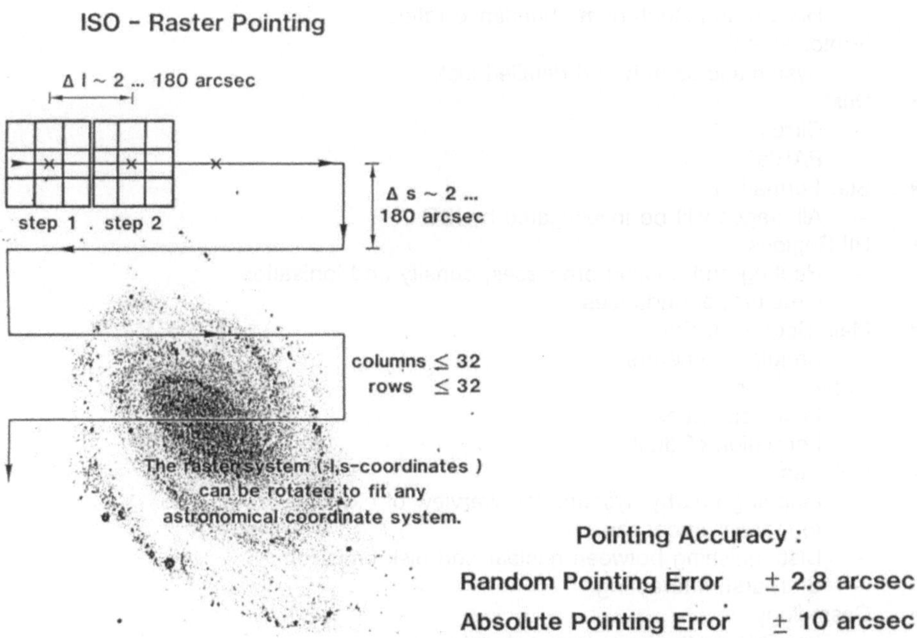

Fig. 15. Extended objects can be imaged by a raster pointing of ISO. The figure shows ISOPHOT's C-100 camera (100 μm) as an example.

5. The Scientific Programme

The majority of ISO's observing time will be available to the community; however, 35% of ISO's time is reserved for the instrument-building PI groups (6.25% each), the mission scientists (5% in total) and the observatory team. A Central Programme, addressing all important aspects of ir astronomy (tab. 2), will be carried out in the guaranteed time. This coordinated effort makes sure that (i) complete sets of data are being sampled on agreed sources, (ii) unbiased samples of sources are selected, (iii) no observational overlaps occur and (iv) the scientific key questions are addressed. Detailed proposals are being worked out now and, the full Central Programme will be published in October 1991. The purpose of publishing the Central Programme is twofold: to inform the scientific community about ISO's capabilities as illustrated by hundreds of observations involving all of ISO's

Table 2. Summary of observational highlights

- Comets
  - Gas/dust composition and evolution
  - IR tails and trails
- Giant Planets
  - Helium and Deuterium abundance ratios
- 'Protoplanets'
  - Systematic search and detailed look
- Dust
  - Cirrus
  - PAH's
- Star Formation
  - All stages will be investigated by ISO
- HII Regions
  - Heating and cooling processes, density and ionisation structure, abundances
- Main Sequence Stars
  - Angular diameters
- Evolved Stars
  - Mass loss rates
  - Formation of dust
- Galaxies
  - Mapping nearby systems for overview of e.g. star formation processes
  - Distinguishing between nuclear and disk emission
  - Starburst, interacting
- Cosmology
  - Invisible matter, e.g. brown dwarfs
  - IR galaxies as standard candles

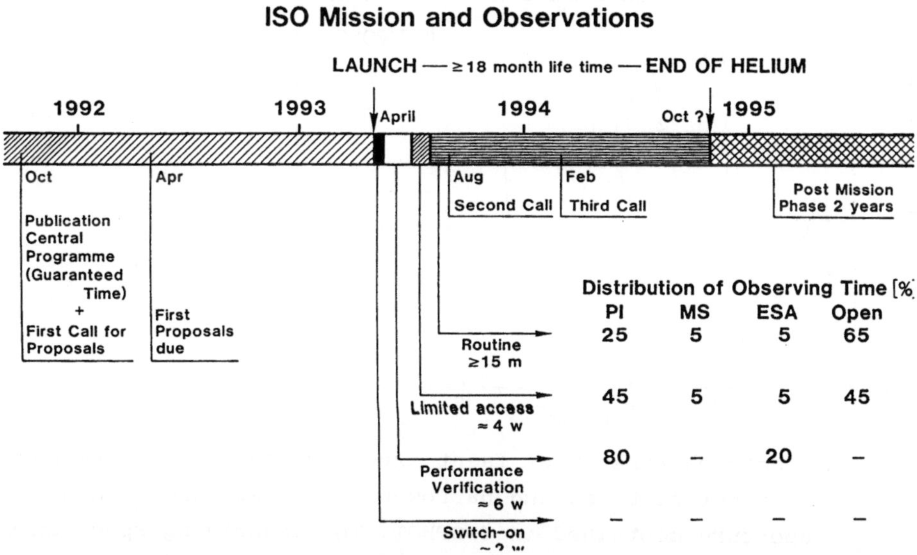

Fig.16. Prior to the 15 month routine observation there is a one month "limited access" phase where the available time is equally shared between the 'guaranteed time observer' and the community. This phase will be devoted to the highest priority observations.

instruments; and to reserve these proposed measurements for the teams involved in the development of ISO, as a return for a decade of effort devoted to the project.

The largest part, 65%, of the observing time is open to the astronomical community via the traditional method of proposal submission and selection. Three calls for proposals are planned, as indicated in fig. 16; the first one is accompanied by the issue of the Central Programme. As examples taken from the long list of potential observations, two ISOPHOT proposals are illustrated in figures 17 and 18. Firstly, spectrophotometry of cirrus clouds at different locations in the Milky Way is planned. The high sensitivity achievable in the wide beam (180 arc sec) allows obtaining a 14 band spectrum within 500 seconds. Information on dust composition and its physical properties can be achieved and compared to the results of sources and sinks of interstellar dust. The other example is the attempt to detect faint halos above galaxies seen edge-on. If these halos are there and are composed of brown dwarfs, several hours of integration

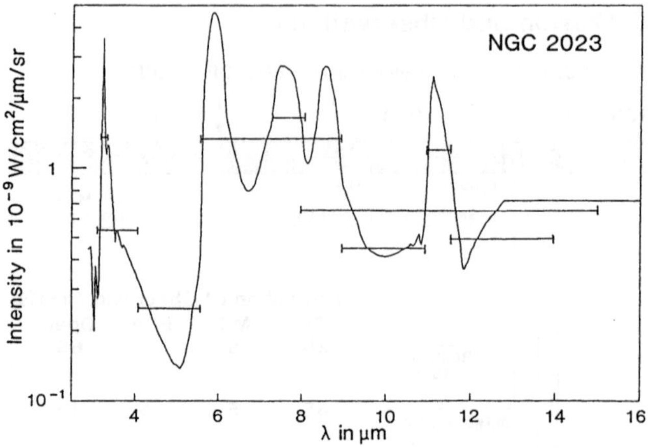

Fig.17. Spectrophotometry of faint extended sources will be possible by a set of broad and narrow band filters and a large aperture contained in ISOPHOT. The underlying spectrum of a bright reflection nebula showing the "unidentified features" assigned now to polcyclic aromatic hydrocarbons (PAH's) was obtained by Sellgren.

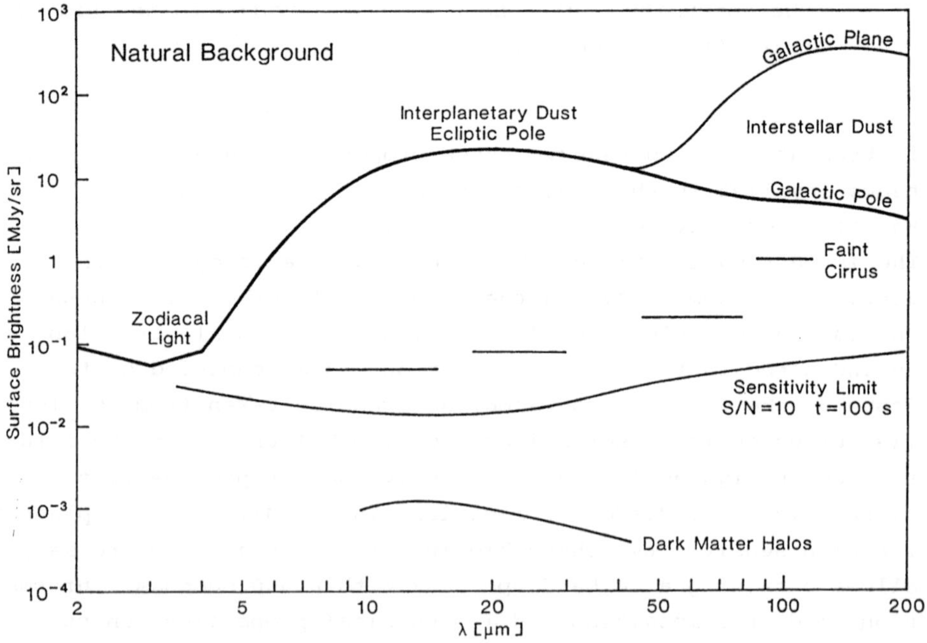

Fig.18. The infrared zodiacal light hampers observations of faint extended sources like the interstellar cirrus and the suspected dark matter halos of other galaxies. The sensitivity limit of ISOPHOT (S/N = 10, 100 s) is plotted.

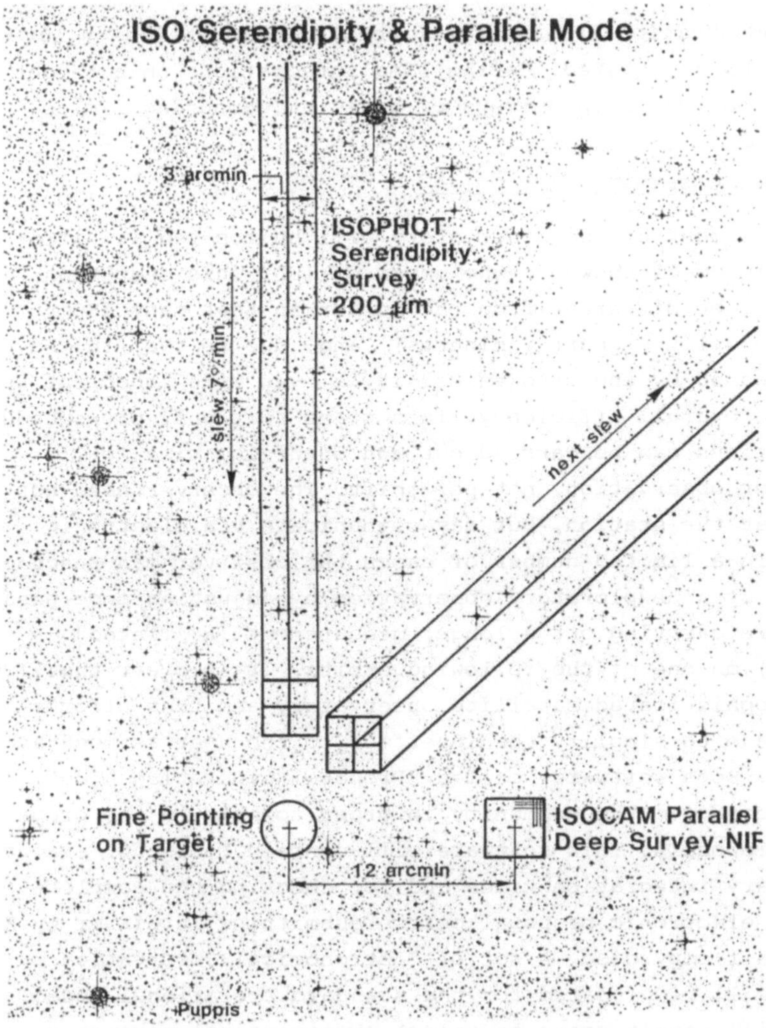

Fig.19. The "dead" time during slews is used by ISOPHOT's 200 μm camera for a serendipedous sky survey. The sensitivity will be ~1 Jy, therefore complementing the IRAS survey towards longer wavelengths.

time with ISO might allow them to be detected. A positive result would be an important clue to the question of the "dark matter" in the universe. Both the observations mentioned require a cryogenically-cooled space telescope and cannot be addressed otherwise. During slews of the satellite, the 200 μm camera will view the sky for a serendipity survey (fig. 19) The sensitivity of ~1 Jy achievable is comparable to the IRAS survey. Depending on the final observational strategy of pointed observations > 10% of the sky can be covered. During long integrations on a target ISOCAM will perform deep surveys of small neighbouring fields.

6. How to Get Observing Time on ISO

The ground segment of the observatory will be located at Villafranca, Spain, well-known to the users of IUE. This operation, however, will be different from the former in that many ISO observations will be short (~minutes), and, due to the limited lifetime imposed by the cryogenics, it is even more important to maximize the overall efficiency. This means that only pre-scheduled observations can be carried out and observing time is allocated per observation, not per shift. The observer need not be present at the station, but may communicate via electronic mail, etc. In a limited number of cases the observer can react on the results of an exploratory observation by adjusting some parameters before carrying out a longer measurement (a "linked" observation) on one of the following orbits. It is clear that ISO will be a sophisticated satellite observatory and thus efforts are being made to simplify the interface to the user as much as is possible. Some software, called "Proposal Generation Aids", will be released to the community in order to help in the preparation of proposals. This software will contain tools for evaluating necessary integration times and for checking the completeness of the electronic proposal before submitting it either by E-mail or on a floppy disc. A set of "Astronomical Observation Templates" will be prepared for the standard operational modes and these will permit users to select observational parameters in a simple way. After the observation, all observers will receive a standard set of data product including raw and processed data, calibration files and some extracted scientific results.
Figure 20 also outlines a rough timescale for the availability of the data after an observation. Preliminary data, mainly for evaluation of linked observations will be supplied within 1 day. Because the calibration of the ir detectors changes along the 24 hour orbit passing the radiation belt, calibrated data may not be available until 1 week after the observation, however, the goal is 1-2 days.
One year after the observer has received his data they will become public and available to all on request to the archive. It is

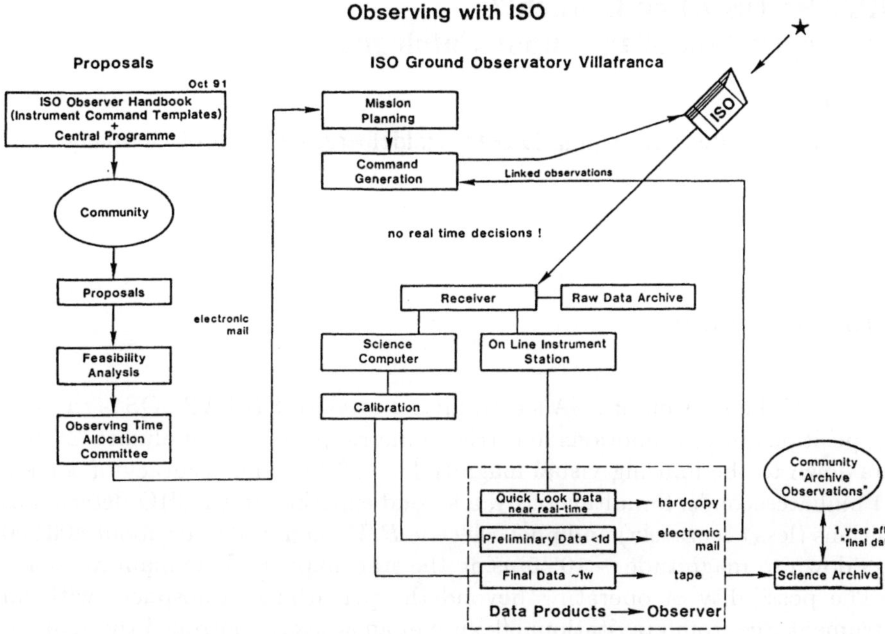

Fig.20. Simplified flow diagrams showing the route from a successful proposal to the distributed data products. Note that for efficiency reasons only prescheduled observations are possible.

to be expected that "archive observations" will extend ISO's lifetime to many times the cryogenic lifetime.

Acknowledgement: C. Cesarsky, Saclay, L. Haser, Garching and P. Clegg, London, supplied us with material on their experiments (CAM, SWS, LWS). Ch. Hajduk and E. Ackermann, Heidelberg, assisted us in preparing some of the diagrammes. The German contribution to ISO and its experiments (PHT, SWS) is funded by the BMFT, Bonn and the DLR, Porz.

HIPPARCOS After Launch!? The Preparation of the Input Catalogue

H. Jahreiß

Astronomisches Rechen-Institut, D-6900 Heidelberg, Fed. Rep. of Germany

1. Introduction

The goal of the mission of ESA's astrometric satellite HIPPARCOS is to measure positions, proper motions and trigonometric parallaxes of almost 120 000 stars down to the limiting visual magnitude $\sim 12^m.5$ to an accuracy of some 2 to 4 milliarcseconds. Simultaneously, a second experiment TYCHO determines positions (less accurately) and photoelectric B, V magnitudes for about 400 000 stars down to magnitude $\sim 10^m.5$ using the star mapper of the main mission.

The possibility of operating beyond the perturbing atmosphere with an instrument free from gravitational flexure in an actively controlled thermal environment makes the HIPPARCOS mission so unique and the expected results so exceptionally much better in quantity and quality than all that was received up to now from ground-based observations. For example, FK5 is at present the only ground-based catalogue with better proper motions ($\pm$ 0.7 milliarcsec per year), but only for 1535 stars brighter than $\sim 6^m.5$, and the *New General Catalogue of Trigonometric Stellar Parallaxes* (van Altena and Lee [1]) lists trigonometric parallaxes for about 7 700 stars with errors in the order of 15 milliarcsec, in other words, approximately half of the stars have parallaxes that are smaller than their standard error. Furthermore, the reduction to absolute remains still questionable, whereas HIPPARCOS would provide absolute parallaxes!

Satellite characteristics, payload, operational principles and data reduction have already been described several times, see for example [2] and [3] or in greater detail [4] and [6]. Therefore, in chapter 2 only the key features will be explained.

HIPPARCOS was successfully launched by Ariane 4 on 8 August. But, the failure of the Apogee Boost Motor (ABM) to fire prevented it reaching its foreseen geostationary orbit. The highly elliptical orbit in which HIPPARCOS remains required a revision of the nominal mission and puts some severe constraints on the planned mission lifetime. But, at present, reliable estimates on the outcome cannot be made. In chapter 3 some possible results of the revised mission are mentioned.

The finite mission duration in connection with the scanning principle applied required a careful selection of the stars to be observed and a precise *a priori* knowledge of the stars' position and magnitude. Special emphasis had

Reviews in Modern Astronomy 2 Editor: G. Klare

to be put on all types of doubles, variables and stars in crowded regions, not to forget minor planets and other solar type systems. The preparation of all this was the task of the Input Catalogue (INCA) Consortium. The difficulties encountered and the final results achieved are described in chapter 4.

2. The Nominal Mission

The satellite's main characteristics are summarized in Table 1. The total launch mass includes spacecraft (400 kg), payload (210 kg) and the ABM propellant (463 kg). The latter contributes still to the mass of the satellite due to the failure of the ABM ignition. The spin rate corresponds to 11.25 revolutions per day or - in the present orbit - five revolutions per orbital period.

Table 1. The main characteristics of the satellite

total launch mass	1140 kg
power requirements	295 W
uplink data rate	2 kbits s^{-1}
downlink data rate	24 kbits s^{-1}
satellite orbit	(geostationary)
inclination to Sun	43°
spin rate	168.75 arcsec s^{-1}

Table 2 gives an overview of the payload characteristics (see also Fig. 1).

A special optical telescope having two distinct fields of view is one of the key features of the payload. A beam combiner superimposes the incoming light on the spherical mirror. In the focal plane the regular grid of the primary detection system modulates the light signal due to the spin motion of the satellite. A transit of a star lasts about 19.2 seconds, and some four to five program stars are always present in the two fields of view. The modulated signal is sampled by an image dissector tube (IDT). The IDT possesses an instantaneous field of view ($\sim$ 38 arcsec diameter) that can be directed at any position in the field of view. By computer control this instantaneous field of view can be switched rapidly to any one star present. Since it can follow only one star at one time, the measurements of the relative positions are not strictly simultaneous.

The two fields of view are separated by a basic angle of about 58°, therefore, not only small-field but also large-field angular measurements can be carried out at the same time. Due to the satellite's rotation - one revolution in just over two hours - each program star is first be observed in the preceding and then in the following field of view. The scanning principle is shown in Fig. 2.

Furthermore, because the axis of rotation changes, each star can be subsequently compared with other program stars in about the same 58° distance but under different position angles. Due to the satellites motions the complete celestial sphere will be scanned within about three months, and in the course

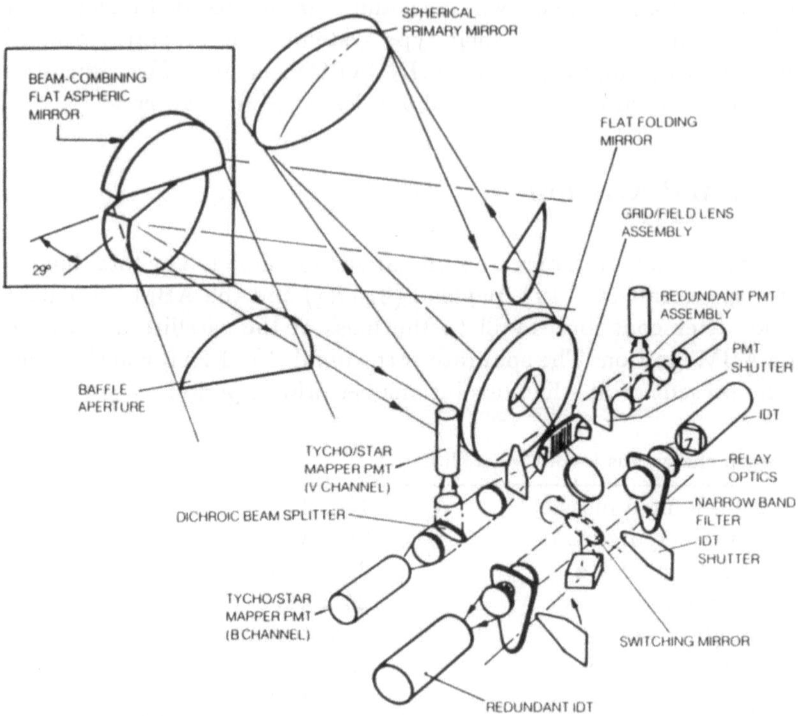

Fig. 1. A schematic view of the payload: Beam combiner, spherical primary mirror and the assembly in the focal plane with the photomultipliers (PMT) of the TYCHO mission and the image disector tube (IDT) of the main mission are explained. From M. A. C. Perryman: *Ad Astra HIPPARCOS*, ESA BR-**24**, 1985

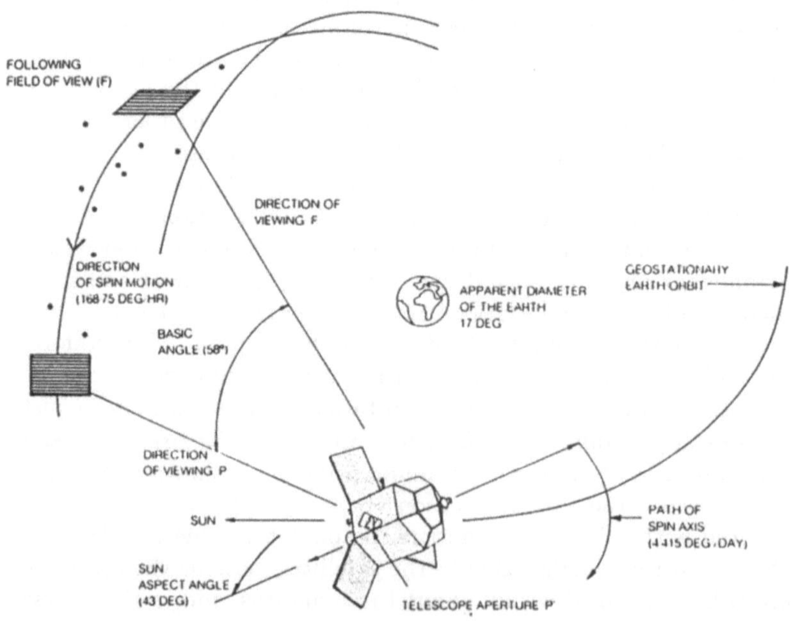

Table 2. The payload characteristics

optics:	
telescope configuration	all-reflective Schmidt
field of view	$0.^{\circ}9 \times 0.^{\circ}9$
separation between fields	58°
diameter of primary mirror	290 mm
focal length	1 400 mm
scale at focal surface	6.8 μm per arcsec
mirror surface accuracy	$\lambda/60$ rms (at $\lambda = 550$ nm)
primary detection system:	
modulating grid	2 688 slits
slit period	1.208 arcsec (8.2μm)
detector	image dissector tube
photocathode	S20
scale at photocathode	3.0μm per arcsec
sensitive field of view	38 arcsec diameter
spectral range	375-750 nm
sampling frequency	1 200 Hz
star mapper (TYCHO) system:	
modulating grid	4 slits perpendicular to scan
	4 slits at ±45° inclination
detectors	photomultiplier tubes
spectral range B_T	$\lambda_{eff} = 430$ nm, $\Delta\lambda = 90$ nm
spectral range V_T	$\lambda_{eff} = 530$ nm, $\Delta\lambda = 100$ nm
sampling frequency	600 Hz

of time a dense net of relative angular distances connecting the program stars will be established in spite of the fact that only one-dimensional measurements are obtained.

The *star mapper* in the focal plane alongside the primary detector grid measures the transit of ~ 40 000 reference stars selected from the Input Catalogue for their high positional accuracy. These measurements are for real time attitude determination (RTAD)of the satellite. Between the transits of reference stars the attitude is maintained by means of gyroscopes. Jets of cold gas are used to bring the satellite back to its nominal attitude. They are actuated almost every 400 seconds.

The star mapper (there is a redundant second one) consists of a non-periodic modulating grid (see Table 2), and is additionally used in the TYCHO experi-

Fig. 2. The measurement principle of HIPPARCOS. As the satellite sweeps out a great circle across the sky, stars continually enter and leave the fields of view of the telescope. All stars simultaneously present in the combined fields of view have their relative separations along the direction of the scan measured. As the spin axes changes the complete sky will be covered after about three months. From M. A. C. Perryman: *Ad Astra HIPPARCOS*, ESA **BR-24**, 1985 ment. For this purpose two photomultipliers measure the light transmitted by the star mapper grid in two different spectral ranges corresponding roughly to B and V in the Johnson system.

The data reduction of the main mission will be carried out by two consortia: FAST and NDAC. Both have chosen a three-step approach.

1. The relative phase measurements taken over a period of about 12 hours - corresponding to six revolutions - are projected on one *reference great circle*, i.e. a least-square solution yields the abscissae of about 2 000 stars from about 70 000 phase estimates.
2. The abscissae for a *well-behaved* subset of some 40 000 stars are combined in a subsequent least-square process into a single global system.
3. The five astrometric parameters of each star are determined by a least-square solution of its abscissae on different reference great circles, which also provide different epochs of observations.

3. The Revised Mission

Since also the last attempt - at the end of August - to ignite the Apogee Boost Motor turned out to be unsuccessful, it was decided on 4 September to abandon the original mission and to proceed with a revised mission. From 7 to 11 September the on board hydrazine was used to raise the perigee of the highly elliptical transfer orbit from 250 km to about 500 km, the apogee remained at about 36 000 km. On 12 September the three solar panels and the fill-in antenna were successfully deployed, and, hopefully, the remaining satellite commissioning and payload calibration will be likewise successful, so that at the end of October the scientific data acquisition can start.

But, even then, the highly elliptical orbit has severe consequences. Obviously, the quantitative and qualitative outcome of the mission depends strongly on the mission lifetime, which is mainly governed by the power requirements. The power supply by the three solar panels will be reduced for two reasons.

1. In its present orbit the satellite passes the high energy radiation in the van Allen belt. This causes a much faster degradation of the solar cells than expected in the geostationary orbit. Yet, a prediction how fast this will reduce the power supply can only be given after some more weeks.
2. In the period from 2nd February to 9th April 1990 and again starting on Friday the 13th July the satellite passes eclipses lasting longer than 60 minutes. With respect to its present 10h 39m orbital period this is much more serious than for the nominal 24h orbit.

To overcome these power constraints the satellite must be put in a hibernation phase, i.e. at least the payload must be switched off.

A further drawback lies in the data acquisition. Contrary to a geostationary orbit more than one ground station is necessary to receive a not too limited data recoverage. First simulations showed the initially destined Odenwald station would cover only 31.4 per cent. With the addition of two further stations, Perth (Australia) and Kourou (French Guiana), the coverage can be raised to 81 per cent. The latter station will be equipped for high rate data reception and transmission to ESOC in Darmstadt at the end of October. Yet, due to the fact that the satellite cannot observe in too great vicinity of the Earth (minimum height about 10 000 km), the really usable observing time recovered by the ground stations will be only about 70 per cent.

One possible outcome of the mission if the satellite survives at least the first hibernation phase is presented in Table 3 compared with the predicted results for the nominal mission. The accuracy will drop by more than a factor of 10. Yet, much more serious, parallaxes and proper motions cannot be expected. Though the outcome would be at least a positional catalogue with an accuracy much better than can be achieved from ground-based measurements one would be far from the initial goal, with its enormous impact on all parts of astronomy.

Table 3. Results of the nominal and a revised half year mission. From J. Van der Ha and K. Clausen: *HIPPARCOS Recovery Mission Requirements*, priv. communication, September 1st, 1989

mission product	nominal mission	revised mission
HIPPARCOS		
Number of stars	120 000	120 000
number of obs. per star	80	2 (?) to 8
typical accuracy of pos.	0.002 arcsec	0.02 to 0.05 arcsec
parallaxes	0.002 arcsec	none expected
proper motions	0.002 arcsec y^{-1}	none expected
TYCHO		
Number of stars	$\sim$ 400 000	$\sim$ 200 000
typical accuracy of pos.	0.03 arcsec	0.1 arcsec
accuracy of B_T and V_T	0.05 mag	0.05 mag

The satellite is now about 460 kg heavier and its approach to the Earth during perigee may cause not only troubles to RTAD due to perturbing torques, but the whole control of the satellite may be more difficult. Under the assumption that no further unexpected problems arise it may even be possible that HIPPARCOS operates during its planned 2 1/2 years. At least 18 months would be necessary to get the five astrometric parameters - positions, proper motions and trigonometric parallax - properly. Two and a half year of operating would mean a two times smaller accuracy than with the nominal mission. To reach the full accuracy of the nominal mission a lifetime of at least 3 1/2 to 4 years would have been necessary.

For quite a lot of important proposals even the outcome of a possible two and a half year mission would still be insufficient. Therefore, the feasibility of a second HIPPARCOS mission in the near future is already under investigation. The costs for the construction of a HIPPARCOS II were estimated to be about one third of the present satellite. It could be built within three to four years. A first decicion on such a project may be taken in November by ESA's Science Program Commitee.

4. The Preparation of the Input Catalogue

To give the HIPPARCOS mission the greatest possible weight all astronomers - regardless of their main research fields - should have the opportunity to participate in the composition of the observational program. For this purpose an invitation for proposals was issued by ESA in 1982 to the whole astronomical community. Altogether 214 proposals were received and supplemented with additional proposals closely related to satellite operation and data reduction. Summing up all proposals yields altogether more than 500 000 objects - obviously with many redundancies. All these objects had to be investigated by the INCA consortium to determine the final 118 321 stars of the HIPPARCOS Input Catalogue. The manner in which this was achieved is described in full detail in [5] where also a complete reference list of all papers related to the work can be found. In the following some insights into this work will be given.

4.1 The INCA Data Base

The first step towards the Input Catalogue was the generation of the INCA data base. At the very beginning the redundancies must be removed as far as possible, i.e. a set of all cross-identifications must be established. For this purpose SIMBAD the data base of the Centre de Données Stellaires (CDS) in Strasbourg was used as *master catalogue*. As a result almost 90 per cent of all stars could be identified immediately, and during progress of the work it turned out that in total about 215 000 stars were proposed for observation by HIPPARCOS.

Since the original mission was designed to observe about 100 000 objects a selection depending on magnitude and scientific importance of the objects had to be carried out. Apart from this the *survey* was defined containing about 55 000 objects down to a limiting magnitude $V \approx 8^m$ depending slightly on colour and galactic latitude of the stars.

Investigations on the amount of easily accessible information revealed that a large portion of INCA stars had still insufficient astrometric and/or photometric measurements. The positional accuracy required at the epoch of observation (about 1990) is $\pm 1\overset{''}{.}5$ rms and the maximal acceptable error in magnitude $\pm 0\overset{m}{.}5$. Therefore, the INCA data base had to be updated by continuously adding and correcting identifications, astrometric and photometric informations and further information, for example for double stars or variables. Additional ground-based astrometric and photometric observations were organized to supply at least the high priority stars with the necessary information.

4.2 Mission Simulations

A selection committee was nominated by ESA to judge the scientific value of the incoming proposals. This committee established a hierarchy of priorities ranging from 1 (highest) to 5 (lowest scientific importance) and made additional remarks on the feasibility of the proposals received. This served as a principal guide for inclusion of an object in the Input Catalogue.

The mission duration is limited and the total observing time necessary to get the required accuracy depends strongly on the magnitude of the star. Compared to a star with $H_p = 7^m$ (the HIPPARCOS magnitude is defined by $H_p = V + 0.38(B - V) - 0.13(B - V)^2$) a star with $H_p = 10^m$ needs almost twice as much observing time and a star at the limiting magnitude $H_p = 12\overset{m}{.}7$ even more than four times as much.

A certain star passes the field of view within 19.2 sec due to the spin motion of the satellite, and it has to share this available observing time with some four to five other stars passing the superposed field of view simultaneously. As a first approach to star selection the concept of *pressure* was developed, i.e. a factor was defined for each proposed star that is proportional to the total time required to observe the star itself and all its neighbours within a radius of $0\overset{\circ}{.}6$ having higher priority. The star was retained if its pressure factor turned out to be smaller than a predefined limit depending on the highest priority assigned by the selection committee (a star may be included in several proposals).

For the observing program selected in this way simplified global simulations were carried out under application of the nominal scanning law, determining a *yield* factor for each star. The yield is proportional to the total observing time for an individual star divided by the number of times the star crosses the field of view. For yield 1 the star achieves its nominal accuracy. Yields significantly larger than 1 allow, in principle, the introduction of more stars into the observing program, and stars with a yield significantly smaller than 1 should be removed.

Since these computerized simulations may produce undesired results, where several high priority stars compete with each other, efforts were made to correct this by INCA *tuning*. Stars proposed only for statistical purposes were tuned down, whereas individually interesting stars in small proposals were tuned up. For example, the first version of the Input Catalogue (IC1) contained only 86 per cent of the proposed nearby stars, whereas in successive versions more than 96 percent could be retained.

Detailed simulations requiring a very large amount of CPU time were only performed on subsamples of stars. Selecting a set of test areas at various regions of the sky with different star densities it should be possible to test the diverse types of observing situations. The experiences out of this and the permanent updating of the INCA data base required several iterations, and the finally delivered Input Catalogue was IC5.

4.3 Astrometric Data

Though SIMBAD contains information on a few valuable astrometric catalogues a real astrometric data base does not yet exist. Therefore, at ARI the major astrometric catalogues were examined for their suitability to provide the required positional accuracy for the HIPPARCOS program stars. A hierachical order of quality was established, and in collaboration with the CDS an extensive cross-index was created allowing the compilation of a first version of the Catalogue des Données Astrométrique (CDA) [7] by selecting for each star the catalogue entry with greatest weight and transforming all stars - as far as possible - into the system of FK4.

Some important astrometric catalogues used are presented in Table 4. Column 2 gives the number of stars contained in the catalogue, column 3 the central epoch and column 4 the mean positional error at the central epoch. Combining these with the proper motion errors in column 5 the positional errors at epoch 1990 were calculated (column 6). PPM is a good example of what can be achieved by adding more - in this case older - observations to an already existing catalogue, AGK3, depending only on two positions with an epoch difference of about 30 years. Not only could the accuracy be improved but also the reliability, which is even more important.

Table 4. Some important astrometric catalogues

Catalogue	N	cent. epoch [years]	ε_{pos} [$''$]	ε_{μ} [$''$/cen]	ε_{90} [$''$]
FK5	1 535	1949	0.019	0.07	0.04
N30	5 268	1916	0.068	0.51	0.4
AGK3RN	38 627	1944	0.085	0.48	0.25
GC ($m > 5$)	31 860	1900	0.16	1.2	1.1
SAO	258 942	1925	0.17	1.3	0.9
AGK3	181 581	1943	0.13	0.9	0.5
PPM	181 581	1930	0.093	0.44	0.28

Later versions of the CDA were improved by permanent updating, i.e. error correcting, addition of recently available catalogues and new observations of INCA stars. Furthermore, the selection method was refined in trying to estimate individual errors in positions and proper motions and retaining the best ones, [8] and chapter 9 in [5]. In the past, astrometric catalogues were usually produced for reference purposes. It was tried to achieve a regular distribution on the celestial sphere and to avoid stars causing observational difficulties, as, for example, double stars or variables. So, from the very beginning it was evident that a large portion of HIPPARCOS program stars, proposed for astrophysical reasons could not be found in any astrometric catalogue.

Indeed, the first cross-identification in 1984 of the then 330 000 CDA stars with the 215 000 proposed stars in the INCA data base showed that more than 30 000 INCA stars had no reliable positions for epoch 1990. This number reduced to 23 000 when a preliminary version of CPC2 - a photographic catalogue covering the southern hemisphere at a mean epoch 1968 - became available. Yet, at least for the high priority stars observational programs were inevitable.

There were two options, plate measurements or meridian circle observations. Plate measurements allow also the investigation of the neighbourhood of each star and can be easily repeated. Since there was no operating meridian circle this method was chosen for the sky south of $\delta < -17.^\circ 5$. The ESO Schmidt plates having an approximate epoch around 1976 were measured to an accuracy of about ± 0.3 arcsec. Identifying the stars with high proper motions - the latter could be got for example from Luyten's catalogues - should be sufficient to meet the required accuracy at epoch 1990. Since tests proved that in certain zones the main source catalogue SAO was not only quite inaccurate but also very unreliable the decicion was made to measure all proposed stars. In total about 135 000 measurements were obtained for some 85 000 program stars. In addition, 18 000 *parasitic* stars were measured, i.e. stars with $B < 15^m$ found within a radius of 25 arcsec around a program star. The latter information was important to cope the *veiling glare* effect (see chapter 4.5). Furthermore, stars in open clusters were measured on plates taken especially for this purpose.

North of $\delta = -17.^\circ 5$ only POSS plates were available. But due to their *old* epoch around 1954 the same strategy wouldn't work, because also smaller proper motions must be taken into account, and for non-astrometric stars these are usually unknown. At that time the Guide Star Catalogue (GSC) was not yet released. So, the two automatic meridian circles in Bordeaux and on La Palma were used to observe the high priority stars north of $\delta = -20^\circ$ providing positions for about 8 000 stars to an accuracy better than ± 0.2 arcsec at an epoch around 1986.

Early this year there remained still a few hundred high priority stars - variables, faint K- and M-dwarfs and high-proper motion stars - which couldn't be measured due to identification problems or very poor positions ($\pm$ 1 arcmin) supplied by the proposers. Most of these could be saved for the mission by identification with the Astrographic Catalogue and/or with the help of finding charts prepared from the recently completed GSC.

4.4 Photometric Data

Photometric data for the proposed stars were needed for various reasons. First of all, the allocated observing time to reach the required accuracy depends very strongly on the stars magnitude. Furthermore, the magnitude-limited survey had to be defined and, last but not least, a large amount of ground-based photometry is necessary to reduce and calibrate during the mission the incoming satellite data: H_p for the main mission and B_T, V_T for the TYCHO experiment.

In examining the available photometric data, which are systematically compiled at the Lausanne and Geneva observatories, at the beginning of the work on the Input Catalogue, it turned out that for about 16 700 proposed stars the information was incomplete or unreliable. Since these stars were mainly faint ones, it was very unrealistic to get accurate photometry in time, i.e. a pre-selection was necessary. Restriction to priority 1 stars and taking into account the first simulations reduced the number of *non-photometric* program stars by a factor of three to about 5 150 stars with the largest portion in the southern sky. Due to observing campaigns organized by the INCA consortium the number of non-photometric INCA stars could be reduced to 395 in May 1989.

4.5 Double Stars, Variables, and Other

During preparation of the Input Catalogue double and multiple stars, stars in very dense areas, and variable stars needed special attention due to the observing mode and the characteristics of the detector system. The presence of a star close to a program star may cause a perturbation depending on magnitude difference and separation.

Numerical studies were carried out to investigate the possible results, and finally it was decided to distinguish essentially three ranges of separation. For separation less than 10 arcsec the two stars can be merged to one entry and the photocenter will be observed. For separations in the range 10-30 arcsec and some given range of magnitude difference an alternating observation of the two components is required to get a precise determination of the astrometric parameters of the brighter component.

For separations larger than 30 arcsec the perturbing effect *veiling glare* can be regarded as an additional term in the total expected error on the signal received from the program star. Depending on the size of this additional term three possible cases must be distinguished:

(1) The observation of the program star is simply impossible due to the fact that it is too near and/or too faint with respect to the perturbing star.
(2) The perturbing star must be included for observation to correct for veiling glare. This caused the addition of about 70 *bright* stars to the Input Catalogue.
(3) The perturbation can be neglected.

To cope with these problems a considerable amount of work had to be put into the preparation of the double star data. The existing compilations such as the *Index Catalogue of Double Stars* (IDS), providing only the first and last observation of double star measurements, were not appropriate for the present purpose. Therefore, the *Catalogue of the Components of Double and Multiple Stars* (CCDM) was created (see chapter 12 in [5]). It combined the information from the IDS with accurate positions mostly from CDA and contains one data record for every component. This allows - as far as possible - a precise prediction

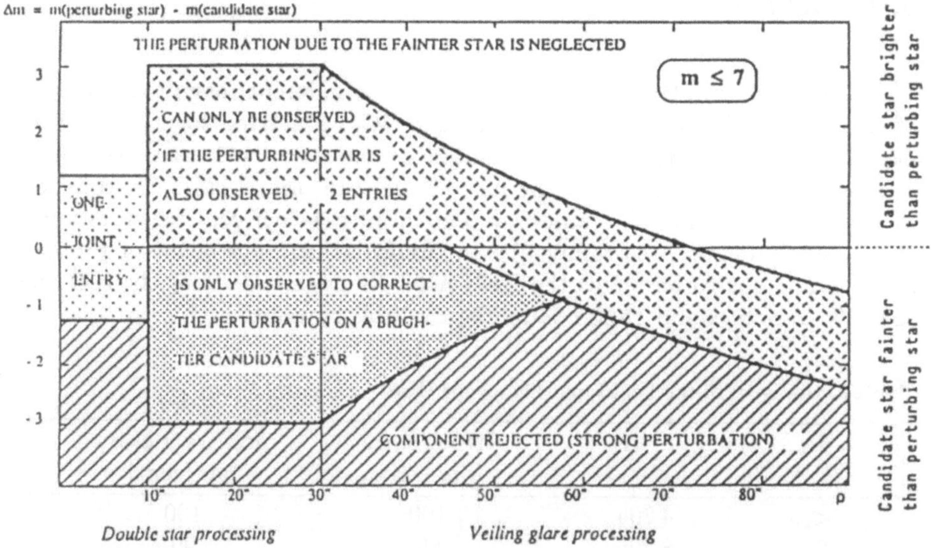

Fig. 3. The principle of double stars and veiling glare processing. Presented is the case for stars brighter than magnitude 7. An analogeous scheme applies to fainter stars. From C. Turon: in Proc. Sitges Coll. *Scientific Aspects of the Input Catalogue Preparation*, eds. J. Torra and C. Turon 215 (1988)

of the 1990-position. The compilation of the CCDM required not only a lot of additional observations - astrographic and photometric - to complete the necessary informations, but also the use of all already known measurements of some thousands of double stars to compute the relative positions for epoch 1990. The investigation of the ESO plate measurements in the southern hemisphere (see chapter 4.2) revealed almost 9 700 new systems with separations from 2 to 30 arcsec. Unfortunately, in the northern hemisphere a similar systematic search for unknown components could not be performed. During updating the CCDM a considerable list of inconsistencies could be removed in the INCA data base as well as in the CDA.

Variable stars, which are of great scientific importance, caused problems from the beginning. It was quite hard to identify these stars in the sky or on plates to get better positions or photometry. Since precise ephemerides must be provided to optimize the available observing time or to know the period when they are not observable at all.

The same applies to the 63 minor planets and two planetary satellites included in the Input Catalogue to get a link to the dynamical reference system. On the other hand links to an extragalactic reference system and to a radio reference frame will be established by the observations of carefully selected radio stars or stars near faint extragalactic sources. Since extragalactic point sources are too faint be observed directly by HIPPARCOS the interpolating measurements will be performed by the Hubble Space Telescope.

5. Conclusions

In Table 5 the number of stars (column 2) for different magnitude intervals of the final version of the Input Catalogue (IC5) are given together with the percentage of all proposed stars (column 3) and priority 1 stars (column 4) retained. It demonstrates that the percentage of priority 1 stars could be kept relatively high even at the faint end. The drop to 44 per cent for stars fainter than magnitude 12 is certainly due to the fact that quite a lot of proposed stars turned out to be below the HIPPARCOS magnitude limit. In IC5 there are 5834 variables and 15 955 double or multiple systems.

Table 5. Statistics of the final Input Catalogue (IC5)

H_p	N	INCA stars [per cent]	priority 1 stars [per cent]
< 6	4 204	100	100
6-7	8 505	99	99
7-8	22 241	92	98
8-9	41 100	74	96
9-10	29 402	41	91
10-11	9 314	26	83
11-12	2 909	29	86
≥12	646	12	44

The enormous improvement in ground-based astrometric, photometric and double star data during preparation of the Input Catalogue are of importance far beyond the HIPPARCOS mission - some 60 per cent of the relevant data in the SIMBAD data base has been updated. Therefore, the main content of the INCA data base will be published in two parts about eighteen months after launch of the satellite, i.e. in early 1991. The first part for which also a printed version will be established contains all 118 321 stars of the Input Catalogue itself including some annexes with information on duplicity and variability. The second part compiles the remaining stars proposed for which less complete information is available. But, nevertheless, they should also be of interest to the astronomical community and will be provided in machine-readable form.

Obviously, HIPPARCOS cannot provide the expected results. We may have to be content with much less accurate data, or we even may have to wait some more years for a second HIPPARCOS mission. Nevertheless, the progress achieved with the HIPPARCOS Input Catalogue can best be described by the words of Adriaan Blaauw [9] in his concluding remarks at the Aussois meeting on the *Scientific Aspects of the Input Catalogue Preparation*: 'Few, if any, satellite projects have produced already before launch so many things useful to astronomy as HIPPARCOS has, and this through the work of many investigators in the context of the INCA project. There is, for one thing, what I might call the general cleaning effect: I hope nobody will mind if, for a moment, and most disrespectfully, I call HIPPARCOS the biggest vacuum cleaner astronomers have ever seen.'

References

1. W. F. van Altena and J. T. Lee: in *Star Catalogues: A Centennial Tribute to A. N. Vyssotsky*, eds. A. G. Davis Philip and A. R. Upgren, L. Davis Press Schenectady, New York, 83 (1989)
2. M. A. C. Perryman: Nature **340**, 111 (1989)
3. U. Bastian: Mitteilungen der Astronomischen Gesellschaft **68**, 81 (1987)
4. The HIPPARCOS Mission: Pre-launch Status, Volume I. The HIPPARCOS Satellite, eds. M. A. C. Perryman, H. Hassan ESA SP-**1111** (1989)
5. The HIPPARCOS Mission: Pre-launch Status, Volume II. The Input Catalogue, eds. M. A. C. Perryman and C. Turon ESA SP-**1111** (1989)
6. The HIPPARCOS Mission: Pre-launch Status, Volume III. The Data Reductions, eds. M. A. C. Perryman, L. Lindegren, C. A. Murray, E. Høg, J. Kovalevsky ESA SP-**1111** 1989
7. U. Bastian and T. Lederle: in The European Astronomy Satellite HIPPARCOS, *Scientific Aspects of the Input Catalogue Preparation*, ESA SP-**234** 185 (1985)
8. H. Jahreiß: in Proc. Sitges Coll. *Scientific Aspects of the Input Catalogue Preparation*, eds. J. Torra and C. Turon 289 (1988)
9. A. Blaauw: in The European Astronomy Satellite HIPPARCOS, *Scientific Aspects of the Input Catalogue Preparation*, ESA SP-**234** 267 (1985)

The Cassini/Huygens Mission

W.H. Ip

Max-Planck-Institut für Aeronomie,
D-3411 Katlenburg-Lindau, Fed. Rep. of Germany

A brief description is given of the Cassini/Huygens project involving a Saturn Orbiter and a Titan Atmosphere Probe. This mission, with NASA providing the Cassini Orbiter and ESA the Huygens Atmospheric Probe, will address a wide range of fundamental issues in the origin of the solar system and planetary atmospheres.

1. Introduction

With the close encounter of the Voyager 2 spacecraft with the Neptunian system on August 24, 1989, the era of fast reconnaissance of the outer planets (except for the Pluto-Charon system) via flyby observations will reach an end. The next phase of outer solar system exploration, characterized by long-term orbiter observations, would await the launch in October, 1989, of the Galileo spacecraft towards the Jovian system. The Galileo mission, which is a joint NASA-BMFT project, consists of a two-year orbiter mission plus the delivery of an atmospheric probe to Jupiter. The date of Jupiter Orbit Insertion (JOI) is targeted to be in December 1995. The return of the scientific data from the Jupiter probe and the Galileo Orbiter are expected to increase dramatically our understanding of the atmospheric structure and composition of this giant planet, its magnetospheric system dominated by the SO_2 gas emitted from Io, and the Galilean satellite system.

As a second step to this major effort in the scientific investigation of the origin of the solar system and the origin of the life, the National Administration of Space and Aeronoautics (NASA), the European Space Agency (ESA) and the Bundesministerium für Forschung und Technologie (BMFT) are currently engaged in a program of space missions to a comet and to Saturn and Titan. The first part of the program is the CRAF (Comet Rendezvous and Asteroid Flyby) mission to comet Kopff and asteroid Hamburga. In this joint venture of NASA and BMFT, the BMFT would provide the retropropulsion engine for the spacecraft. The other part of this ambitious program is the Cassini Saturn Orbiter/Huygens Probe mission to Saturn and Titan. In this joint effort between the U.S. and Europe, NASA would provide the Saturn Orbiter and ESA the Huygens atmospheric probe to Titan. Both the CRAF spacecraft and the Cassini Orbiter share common heritage of the Mariner Mark II spacecraft which are being designed at the Jet Propulsion Lab (JPL). After the selection of the Cassini/Huygens mission by the Space Program Committee (SPC) of ESA as the next new space project in Dec. 1988, the approval of the Mariner Mark II program by the US Congress is still pending. There is every reason to believe that a positive decision will be forthcoming.

The present arrangement between ESA and NASA is that the Announcements of Opportunity (AO) should come out in October, 1989. The experimental proposals

Reviews in Modern Astronomy 2 Editor: G. Klare

for the Saturn Orbiter would be evaluated by NASA and those for the Huygens Titan Probe by ESA. In the selection process, there would be participation by ESA and NASA representatives on both panels. The final selection of instruments is expected to be made in October, 1990. In the following, the scientific objectives and mission designs for the Cassini/Huygens mission as reported in the Cassini Phase A Study Report (SCI(88)5) will be briefly summarized. This joint ESA-NASA study has been supported by a science team and the engineering staffs from ESTEC, ESOC and JPL over a period of five years.

2. Titan and Saturn

In Galileo's telescopic observations of solar system objects, Saturn appeared as the most perplexing planet because of its variable appearance. A lot of scientific debates had raged centering around Saturn's variable configuration at this starting point of planetary astronomy. It was Huygens in 1665 who found the solution to the problem by suggesting that Saturn was surrounded by a ring system. (His discovery of Titan is now honored by the naming of the Titan atmosphere probe as the Huygens Probe.) The Observations by the Voyager spacecraft showed that the Saturnian ring system actually consists of a myriad of narrow ringlets with widths ranging from a few km to a few tens km. Many of these fine structures, including the twisted F-rings and the narrow rings embedded in the Cassini division, are not yet understood. The ring system as a whole is subject to various dynamical effects such as the gravity waves excited by the neighbouring satellites, the bombardment of interplanetary meteoroids, and the electromagnetic coupling with the planetary magnetosphere. The corresponding time scales for the dispersion of the rings have been estimated to be as short as $10^7 - 10^8$ years. The detailed observations afforded by the Cassini mission should permit definite answers to many of these outstanding questions.

Other unique features of the Saturnian system include the axial symmetry of the planetary magnetosphere, the plasma interaction of the icy satellites and ring particles with the huge magnetosphere, the plasma interactions of Titan's atmosphere and the very complex geological structures of the icy satellites. For example, only when we return to Saturn with the arrival of the Cassini spacecraft would we be able to study the exact cause of the very different albedos on the leading and trailing sides of Iapetus. At the same time, the possibility of re-surfacing of liquid water on Enceladus would be thoroughly investigated by a collection of advanced instruments on the spacecraft with unprecedented spatial resolution.

The Voyager 1 encounter with Titan has produced a most intriguing picture of this gaseous satellite. Its thick atmosphere with a surface pressure of 1.5 bars is composed mainly of nitrogen and methane. The photolysis of the methane gas has led to the escape of dissociated hydrogen gas and the formation of complex organic hydrocarbon molecules and heavy polymers in the atmosphere. The orange colored aerosol particles so formed are distributed in an opaque layer obscuring the satellite surface from direct imaging. A continuous photolysis in this nitrogen-rich atmosphere could lead to the presence of a km-thick ocean made up of CH_4 and C_2H_6 liquids at Titan's surface. The radar instrument and the microwave spectrometer/radiometer experiment onboard the Cassini spacecraft will address the issue of global ocean. During close flybys of Titan, remote sensing instruments such as the far infrared experiment, the ultraviolet spectrometer, and the microwave spectrometer/radiometer experiment will study the atmospheric structure and composition in great detail. The heart of the Titan science, however, lies in the atmospheric descent observations of the Huygens Probe which is to be described below.

3. The Cassini Orbiter Mission Profile

According to trajectory studies, the optimal launch date is in April, 1996 after which a period of 6.8 years of interplanetary cruise and manuevers will follow before Saturn arrival in October 2002. A close flyby of an asteroid (66 Maja is the current favorite) sometimes in 1997 is planned. The Jupiter flyby in 2000 will allow the exploration of a magnetospheric region not accessible to Voyager and Galileo. In addition, the interplanetary cruise phase will provide ample opportunities for study of the heliosphere up to a distance of 10 AU.

Upon Saturn arrival, the spacecraft will make a burn at the first pericone to enable Saturn Orbit Insertion (SOI). During the first inbound trajectory, a close flyby of Titan (up to an altitude of about 1500 km) could be scheduled such that new and important data on the atmospheric structure of Titan can be obtained. A number of uncertainties remaining in our post-Voyager knowledge of Titan's atmosphere (i.e., temperature and wind speeds at high altitude)thus could be eliminated. In this manner, the descent profile of the Hygens Probe may be further optimized.

In the design of the mission, there would be a number of trade-offs to be investigated such that different scientific objectives can be optimized. For example, to be studied would be the opportunity of ring study during the SOI phase at which time the ring system may be observed with the highest resolutions; and the plasma environment in the vicinity of the rings could be surveyed by the particles-and-fields experiments. Also, the satellite and magnetosphere tours might be modified in the course of the mission as the atmospheric environment of Titan has become better known for the purpose of trajectory calculations. As Titan is the only satellite which can provide enough gravitational assistance for shaping the trajectory of the Cassini Orbiter, in the course of the mission there would be about 30 or more close encounters with Titan down to an altitude of about 1000 km. During the nominal mission of four years, the orbit of the spacecraft would be gradually pumped up such that at the end of the mission, the orbital inclination of the spacecraft would reach 80° or higher. These high-inclination orbits are required to investigate the auroral zone and cleft region of the polar ionosphere. These high-latitude regions are believed to be the sites of strong plasma outflows as well as plasma turbulence and wave generations unique to the Saturnian magnetosphere.

At SOI, the spacecraft will be inserted into a highly elongated orbit with a period of around 3 months. At its first inbound leg, the spacecraft will release the Huygens Probe twelve days before the Titan overfly. The Orbiter will be subsequently deflected and delayed in such a way that the spacecraft will reach the closest approach of Titan at around 1500 km and about 3 hours after the atmospheric entry of the Huygens Probe. The main function of the Cassini Orbiter in this phase is to act as a radio-relay station to transmit the data to Earth via NASA's Deep-Space Network (DSN).

4. The Huygens Probe Mission Profile

The initial entry speed of the Probe would be about 6 km/s. Its 3.1 m diameter aerodynamic decelerator will slow down the descent speed to subsonic value at an altitude of < 175 km. The deployment and staging of the parachute system would allow a total descent time of 2-3 hours facilitating scientific measurements by the Probe instruments. At surface impact, the terminal speed would be on the order of a few m s^{-1}. It is thus possible that certain post-impact measurements could still be made before the loss of radio contact with the Orbiter. The design of the Huygens Probe, however, would not guarantee its survival after impact at either solid or liquid surface.

During the atmospheric descent, the instruments on the Huygens Probe will be operating to gather important chemical and physical information pertinent to the structure and composition of Titan's atmosphere. For example, the gas chromatograph/neutral mass spectrometer (GC/MS) will measure the atmospheric composition as a function of altitude and in combination with the aerosol collector and pyrolyser (ACP), will provide chemical analysis of the aerosols. Composition measurements of selected gas species (i.e., $CH_4, CO, CO_2, C_2H_2, C_2N_2, C_3H_4, C_3H_8$, etc) with high precisions would be made by the probe infrared laser spectrometer (PIRLS). The descent imager/infrared spectral radiometer (DI/SR) would allow in-situ measurements of the atmospheric and cloud structure plus imaging of the surface features. Doppler radio tracking of the Probe from the Orbiter should provide ground-truths for zonal wind profile when compared with measurements from the microwave spectrometer/radiometer on the Orbiter. Last but not least, the accelerometers placed in the Huygens Probe would yield very interesting information on the surface properties at impact. In the event of landing on a liquid surface, the possible incorporation of a science package designed to sample physico-chemical nature of the liquid material would give us a first glimpse of the state of the Titan Ocean.

5. Summary

The Cassini/Huygens mission to Saturn and Titan is one of the most ambitious planetary missions ever planned. Its wide scope of scientific investigations covers almost all fields of planetary sciences and space physics. The international cooperation fostered by NASA, ESA and BMFT in the framework of the Mariner Mark II Program certainly will lay the foundation for future research in the areas of the origin of solar systems, the formation of planets and ring systems, the origin and evolution of planetary atmospheres, and pre-biotic chemistry in primordial atmospheres.

Plans for High Resolution Imaging with the VLT

J.M. Beckers

European Southern Observatory, Karl-Schwarzschild-Str. 2, D-8046 Garching b. München, Fed. Rep. of Germany

1. INTRODUCTION

Astronomical telescopes have as their main function the collection of radiation of astronomical objects and its concentration in as sharp an image as possible. In the past most astronomers have emphasized the light gathering function as is evidenced by the common characterization of telescopes by their diameter. Only in specialized subfields, like solar and planetary astronomy, has the angular resolution been a prime characteristic. In the recent renewed interest in building large telescopes both light gathering and angular resolution have received major emphasis since it is clear that also in nighttime observations of faint objects both contribute more or less equally to the quality of the results. Improvements in telescopes, seeing management of the telescope, building and site, and sophisticated manipulation of the incident radiation by adaptive optics, interferometry and other techniques have and will result in much improved astronomical observations and hence a better understanding of the universe.

One can divide the benefits of higher angular resolution in three classes. The first is the *Study of the Structure of Astronomical Objects*. It applies to the morphology of extended sources like stellar surfaces and envelopes, regions of star and planetary system formation, resolving binaries and star clusters, resolving gravitationally lensed QSO's, resolving stars in other galaxies, studying the morphology of distant, large z, galaxies, and the evolution of supernovae and their early remnants. In this use of high angular resolution a qualitatively different benefit is obtained then that of the larger collecting area so that it is hard to say which is more determining for the merit of the telescope without identifying the scientific objective. That is not the case for the following two benefit classes.

The second and third benefits of higher angular resolution applies to point or pointlike sources (size comparable to point spread function PSF). The second refers to the *Improved Spectral Resolution* which results from the ability to narrow the spectrograph slit. As is well known, it is hard and expensive to build grating spectrographs with high spectral resolution to mate with telescopes with large apertures (diameter D). They require large gratings which can only be made by mosaicing smaller ones and they are bulky requiring often optical feeds which are lossy in light transfer. Image slicers are sometimes used to ease these limitations at the cost of loss of ability to study a field of view and some light loss of their

Reviews in Modern Astronomy 2 Editor: G. Klare

own. In practice these problems lead to limited size spectrographs resulting in limited spectral resolution R (= wavelength divided by spectral bandwidth). The smaller image sizes (width d) resulting from the higher angular resolution gives then a proportionally higher R. For example: a R2 grating (tangent blaze angle = 2) of 20 cm width used at an 8 meter telescope results in a spectral resolution of R = 20,000 for a 1 arc sec slit width but R = 200,000 when the slit width is reduced to .1 arc sec. In poor seeing (d $\gg$ 1 arc sec) this would mean an effective loss in telescope light gathering proportional to the slit width because of spill over on the slit. Improving the image quality d therefore gives the same benefit as increasing the telescope area (or D^2).

Thirdly improved angular resolution will lead to *Improved Detection of Pointlike Sources Against a Dominating Background* as is the case at visible wavelengths for faint (V > 21) sources due to sky background and in the infrared because of thermal radiation from either the telescope/instrument or the sky. In this case the signal-to-noise ratio (SNR) in the observations is proportional to D/d so that improving image quality is as important as improving the telescope size. In the case of spectroscopy of these sources the second and third benefits of course add so that the gain in image quality becomes dominating.

In the second section I will review the methods used to improve the angular resolution of large telescopes. Sections 3 and 4 will describe plans to use the VLT with adaptive optics and interferometric imaging techniques to improve image quality, and section 5 with place the VLT plans in broader (past and future) perspective.

2. WAYS TO INCREASE ANGULAR RESOLUTION IN ASTRONOMICAL TELESCOPES

In the following I will discuss the prerequisites for high resolution imaging (site, telescope, building) and I will distinguish between what I will call "geometrical methods", which do not use the wave nature of light in the manipulation of the electromagnetic signal, and "wave optics methods", which do.

2.1 Prerequisites

The prime prerequisite must be the selection of the best available astronomical site in terms of "seeing". Seeing in this context primarily emphasizes the image quality for long exposures (assuming perfect tracking. no optical aberrations and no atmospheric chromatic dispersion), or average wavefront distortions, but should also include the rate of change of the wavefront distortions. Site selection has generally been based on the former. Not all sites are selected on the basis of the results of a site selection campaign. Often seeing and other measurements have been made for "site confirmation": of a site which was selected on the basis of other considerations. Acceptance of those sites then depends on the quality of seeing which is at that moment in vogue, a dangerous procedure because it has become clear that astronomers have often underestimated the inherent optical quality of the atmosphere because of image deteriorations by their telescopes, domes, and other manmade causes.

A site comparison between an ocean site (Mauna Kea) and inland site (Mt Graham) done to select a site for the US National Telescope agreed with Walker's notion[1] about the better quality of the ocean sites. Although subject to some uncertainty in interpretation in absolute values[2] this comparison resulted in estimates for the median image size (FWHM) of .5 arc sec at the ocean site and .9 arc sec at the inland site. Unfortunately there are no acceptable (high and clear) ocean sites (like Mauna Kea and Roque de la Muchachos) in the southern hemisphere (La Réunion appears too cloudy). In the following the VLT site will assumed to have a median atmospheric FWHM of .9 arc sec.

The second prerequisite is to build an observatory on the site which does not negate the inherent quality of the seeing of the site. That means excellent optics and no manmade seeing due to heating in the telescope, dome and building and daytime solar heating and nighttime cooling by radiation. The successful philosophy accepted for the MMT was to eliminate these effects by "seeing management" aimed at creating a uniform thermal environment by reducing as much as possible heating, by insulating surfaces to reducing time constants, by free air circulation in the dome, and by reducing radiation cooling effects. As a result the observatory seeing was reduced to below .1 arc sec as was evidenced on those scarce seeing nights when atmospheric seeing with the individual 1.8 meter telescopes was virtually undetectable. The ESO NTT followed a similar philosophy which resulted in the unparalleled .3 arc sec image which was obtained recently. The MMT was unfortunately never able to give such image quality because of the quality of the optics.

I have distinguished[3] between "good" telescopes and "poor" telescopes on the basis of their inherent image quality (optics and observatory seeing) as compared to the inherent seeing of the observatory site. The latter should be taken as the seeing on excellent seeing nights (eg the ≈ 3 percentile nights which result in about half the FWHM of the median nights). A good telescope is one in which this telescope/observatory image width ($FWHM_T$) is substantially less (eg one half) than that of the atmosphere ($FWHM_A$) on excellent seeing nights, or when expressing both atmospheric and telescope wavefront deformations[3] in the so-called Fried's parameter r_o, in which $r_o^T \gg r_o^A$. The ESO-NTT with its ≈.1 arc sec $FWHM_T$ is an example of a good telescope with a large aperture. It sets the standard for telescopes to come. Specifications for the VLT telescopes call for the same qualification.

2.2 Geometrical Optics Methods

There are a number of ways to obtain better images with the VLT without recourse to image reconstruction techniques by interferometric or adaptive optics means. These so-called "geometrical optics' methods include:

(a) selection of a *good seeing night*. The NOAO and ESO/VLT site comparison campaign showed that one might expect half the median image size for the best ≈ 3 percentile nights. For the assumed VLT site quality that implies .45 arc sec at visible wavelengths and somewhat better (.3 arc sec at 5 µm) in the infrared (see figure 1).

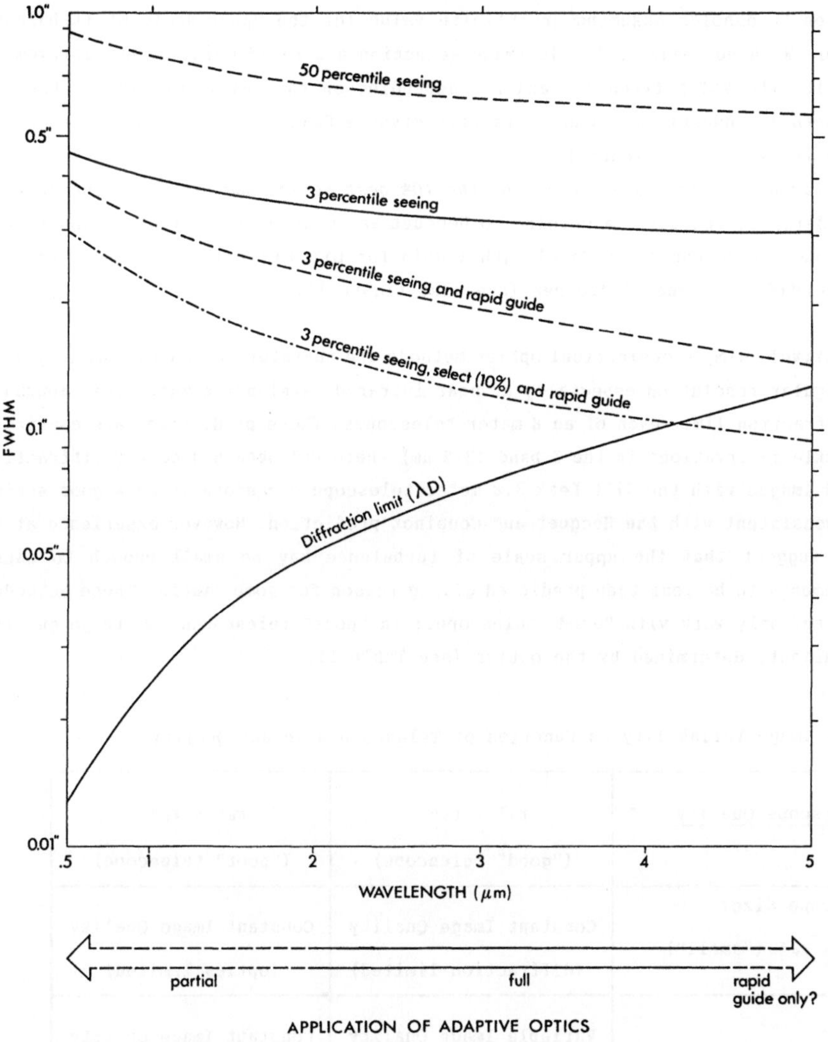

Figure 1: Image quality as a function of wavelength assumed for the VLT site and for various geometrical optics methods used to sharpen the images assuming a purely Kolmogoroff type spectrum for the wavefront corrugations (upper scale of turbulence >> 8 meter). Below abscissa: wavelengths at which full and partial adaptive optics will be applied.

(b) *rapid guiding* while integrating the radiation on the detector. This technique is being used on Mauna Kea[4] and La Silla[5] and can result in substantial image improvements depending on the ratio of telescope diameter to Fried's parameter (D/r_o) and on the spatial spectrum of atmospheric seeing. Hecquet and Coupinot[6] analyzed the improvements which may be expected by using short exposures and different image

selection thresholds assuming an infinite value for the upper scale of turbulence (L >> D). With no image or little image selection and rapid guiding the improvement ranges for the VLT between 1.2 and 2.0 depending on the wavelength (since D/r_o is wavelength dependant). At .5 and 5 μm this gives a FWHM of .3 and .14 arc sec resp. for the VLT (see also figure 1).

(c) *rapid guiding and selection of the 10% best seeing moments* while integrating on the detector increases according to Hecquet and Coupinot[6] the improvement factor to 1.5 and 3.0 depending on wavelength (again for L>> 8 meter). At .5 and 5 μm this gives a FWHM of .3 and .1 arc sec (see also figure 1).

Relatively simple geometrical optics methods are therefore very powerful in giving high angular resolution especially at near infrared wavelengths where they approach the diffraction limit even of an 8 meter telescopes. These predictions are confirmed by speckle observations in the K band (2.2 μm) where we[7] obtained nearly diffraction limited images with the Kitt Peak 3.8 meter telescope occasionally on a good seeing night consistent with the Hecquet and Coupinot prediction. However experience at La Silla[5] suggest that the upper scale of turbulence may be small enough to cause improvements to be less then predicted giving reason for some caution.These methods, of course, only work with "good" telescopes; in "poor" telescopes the image quality is invariant, determined by the optics (see Table I).

TABLE I Image Variability as Function of Telescope Size and Quality

Telescope Quality	$r_o^T > r_o^A$ ("good" telescope)	$r_o^T \leq r_o^A$ ("poor" telescope)
Telescope Size: $D \leq r_o^A$ ("small")	Constant Image Quality (diffraction limited)	Constant Image Quality (optics limited)
$D > r_o^A$ ("large")	Variable Image Quality (image selection dependant)	Constant Image Quality (optics limited)

2.3 Wave Optics Methods

To obtain even higher angular resolution with the VLT it is necessary to go to interferometric techniques. These include:

(a) *Speckle Imaging* in which diffraction limited images are obtained with the individual telescopes after detection by analysis of many atmospherically distorted, speckled images of the object. This technique is rapidly reaching a substantial level of maturity in which important astronomical results are obtained. Techniques now being developed for 4 meter class telescopes will directly be applicable to the VLT.

This form of high resolution imaging results of course only in the first class of benefits (section 1), it does not help spectroscopy or the detection of objects against a bright background. On the other hand it is a well developed technique which will result with the VLT in 1/80 arc sec resolution at .5 μm, well beyond what is possible with geometrical techniques. Also it is likely not to be replaced soon by other techniques (adaptive optics, see below) because of their cost and because of the absence of ways to sense the wavefront at short wavelengths.

(b) *Adaptive Optics* in which the wavefront is restored before detection by means of a rapid wavefront sensor acting on the object itself or on a nearby object within the so-called isoplanatic patch and wavefront correction by an adaptive mirror. A number of project are now underway to implement adaptive optics on astronomical telescopes[8,9] and it is expected that some of these systems will be operational soon. The VLT will use adaptive optics designed to function well at near infrared wavelengths (2 μm) and above to give diffraction limited images for the individual telescopes. It will give all three benefits of high angular resolution: morphology, high resolution spectroscopy and improved detection against a bright background. In addition it will greatly benefit interferometric imaging with all telescopes since the concentration of all radiation in one Airy disk in each telescopes will increase the sensitivity of the interferometric array tremendously even at shorter wavelengths where the partial effect of adaptive optics will be to give a speckle image which includes one very bright speckle. The principles of astronomical adaptive optics have been discussed elsewhere[10,11,12] and will not be repeated here. In section 3 I will discuss its application to the VLT.

(c) *Interferometry* using the VLT as an interferometric array, the optical analog of radio interferometric arrays. Experiments in Europe, the USA and Australia have clearly demonstrated our ability to combine the light of widely separated optical telescopes coherently and to detect both the amplitude and the phase of the resulting interference fringes. The VLT array is conceived to make use of this and of the imaging techniques already existing in radio astronomy to achieve angular resolutions more than an order of magnitude higher than can be achieved with the individual 8 meter telescopes (≤ .001 arc sec). Together with the adaptive optics in each of the telescopes, including the 2 meter class "roving" telescopes which aim at better filling the (u,v) plane, this will be achievable on faint objects. Section 4 will go into more detail on the VLT interferometric mode. Like speckle imaging interferometric imaging results however only in the first benefit of high resolution imaging (morphology) and not in the other two (spectroscopy,detection against sky background).

3. PLANS FOR THE INCORPORATION OF ADAPTIVE OPTICS IN THE VLT

The VLT will incorporate both "active" and "adaptive" optics. It is now convention to refer to the relatively slow control (< 1 Hz) of the primary mirror shape and the position and tilt of the secondary mirror as *active optics*. It is intended to remove drifts in the telescope optical quality caused by eg thermal and gravitational effects to make it a "good" telescope as defined in section 2.1. The goal of the *adaptive optics* is to remove the much faster changes in the wavefront caused by atmospheric seeing and thus effectively making the VLT telescopes behave like "small, good" telescopes, in the sense of the definitions given in Table I, giving always diffraction limited imaging. Rapid wavefront sensing will be done with a Hartmann-Shack sensor using either natural stars or artificial stars[13]. Because of photon flux limitations and of detector noise a significant sky coverage is limited to the longer wavelength only (≥ 2 μm when using natural stars, ≥ 1 μm when using artificial stars) where the larger r_0^A values allow a coarser and slower sampling of the wavefront and where the size of the isoplanatic patch is larger. For .9 arc sec seeing at visible wavelengths (.5 μm) D/r_0^A equals 31 and 13.6

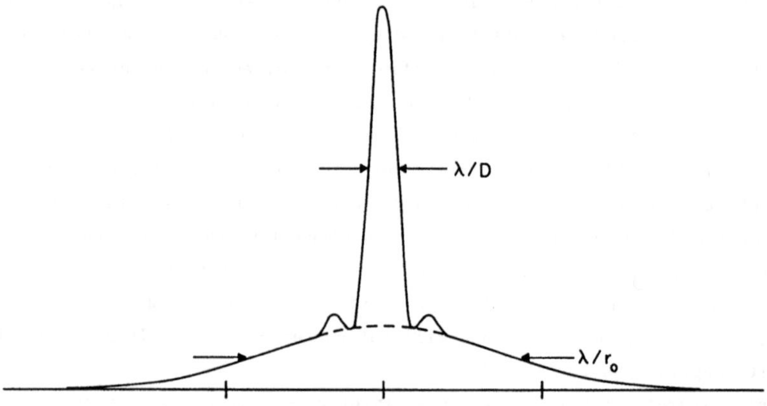

Figure 2: Point spread function for partial adaptive optics system. The spike at the center approximates the diffraction profile of the telescope, the broad background corresponds to the original seeing disk.

at 1 μm and 2 μm wavelength respectively so that about 980 and 190 spatial adaptive elements are needed for a well working adaptive optics system, and 1/4 of that for a system which works at the 3 percentile .45 arc sec seeing level. Current plans for the VLT adaptive optics call for a 16x16 element square actuator array resulting in about 200 elements inside the pupil allowing it to work well at median seeing conditions down to 2 μm and in excellent seeing conditions (using artificial stars when available, or on bright stars only) down to 1 μm.

However, even at shorter wavelengths or under worse seeing conditions the adaptive optics system is expected to improve the image quality. Computer models of the functioning of adaptive optics have shown that the average point spread function (PSF) of such "partial" adaptive optics system consists of the sum of two profiles, one being the diffraction profile of the telescope (approximately the Airy disk), the other being the original uncorrected image profile (see figure 2). The performance of a partial adaptive optics system can be characterized by the fraction of the energy present in the diffraction spike. This fraction Σ is approximately

TABLE II Modified Strehl Ratio Σ as a Function of Wavelengths and Seeing

Wavelength (μm)	Band	median seeing	excellent seeing
5.0	M	98% (1000)	99% (500)
3.4	L	95% (1000)	98% (600)
2.2	K	85% (900)	96% (900)
1.65	H	73% (840)	93% (900)
1.25	J	58% (750)	85% (890)
0.90	I	33% (620)	71% (810)
0.70	R	14% (380)	57% (740)
0.55	V	4% (170)	40% (630)
0.44	B	---	20% (440)
0.365	U	---	11% (340)

equal to the so-called Strehl ratio S (which is the central intensity of the PSF expressed in units of the central intensity of the perfect, undisturbed diffraction profile). Table II lists estimates for Σ for different wavelengths for the planned VLT adaptive optics system (working perfectly) and for the assumed median and excellent (3 percentile) seeing. In parenthesis is listed the approximate contrast of the center of the diffraction spike with respect to the seeing profile background. It should be noted that this contrast remains very high even when the adaptive optics is functioning far from its ideal condition. The reason for that is that at the shorter wavelengths where this occurs the narrowing of the diffraction disk and the slight increase in width of the seeing disk offsets to a considerable extend the decrease in Σ. It should also be noted that the partial mode is very powerful for some applications like interferometry and some spectroscopy and morphology. At the R band, for example, the width of the diffraction part of the PSF equals .02 arc sec, 30% of that of the Hubble Space Telescope (HST), whereas the energy in the spike is more than 1.5 and 6 times that of the HST respectively. The partial adaptive optics mode is however limited in its use because of the presence of the bright seeing halo around the diffraction spike and because the variations in seeing will cause major changes in Σ making photometry difficult.

At this moment the main technology issues which face the implementation of adaptive optics are the need to make current astronomical prototypes[9] in Europe, China and the USA work (military systems are rumored to be functioning routinely), to use these to optimize their control algorithms, and to fabricate the 256 actuator flat mirror needed for the VLT. In the longer term one might consider incorporating Hartmann-Shack wavefront sensors working in the K band which would significantly increase sky coverage in at low galactic latitudes[10,11], developing adaptive secondary mirrors which would reduce light losses and thermal emissivities because of the absence of the need to reimage the pupil, and the testing and incorporation of the artificial star concept.

4. PLANS FOR THE INCORPORATION OF INTERFEROMETRY IN THE VLT

4.1 Background

By its very layout the VLT presents itself as the optical equivalent of a radio interferometer. The VLT has the advantage over other ad hoc proposed interferometer plans using for example the large telescopes on Mauna Kea in that it from the beginning has been planned to work as an optical interferometric array and hence can be optimized for it. At this moment the implementation of the VLT Interferometric Mode is being defined with the help of a VLT Interferometry Panel consisting of experts in the discipline.

It used to be the case that the angular resolution of optical telescopes far exceeded that of radio telescopes even though the former were not diffraction limited. The advent of angular interferometry, which originated with Michelson's experiments in the optical wavelength region, in radioastronomy radically changed that. Currently radio astronomy notwithstanding its use of longer wavelengths, but also because of its use of longer wavelengths, has bypassed optical astronomy achieving angular resolutions with very long baseline interferometry of a milliarcsecond and less. The recent successes of radio astronomy has challenged the optical experimental astronomy community to implement the same at its shorter wavelengths. Starting with Antoine Labeyrie's experiments at CERGA it has grown to include other successful arrays in the USA and Australia. These have shown that effective light interference at optical wavelengths can be achieved, even though the short, even visible wavelengths and the atmospheric seeing make it a challenge to maintain fringe contrast. The large apertures of the individual MMT and CERGA/GI2T telescopes have shown that this is possible even in the so-called multispeckle mode ($D >> r_o^\lambda$). Important astronomical results have already been obtained with these arrays in their measurements of stellar diameters, binary stars, and in astrometry. Next is the challenge of astronomical high resolution imaging requiring three and more telescopes and the application of the sophisticated imaging algorithms developed already in radio astronomy to the optical observations. The technical and analytical pieces needed to do optical interferometric imaging at the milliarcsecond level are there, the requirements and tolerancing of the components of optical interferometric arrays are well understood, and the VLT initiative offers the challenge to do it at a very high sensitivity level.

In comparing optical and radio interferometry it is important to stress both the similarities and differences of the techniques. The similarities lies in their concepts and in the use of their observations of fringe contrast and phase to make images of the sky. Technically, because of their very different wavelengths/ frequencies, the ways and problems encountered to obtain these fringes/phases are very different. Tolerances for vibrations and surface ("optical") quality are much more stringent at optical wavelengths. Atmospheric seeing results in a very short time constant (≈ .01 sec) for the phase variations at visible wavelengths. Probably the biggest difference lies in the need the mix the electromagnetic radiation of the elements of the array before detection at visible wavelengths whereas by the use of local oscillator/heterodyne techniques in radioastronomy the radiation is detected at the individual telescopes. The IF frequency signal can then be electronically amplified and mixed with the IF frequencies of the other telescopes without loss in signal-to-noise due to signal splitting. In optical interferometry the original radiation has to be split in as many ways as there are other telescopes resulting in a loss in signal-to-noise in the signal of telescope pairs which is only partly offset by the gain in the number of baselines (the exception to this is the IR heterodyne array build by Townes and his collaborators which employs radio methods using a laser as local oscillator. This, however, at a cost of usable spectral bandwidth, and hence sensitivity). At optical wavelengths the gain of going to an interferometric array with many elements is therefore not as large as it is at radio wavelengths so that the VLT with its 6 to 8 telescopes is a reasonable choice. This disadvantage of optical vs radioastronomical interferometry is however offset by the ability of optical astronomy to use area detectors equivalent to "multiple feeds" in radioastronomy containing as many as a thousand or more Airy disk elements. Wide field-of-view optical interferometry is therefore possible and will be of interest for a number of astronomical objects.

4.2 The VLT Interferometric Mode

The VLT interferometric mode as described in the VLT proposal[14,15,16] combines the four large stationary 8 meter telescopes with two movable auxiliary telescopes with apertures in the 2 meter class. The baseline and configuration of the large telescopes is largely site dependant but will probably be in the 100 - 150 meter range and approximately linear. The exact location of these telescopes is presently being defined within these site constraints to give optimum (u,v) plane coverage as well as best image reconstruction conditions. Optimum (u,v) plane coverage with the large telescopes does not mean full (u,v) coverage because of the stationarity of the telescopes and their linear configuration. The function of the movable auxiliary telescopes is to give nearly full (u,v) plane coverage. To enhance this latter capability plans are being developed to add as many as 2 more movable 2 meter class telescopes to the array funded by individual ESO member countries. The resulting "small" array of four movable 2 meter class telescopes will be a powerful interferometric facility in its own right which will be available for interferometric imaging 100% of the time. The full "large" array which combines the smaller with some or all of the large telescopes will be available only part time because of other heavy demands on the 8 m telescopes, the amount of time depending on the success of interferometric imaging with the VLT and the need for adding the large telescopes to achieve the scientific objectives.

At this moment the final layout of the VLT interferometric mode is being defined by ESO with the aid of the VLT Interferometry Panel. This definition includes:

(i) The location of the 8 m telescopes within the restraints imposed by the still to be chosen site for the VLT (and the available budget !). Conflicting desires for maximum (u,v) plane coverage by using non-redundant array configurations and the optimization of sensitivity and dynamic range in the observed points in the (u,v) plane by using redundant array configurations have to be resolved.

(ii) The size and design of the movable auxiliary telescopes and their mode of transportation. Budget will dominate the determination of the size. The telescope mounts will be of the same type as that of the large telescopes (Alt-Az) to simplify the control of the relative image, pupil, and polarization frame of reference orientations between the unit telescopes of the array. Telescope transportation will be either in the so-called "hopping" mode in which the telescopes are moved from one location to another, using delay lines to adjust pathlengths, or in the continuously moving mode where delay lines can be avoided[17].

(iii) The form in which the beams will be combined for *on-axis operation* (field-of-view equal to the Airy disk). For efficient fringe formation for on-axis operation it is necessary to closely control the polarization properties of the different legs of the array, not just the retardation effects and partial polarizations due to off-normal reflections but especially the direction of the polarization frame of reference which can easily cause a destruction of fringe contrast[18,19]. This, as well as the concomitant restrictions on pupil and image orientation needed for wide field-of-view operation (see below), places stringent restriction on the optical design.

The layout of the beamcombination in the image mode will be similar to that proposed by Merkle[20] as shown in figure 3. It will include delay lines using a Cat's Eye reflector, optics which relay the light to the beamcombining optics (probably a modest size telescope) and which images the pupils near the beamcombining optics, an image tracker, a pupil tracker and a fringe tracker (not shown). Because of the

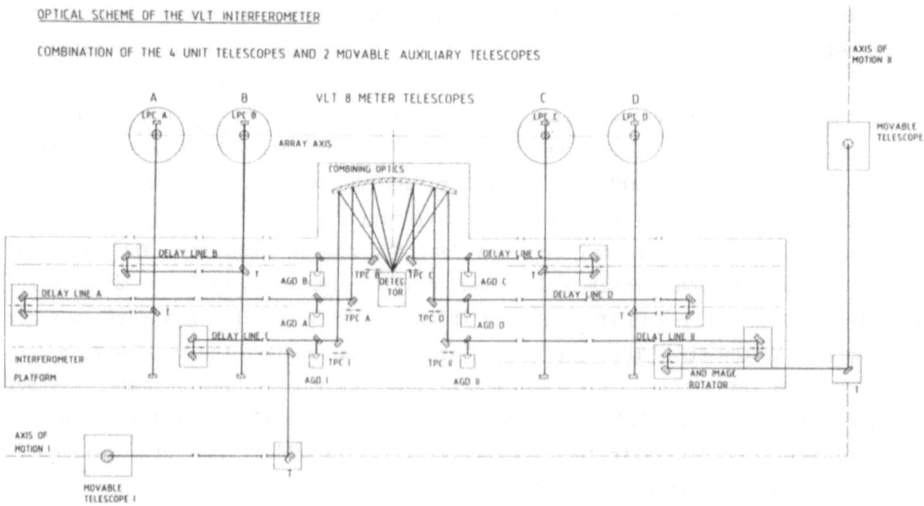

Figure 3: Coherent beamcombination of the VLT as proposed by Merkle[20]. AGD = acquisition and guiding unit, LPC = longitudional pupil control, TPC = transversal pupil control and T = image tracking mirror. Not shown is the fringe tracker (FT).

large number of optical surfaces involved in the beamcombination it is essential that they all be of very high quality (since optical quality is one of the other main causes for fringe contrast decrease) and of high reflectivity/throughput. By having separate optics for the long (> ≈ 600 nm) and short (≈ 400 to ≈ 600 nm) wavelength regions it is expected to maintain reflectivities better than 98% each in the protected, possibly evacuated, relay optics environment.

(iv) The design of a *wide field-of-view* beamcombiner. This requires the preservation of the varying pupil configuration of the VLT array[21,22] at all pointing directions in the sky. A wide (few arc seconds) field-of-view means that either the array stays phased (the white-light fringe stays with the image as it is moved through the FOV, the so-called "Phased-FOV") or coherent (fringes remain visible as the pathlengths vary by less than the coherence length, the so-called "Coherent-FOV"). A wide FOV is desirable when tracking fringes on a nearby object or when studying multiple or simple extended objects. To implement it requires the design of the optics to preserve the pupils in detail, their (variable) configuration and sizes, their orientations and their distances to the beamcombiner,

(v) The analysis on the tolerances on mechanical and optical performance of the array components. The exacting nature of the optical quality and the polarization requirements were already mentioned. Other factors which contribute to fringe contrast decrease are exposure time, vibrations, instrument drifts, improper phasing/ coherencing, inadequate detectors, etc.

(vi) The way to best combine the radiation from the large and small telescopes. When using adaptive optics (essential to achieve maximum sensitivity) the ≈ 16 x larger amount of radiation collected by the larger telescopes is put in an image ≈ 16 x smaller in area so that the interference becomes very ineffective in the overlapping area in an image plane where equal image scale, and hence relative pupil size, is preserved. Optimum interference occurs when the diffraction discs are matched in size in which case the equivalent interferometer aperture of a dual telescope array becomes equal to their geometrical mean[23] (or 4 meter for a 2 and 8 meter telescope). That however means that all exit pupil size must be the same thus destroying the wide FOV capability when using an optimal mixture of large and small telescopes.

The definition of the VLT interferometric mode is expected to be complete within a year. It is urgent because the final design of the large telescopes, an essential component of the VLT interferometric mode, is in progress and because the site work will start soon after the site selection in 1990. An extended phase A design of the auxiliary telescopes and the delay lines is being started and should lead to the construction in a few years.

5. HIGH RESOLUTION IMAGING WITH THE VLT IN A BROADER PERSPECTIVE

The VLT combines in it a number of most remarkable features. As a collection of four individual very large telescopes, which will each be outsized only by the 10 meter Keck telescope, they will provide astronomers with a multiple capability for doing state-of-the art groundbased astronomy expanding their earlier lightgathering capability of 4 meter class telescopes by a factor of four. By combining the four telescopes in the so-called incoherent mode the lightgathering capability will be close to that of an equivalent 16 meter aperture telescope. By optimizing individual telescopes for different wavelength regions it may be possible to increase efficiency as will be the plan to assign different instrument complements to different telescopes thus reducing losses in observing time which often accompany instrument changeovers.

The NTT performance has demonstrated that active optics holds every potential to make the individual VLT telescopes "good" telescopes, in the sense described in section 2, enabling them to give remarkably high angular resolutions with geometrical optics methods approaching the 8 meter diameter diffraction limit at 5 µm. Adaptive optics will extend this wavelength to 2 µm and below in the case of partial adaptive optics or by the use of artificial stars. Speckle interferometry techniques will reach down to the diffraction limit of the individual telescopes at even shorter wavelengths. High angular resolution will therefore, in addition to their large collecting areas, be a major feature of the individual VLT telescopes. These techniques to give high angular resolution with large telescopes is relatively a recent development as demonstrated in figure 4. Historically the increase in angular resolution may therefore turn out to be the dominating technical feature in the development of groundbased astronomical capability in the near future. I like to define as a figure of merit (FOM) for astronomical telescopes $D^2.d^{-2}.QE$. Table III lists the development of the FOM for single telescopes with time, spectacular in the 1970's because of the introduction of CCD arrays and in the present because of the increasing collecting area but especially because of the gains in image quality.

The gain in all parameters in the long term future (>> 2000) has to level off, however, as the sizes of individual telescopes level off as the result of engineering and fiscal limitations. As was the case in radio astronomy one then might expect the development of high angular resolution imaging with interferometric arrays from the ground as well as from space to lead the future. It is therefore especially this feature of the VLT that may be the most remarkable one from the long term development of astronomy perspective.

TABLE III Development of Telescope Parameters with Time

Year	D(cm)	d(")	QE(%)	Nr Pixels	FOM
1610	5	15	3[a]	≈ $3x10^6$	$3x10^{-1}$
1700	22	4	3[a]	≈ $3x10^6$	$9x10^1$
1800	120	3	3[a]	≈ $3x10^6$	$5x10^3$
1920	250	1.5	0.2[b]	10^7	$3x10^3$
1960	500	1.0	1[b]	10^8	$3x10^5$
1980	600	1.0	80[c]	10^5	$3x10^7$
2000	1000	0.02	80[c]	$4x10^6$	$2x10^{11}$
>2000	2000	0.005	80[c]	10^8	$1x10^{13}$

[a] = eye ; [b] = photography; [c] = CCD and CCD mosaics

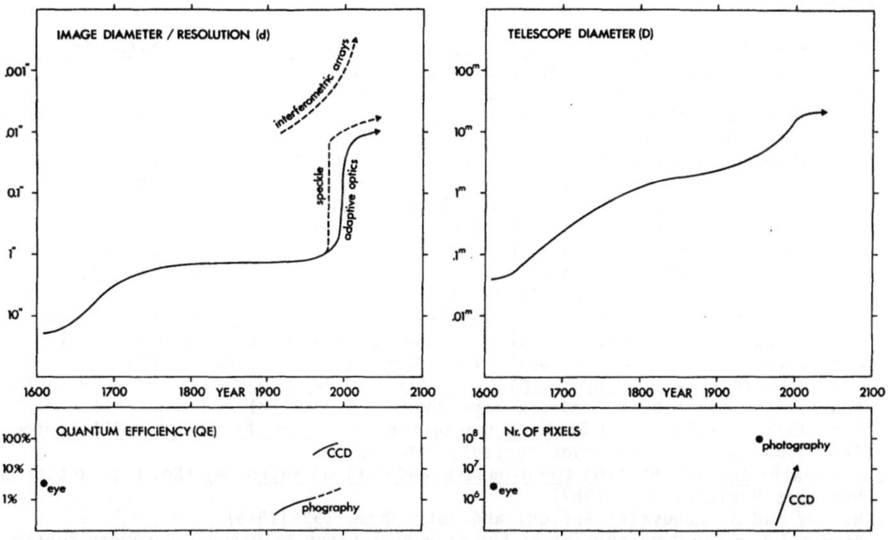

Figure 4: Historical and anticipated future development of ground based optical telescope parameters from the year of the invention of the telescope by Lippershey into the 21st century. Upper left: The evolution of the FWHM (d) of the telescope point spread function (full line) and the angular resolution using interferometric image recovery techniques (dashed lines). The smallest image diameter achievable with individual telescopes is expected to level off at about 20 meters diameter as individual telescopes reach their technical and economical size limit (upper right). Coupled strongly with the telescope parameters in terms of astronomical capability is the quantum efficiency (lower left) and pixel capacity (lower right) of detectors.

6. CONCLUSION

As the angular resolution in astronomical observations keeps on increasing with the use of high quality groundbased telescopes like the ESO NTT, with the further application of interferometric imaging and with the use of the Hubble Space Telescope, one might expect the interest in even further improvements in astronomical imaging to increase. The VLT will provide an exceptionally powerful capability with its combination of good imaging, adaptive optics, and interferometric imaging using its array configuration. Its implementation, like any pioneering venture, poses a challenge to those involved in its implementation. However, there are no pieces of the VLT system which appear mysterious or unsurmountable. The major challenge will be in the systems engineering to make all pieces work together with the required operational reliability.

REFERENCES

1. M.F. Walker: Identification, Optimization, and Protection of Optical Telescope Sites (Eds. R.L. Mills and O. Franz), 128 (1986)
2. H.M. Martin: Publ. Astron. Soc. Pac. 99, 1360 (1987)
3. J. M. Beckers: Solar and Stellar Granulation (Eds. R.J. Rutten and G. Severino, Kluwer Academic Publishers), 55 (1988)
4. L.A. Thompson and H.R. Ryerson: SPIE Proceedings 445, 560 (1984)
5. F. Maaswinkel, F. Bortoletto, S. d'Odorico and G. Huster: Instrumentation for Ground-Based Optical Astronomy (ed. L.B. Robinson, Springer Verlag), 360 (1987).
6. J. Hecquet and G. Coupinot: J. Optics (Paris), 16,21 (1985)
7. J.M. Beckers, J.C. Christou, R.G. Probst, S.T. Ridgway, and O. von der Lühe: ESO Conference and Workshop Proceedings No.29, High-Resolution Imaging by Interferometry (ed. F. Merkle), 393 (1988)
8. J.M. Beckers and F. Merkle: JNLT and Related Engineering Developments (eds. T. Kogure and A. Tokunaga), Astrophysics and Space Science, in press (1989)
9. J.M. Beckers and F. Merkle: SPIE Proceedings 1130, in press (1989)
10. J.M. Beckers: Towards Understanding Galaxies at Large Redshift, (eds. R.G. Kron and A. Renzini, Kluwer Academic Publishers), 319 (1988)
11. J.M. Beckers and L.E. Goad: Instrumentation for Ground-Based Optical Astronomy (Ed. L.B. Robinson, Springer Verlag), 315 (1987)
12. F. Merkle: Instrumentation for Ground-Based Optical Astronomy (Ed. L.B. Robinson, Springer Verlag), 366 (1987)
13. R. Foy and A. Labeyrie: Astron. and Astrophys. 152 (1985)
14. Proposal for the Construction of the 16-m Very Large Telescope, European Southern Observatory, (1987)
15. F. Merkle: J. Opt. Soc. Am., A5, 904 (1988)
16. P. Léna: ESO Conference and Workshop Proceedings No. 29, High-Resolution Imaging by Interferometry, (ed. F. Merkle), 899 (1988)
17. M. Vivekanand, D. Morris and D. Downes: ESO Conference and Workshop Proceedings No. 29, High Resolution Imaging by Interferometry, (ed. F. Merkle), 1071 (1988)
18. W. Traub: ESO Conference and Workshop Proceedings No. 29, High Resolution Imaging by Interferometry, (ed. F. Merkle), 1029 (1988)
19. J.M. Beckers: SPIE Proceedings 1166-37 (1989)
20. F. Merkle: ESO Conference and Workshop Proceedings No. 24, ESO's Very Large Telescope, (eds. S. D'Odorico and J.-P. Swings), 403 (1986)
21. J.M. Beckers: SPIE Proceedings 628, 255 (1986)
22. L.D. Weaver, J.S. Fender and C.R. De Hainaut: Optical Engineering 27, 730 (1988)
23. J.M. Beckers, Diffraction Limited Imaging with Very Large Telescopes, (NATO summerschool proceedings, eds. D. Alloin and J.M. Mariotti), in press, (1988)

A Correlation Tracker for Solar Fine Scale Studies

Th. Rimmele[1,*] *and O. von der Lühe*[2,*]

[1]Kiepenheuer Institut für Sonnenphysik, Schöneckstr. 6,
D-7800 Freiburg, Fed. Rep. of Germany
[2]Institut für Astronomie, ETH Zentrum, CH-8092 Zürich, Switzerland

1 Introduction

A solar feature correlation tracker was designed, built, and successfully tested in a joint effort of the National Solar Observatory (NSO) in Sunspot, USA, and the Kiepenheuer Institut (KIS), Freiburg, Germany. The purpose of the system is stabilizing image motion which is caused by telescope shake and by seeing at the post-focus instruments of the NSO and KIS vacuum tower telescopes in Sunspot and in Izaņa. The tracker system features a matrix diode array as detector, fast digital processors, and an agile mirror as the optical active element. The processor consists of commercial and in-house built hardware. A more detailed description of the system is in press[4].

2 Correlation tracker system overview

The tracking principle is based on a method studied by one of us earlier [2]. A reference picture of the area of interest on the Sun is initially generated. In operating mode, the "live" pictures $I_L(\vec{x})$ produced by a fast matrix detector are compared with the reference $I_R(\vec{x})$ by calculating their cross covariance $CC_{LR}(\Delta)$:

$$CC_{LR}(\Delta) = \int\int I_L(\vec{x}) \times I_R(\vec{x}+\Delta)d\vec{x} = \mathcal{F}^+\Big[\mathcal{F}^-\big[I_L(\vec{x})\big] \times \mathcal{F}^{-*}\big[I_R(\vec{x})\big]\Big]$$

where $\vec{x}$ denotes a two-dimensional spatial coordinate and Δ denotes a two-dimensional spatial lag. $\mathcal{F}^-$ and $\mathcal{F}^+$ denote forward and inverse Fourier transforms, respectively, and the asterisk superscript denotes complex conjugation. Our hardware computes the entire cross covariance function with Fourier transforms, which has a maximum at lag Δ_{max} where both pictures match optimally. The position of that maximum relative to zero lag indicates the relative shift between live and reference images and is taken as an error signal. The signal is applied to a steering device which recenters the live image on the matrix array. A beamsplitter in the optical path makes the stabilized image available to post focus instrumentation.

The correlation tracker consists of the main components described below and in fig. 1. Some components are commercial products; others, like item 2 and 3, were designed and built at the NSO electronics shop in Sunspot.

- A 32 x 32 pixel diode matrix array. The detector can be scanned with a rate of 976 pictures per second.

*Both authors were with the National Solar Observatory (NSO), Sunspot, NM 88349, USA, for most part of this project. NSO is operated by the Association of Universities for Research in Astronomy, Inc., under contract with the National Science Foundation

Reviews in Modern Astronomy 2 Editor: G. Klare

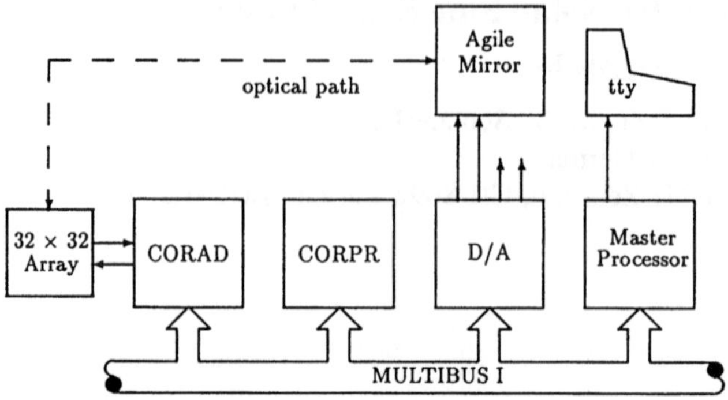

Figure 1: Main components of the correlation tracker system

- A video signal conditioner and digitizer (CORAD, "correcting analog to digital converter"). This component digitizes data from the array detector, performs dark current and gain table corrections, and presents clean data to the correlation processor for further analysis. The component can transfer image data ("snapshots") to the master processor, e. g., dark current and flatfield data.

- A "correlation processor" (CORPR). This processor accesses data from CORAD through a high-speed bus, and computes its Fourier transform with fast, dedicated hardware. For reasons of processing speed, only a 16 by 16 pixel subfield is actually processed by CORPR. The location of this subfield is freely programmable within the 32 by 32 pixel detector field; this feature is used for high-precision scanning. A cross product is generated from the transforms of live data from CORAD and the reference transform, the product is inversely transformed. The resulting cross covariance function is acessible to the master processor.

  Both CORAD and CORPR are controlled by on-board microcode programs.

- A master processor, consisting of a commercial 10 MHz MC68000 - based single board computer. The processor provides the user interface, initializes and supervises hardware and schedules tracking tasks. Mirror drive signals are generated by the master processor which determines the peak position in the covariance functions produced by CORPR.

- A four - channel digital-to-analog (D/A) converter. The D/A receives mirror drive signals, converts and applies them to the steering device.

- A steering (agile) mirror concludes the servo path as the active element. This device was described in [3].

Reference updating is a task that is also performed by the master processor. A snapshot is requested from CORAD every second. The snapshot picture is added to a buffer if its contrast is larger than a selectable value. Reference updating can be made target-dependent this way. After a fixed period, currently 60 s, the old reference is replaced with the new. Tracking is then resumed with the new reference.

3 Performance tests of the correlation tracker system

The correlation tracker system was tested at the 76 cm NSO vacuum tower telescope during observing runs in June and August 1988. We have operated the tracker on the Sun on a variety of structures under seeing conditions that range from excellent to poor. We could lock on granulation everywhere on the disk nearly instantly, moments of severe blurring and regions very close to the limb excluded. We could also lock on sunspot umbrae with light bridges, the inner and outer edges of sunpot penumbrae and regions near the limb where faculae have high contrast. The tracker would lock reliably when the structure had a contrast of 3% rms or more. The tracker would not lock on areas within a few arcsec from the solar limb where the granulation contrast is very low and faculae are absent. Also, it does not establish lock on the solar limb in the direction parallel to the limb. The system reacts very benignly to periods of poor seeing as soon as stable lock has been established. The lock position is normally kept even through moments of severe blur, as long as remnants of the granular pattern remain detectable. Occasionally, the tracker would lose lock during moments of extreme blur when no granulation is visible. The old lock position is immediately recovered as soon as seeing quality improves and the granulation becomes visible again.

CCD pictures covering 12 × 12 arcsec were analyzed to investigate the improvement in image sharpness for long exposures. Bursts of 20 8-ms exposures were recorded on magnetic tape. A diagnostic of the tracker performance is the difference in sharpness of 20-frame averages of uncompensated and compensated pictures, representing 2 s exposures. Fig. 2 shows two sample pairs of average pictures. The increase of sharpness in the corrected image is clearly seen.

We have built and tested a system that stabilizes image motion using any solar small scale structure with sufficient contrast as tracer. The system is designed for the use on a ground-based telescope, which requires stable performance that is largely independent from image sharpness. We have demonstrated the reliable performance of the system under a variety of day-time seeing conditions. We are very satisfied with the performance which we attribute to three of its features: the calculation of the complete covariance function with Fourier transforms, the application of a two-dimensional matrix detector, and a very flexible hardware architecture.

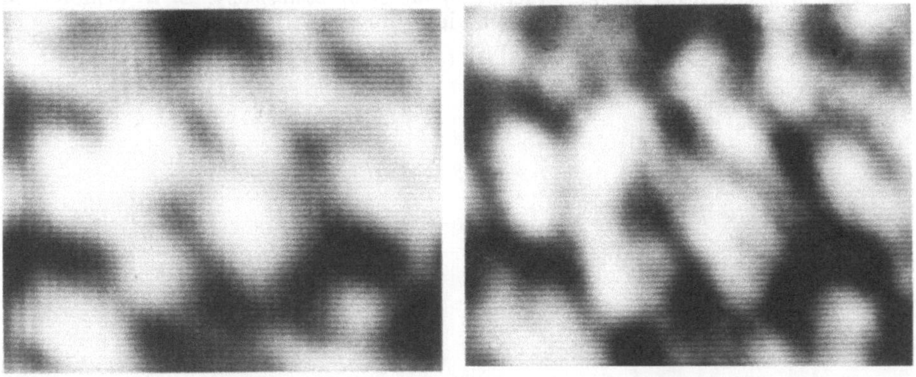

Figure 2: 20 frame averages of uncorrected (left) and corrected (right) CCD pictures

Our approach is computationally expensive in comparison with proposed or already existing correlation tracker systems [1][5]. However, the position of the maximum of CC_{LR} is a linear measure of the live image offset from the target position, it is structure independent and is largely independent of seeing quality[2]. In addition, the maximum can be searched for in the entire field of view of the detector, which constitutes the acquisition range of the tracker, and helps to retrieve the absolute maximum position in case it got lost during the operation of the system.

We have considered and rejected an approach based on linear detectors, as we found that it is very sensitive to motions perpendicular to the scanning directions of the detector and might easily get lost. Another possibility to decrease system complexity would be "hard-clipping" of the image data, i. e., reducing image intensity discrimination to only two levels, 0 and 1. Fast, digital correlators are available for this type of data. We have studied that approach as well and found that with a two-dimensional detector, correlation tracking is easily possible. The penalty is a factor of 3 or so decrease in tracking accuracy, compared with tracking using continuous data.

Acknowledgements: The largest contribution to this project came from Lee Widener who has designed and built CORAD and CORPR and has developed the microcode for the two components. Glen Spence built the detector and the readout logic. Phil Wiborg provided part of the system software and continuous, friendly advice. Dick Dunn was very effective in getting this project organized and under way. We gratefully acknowledge the support that we received from the administrations of NSO and KIS. Part of the work was funded in the early stages by the Deutsche Forschungsgemeinschaft under contract Schr 125/21-1.

References

[1] Edwards, C. G., Levay, M., Gilbreth, C. W., Tarbell, T. D., Title, A. M., Wolfson, C. J., Torgerson, D. D.: 1987, *Bull. Am. Astron. Soc.* **19**(3), 929.

[2] von der Lühe, O.: 1983, *Astron. Astrophys.* **119**, 85-94

[3] von der Lühe, O.: 1988, *Astron. Astrophys.* **205**, 354-360

[4] von der Lühe, O., Widener, A. L., Rimmele, Th., Spence, G., Dunn, R. B., Wiborg, P.: 1989, "Solar feature correlation tracker for ground-based telescopes," *Astron. Astrophys.*, in press

[5] Rayrole, J.: 1987, "French polarization-free telescope THEMIS," in *The role of fine-scale magnetic fields on the structure of the solar atmosphere*, E. H. Schröter, M Vázquez and A. A. Wyller, Eds., Cambridge University Press, 1987.

The Muenster Redshift Project (MRSP)

P. Schuecker, H. Horstmann, W.C. Seitter, H.-A. Ott, R. Duemmler, H.-J. Tucholke, D. Teuber, J. Meijer, and B. Cunow

Astronomisches Institut der Universität Münster,
D-4400 Münster, Fed. Rep. of Germany

1 Introduction

In recent years observational evidence has been accumulating that currently available large-area galaxy redshift surveys (reviewed by Giovanelli and Haynes 1990) do not cover a representative part of the universe. It was found that the small scales reached and the small number of objects observed introduce systematic errors into the derivation of global quantities (galaxy correlation functions, mean galaxy density, luminosity function *etc.*) and of cosmological parameters (Hubble constant, deceleration parameter, density parameter, cosmological constant).

In order to increase significantly the scales and the number of redshifts, the Astronomical Institute Muenster, in 1986, has started the Muenster Redshift Project (MRSP), where redshifts z are measured automatically from low-dispersion objective prism plates. The number of galaxy redshifts per square degree is approximately 250, the scale reached $z = 0.3$, compared to about 2 galaxies per square degree and $z = 0.05$ for currently available large-area surveys. This is a significant growth, gained, however, with the loss of resolution in redshift space: the low dispersion of the spectra gives redshift accuracies of $dz = 0.01$ or $30\,h^{-1}\mathrm{Mpc}$ ($H_0 = 100\,h\,\mathrm{km\,s^{-1}\,Mpc^{-1}}$, $q_0 = 0.5$ throughout this paper). Nevertheless, in most cases the large numbers of objects compensate for the statistical redshift errors, while the derivations of global and cosmological quantities are less affected by small-number statistics, are more representative, and thus lead to more reliable values. The detection of voids on scales $z < 0.02$ is not possible, unless the structures in redshift space are sharpened, using, *e.g.* deconvolution techniques.

2 Procedures and Data

The MRSP data are obtained by automatic reduction of pairs of direct and objective prism Schmidt-telescope photographs. The direct exposures, in the form of film copies of the ESO-SRC atlas plates, and film copies of UK objective prism plates (IIIa-J emulsion, dispersion 246 nm mm^{-1} at H$_\gamma$) are digitized with the microdensitometer PDS 2020 GM^{plus} of the Astronomical Institute Muenster.

Reviews in Modern Astronomy 2 Editor: G. Klare

Each digitized plate ($5.5° \times 5.5°$) contains 20 000 pixels × 20 000 pixels; each object, direct image and spectrum, is stored in a small picture frame. The basic quantities derived from the direct plates are object positions, magnitudes, object types (star, galaxy, elliptical-like or spiral-like galaxy) and position angles of the galaxy major axes. From the objective prism plates galaxy redshifts, spectral types and quasar candidates are obtained. The reduction methods for direct and objective prism plates are described by Horstmann (1988a, b), Schuecker (1988) and Schuecker and Horstmann (1990).

So far, 350 000 galaxies ($m_J < 22^{\rm m}.5$) were detected in 250 square degrees. Of these, 175 000 ($m_J < 20^{\rm m}.5$) have morphological types, and 40 000 ($m_J < 19^{\rm m}.7$) galaxies in 160 square degrees have measured redshifts. By 1991, half a million galaxy redshifts in 2 000 square degrees will be available.

3 Clustering of Galaxies

Clusters of galaxies and filamentary structures are detected on the direct plates as regions of increased galaxy surface density (**Fig. 1**). The most distant cluster identified so far has redshift $z = 0.56$. The average number of rich clusters per square degree is about 0.3. The two-dimensional distributions of elliptical-like and spiral-like galaxies show significant differences (Horstmann 1988b). Elliptical-like galaxies are generally more abundant in regions of enhanced galaxy surface density, in agreement with the morphology-density relation described by Dressler (1980), though rare clusters with a rich spiral population are also found.

In three dimensions, clusters are defined through the coincidence of high surface density and concentration of galaxies in redshift space. When several such clusters are found in close vicinity of each other they constitute superclusters. At least three superclusters are found in the 160 square degree survey field. They have redshifts $z = 0.11$, 0.13 and 0.15. The most prominent concentration, the Sculptor supercluster, lies at $z = 0.11$. The center at $R.A. = 0^h\,42^m$ and $Decl. = -29°$ is defined by at least six rich clusters of galaxies. The filamentary structures between the clusters (**Fig. 1**) are interpreted as bridges of galaxies connecting the members in the nucleus of the supercluster on scales of several to a few tens of Mpc. In **Fig. 2** redshift histograms of the central region of the supercluster core are shown. The smooth superimposed curves are the median distributions calculated from 40 000 galaxies in a volume of $7.2\,10^6\,h^{-3}\,{\rm Mpc}^3$. It can be shown that the median represents a homogenous galaxy distribution (Schuecker and Horstmann 1990). From the figure it is seen that the richest clusters have overdensities of the order of 4 above the median. From the spatial distribution of all galaxies in 160 square degrees it is found that the supercluster extends over $\geq 55\,h^{-1}$ Mpc in the direction of declination, and over $\geq 45\,h^{-1}$ Mpc in both the direction of right ascension and

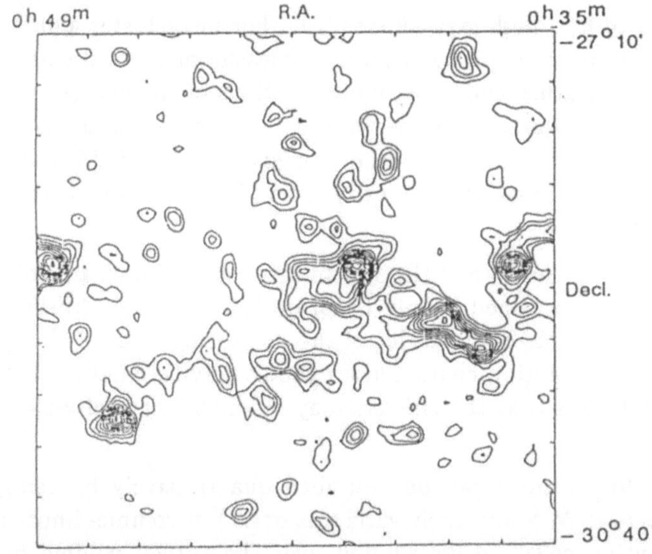

Fig. 1. Galaxy surface density distribution in the nucleus of the Sculptor supercluster at $z = 0.11$, *i.e.* $330\,h^{-1}$Mpc.

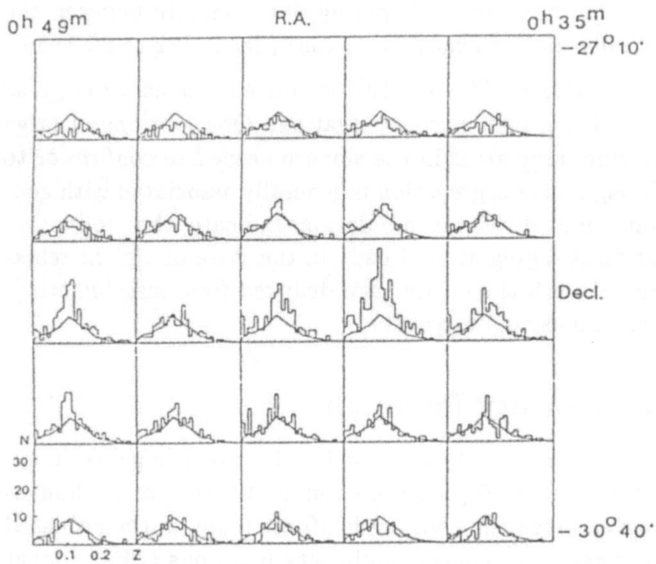

Fig. 2. Distribution of galaxies of the same field in redshift space. The smooth curve represents the median galaxy distribution obtained from a volume of $7.2\,10^6\,h^{-3}$Mpc^3. The apparent peak is a result of the magnitude-limited survey.

redshift. It is expected to find high x-ray fluxes from hot intracluster and intrasupercluster gas in the supercluster core. ROSAT measurements will permit the derivation of the temperature and density distributions of the hot gas and the gravitational potential of the supercluster core within certain limits. With this and the optical brightness of the system, obtained from the MRSP data, it is possible to determine the mass-to-luminosity ratio (M/L) on supercluster scales (Seitter *et al.* 1989a).

The individual clusters in the core of the supercluster show significant subclustering which cannot be explained simply by projection effects of different clusters, with redshift differences of the order of the measuring error. N-body simulations show that subclustering characterizes young gravitationally interacting systems. This indicates that the clusters may be in the state of violent relaxation.

Large-scale density fluctuations can be described quantitatively by using the mean galaxy numbers $< N >$ and their variances $\sigma^2(N)$ in volume- limited subsamples. The counting cells are distributed over the survey volume by tesselation. Application of the theory of gravithermodynamics (Saslaw and Hamilton 1984) yields the b-parameter, defined as $b = 1 - \sqrt{\frac{<N>}{\sigma^2(N)}}$, which is the ratio of excess potential energy over the Poisson case to mean kinetic energy of the system. For the present survey fields, b-values of 0.5 are obtained on scales up to $100h^{-1}$ Mpc, indicating large-scale clustering, intermediate between the Poisson ($b = 0$) and the fully hierarchical ($b = 1$) case (Schuecker *et al.* 1989).

Indications that faint galaxies ($M_J > -19$) do not cluster as strongly as bright galaxies ($M_J \leq -19$) suggest mass segregation. Observations of large numbers of galaxies covering large areas in the sky are needed to confirm or to reject these findings. Though mass segregation is generally associated with gentle relaxation, preliminary runs of N-body simulations indicate, that collapsing clusters show significant mass segregation already in the state of violent relaxation. This is in agreement with the cluster age deduced from subclustering, which remains the unambiguous sign of youth.

4 The Galaxy Luminosity Function

From the basic observational diagram (Seitter 1988), which relates galaxy magnitudes m, redshifts z and numbers $N(m, z)$, no systematic first order changes of galaxy densities at a given magnitude with redshift are found in the unbiased magnitude and redshift ranges, *i.e.* galaxy luminosity functions (LFs) appear to be redshift-independent within the z-interval covered by our observations. This excludes strong global evolution of galaxies on scales $z \leq 0.3$.

The mean LF is calculated from galaxies of different morphological types, from high- and low-density regions, and from different redshifts. The galaxy densities are corrected for field effects on the Schmidt plates, K-dimming and

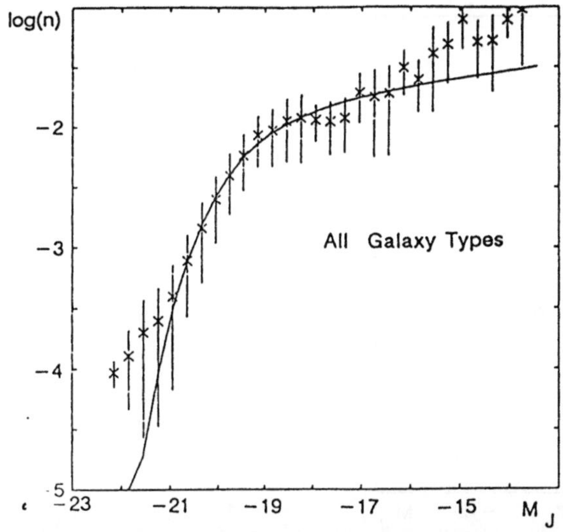

Fig. 3. Mean luminosity function of 40 000 galaxies with superimposed Schechter function. The lines indicate the mean errors.

universal expansion. The combined LFs of all fields (**Fig.3**) show significant substructure, *i.e.* an excess of bright galaxies with $M_J < -21$ (cD-effect?), a dip near $M_J = -17$ and an excess of dwarf galaxies with $M_J > -15$ (Schuecker and Horstmann 1990). Similar structures are found in high density regions, *i.e.* the Virgo cluster (Binggeli *et al.* 1985). It should be noted that only galaxies with $-21 < M_J < -17$ are well represented by a (smooth) Schechter function (Schechter 1976), here with $M_J^* = -19.55$, $\alpha = -1.15$ and $n^* = 0.015$.

For clusters of galaxies (high density regions) the faint end slope is usually $\alpha = -1.25$, whereas for the general field (low density region) smaller values are measured, *e.g.* $\alpha = -1.07$ (Efstathiou et al. 1988). In the MRSP a mixture of high and low density regions is observed. It is thus not unexpected that the faint end slope α of the MRSP sample has a value between those for clusters and field.

Examples for the use of the mean or general LF in observational cosmology are given *e.g.* in Binggeli *et al.* (1985), Felten (1985) and in **Sec. 5**.

Fig.4 shows the LFs for galaxies with apparent ellipticities $E < 0.6$ (crosses) and $E > 0.6$ (circles). They are derived only for galaxies with high reliability redshifts , *i.e.* galaxies for which redshifts from all three redshift measuring methods (Schuecker 1988) are available. The sample includes about half of the total redshift catalogue.

A Gaussian distribution with mean absolute magnitude $< M_J >= -17$ and standard deviation $\sigma_M = 2.5$ is a good approximation for the subsample of galaxies with $E > 0.6$. The high ellipticities of the galaxies suggest that this

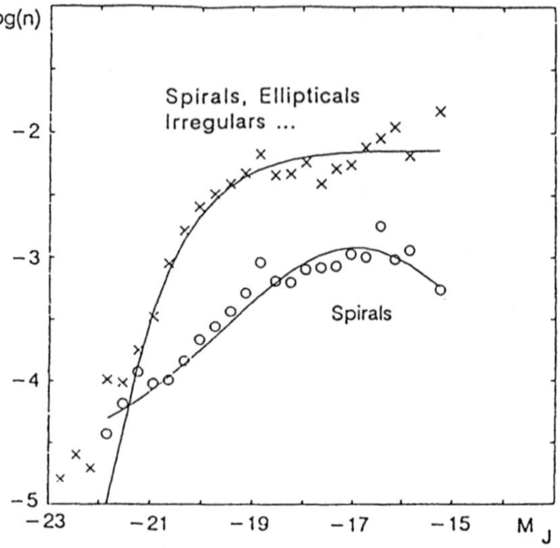

Fig. 4. Luminosity functions for galaxies with high reliability redshifts (s. text) and with apparent ellipticities $E < 0.6$ (crosses) and $E > 0.6$ (circles). Superimposed are a Schechter and Gaussian function, respectively.

sample is dominated by edge-on spirals. For these galaxies internal absorption together with projection effects might be important, but the corresponding corrections are very uncertain and were not applied. The shape of the given LF is representative for the total sample of spiral galaxies only if it is not systematically affected by galaxy orientation.

The LF for galaxies with $E < 0.6$ shows substructures similar to those found in the mean LF (**Fig.3**). Compared to the sample of spirals a significant increase in number density at the faint end is seen. For galaxies with $-21 < M_J < -17$ a Schechter function with $M_J^* = -19.6$ and $\alpha = -0.8$ fits the LF. It is expected that this subsample contains galaxies of all morphological types excluding edge-on spirals.

Fig.4 shows that the LF of spiral galaxies cannot be represented represented by a Schechter function. Also, indication exists that the increasing number densities at the faint end of the mean LF are not caused by spiral galaxies. Possible candidates for faint end populations are dwarf elliptical and irregular galaxies. It is expected that all substructures found in the mean LF might be explained by the superposition of the LFs of galaxies of different Hubble types and luminosity classes (Binggeli et al. 1985). Their detailed analysis must rely on a more sophisticated morphological classification of the galaxies which is in preparation.

5 Density Parameter Ω_0

The density parameter $\Omega_0 = \frac{8\pi G \rho_0}{3 H_0^2}$ represents the present density ρ_0 of the universe normalized to H_0^2. It can be calculated by integrating luminosity densities ℓ_i of various sources i weighted with the corresponding $(M/L)_i$. The most critical step in these calculations is the determination of M/L.

Because of the large present uncertainties in the M/L-values of extragalactic systems, only rough estimates of Ω_0 can be obtained. Here, the value $M/L = 30...50\,h\,(M/L)_\odot$, as derived from pairs of galaxies (Faber and Gallagher 1979), is assumed for all galaxies. With the luminosity density $\ell = (2.3 \pm 0.5)\,10^8\,h\,L_\odot\,\mathrm{Mpc}^{-3}$, obtained from the mean LF, the density parameter for luminous matter is $\Omega_0 = 0.03$.

It is interesting to note that dynamical studies of rich clusters of galaxies yield $M/L = 300\,h\,(M/L)_\odot$ and thus $\Omega_0 = 0.3$. If the tendency of increasing M/L with increasing object size also holds on larger scales, it is expected to find $M/L >> 300\,h\,(M/L)_\odot$ for superclusters. This could bring the density parameter close to $\Omega_0 = 1$. Thus, the detection of the largest structures of the universe and the determination of their sizes, luminosities and masses is not only relevant for the understanding of galaxy formation processes, but also for the analysis of the general dynamical evolution of the universe.

6 Cosmological Parameters H_0, q_0

With MRSP data the Hubble constant is obtained by a statistical method, which compares a well-calibrated LF (measured in our vicinity) with uncalibrated LFs observed at larger distances ($z \approx 0.1$). The values of the Hubble constant, resulting from the use of the Virgo LF for reference, depend only on the distance modulus of the Virgo cluster, not on its redshift, which is affected by local velocity anomalies. The currently favoured values of the distance moduli of the Virgo cluster are $m - M = 31.0$ and 31.7 (van den Bergh 1988). With these distance moduli the Hubble constants $H_0 = 77.3 \pm 1.5\,\mathrm{km\,s^{-1}\,Mpc^{-1}}$ and $56.0 \pm 1.1\,\mathrm{km\,s^{-1}\,Mpc^{-1}}$, respectively, are derived.

The deceleration parameter is determined from the classical redshift-volume test. Preliminary results, obtained with 1 000 galaxy redshifts, yield $q_0 < 0.2$. Details of the determination of cosmological parameters are presented in Seitter *et al.* (1989b). More reliable estimates will be obtained in the near future with the larger MRSP sample now available.

7 High Redshift Objects

High redshift objects offer a new challange because for $z > 0.3$ noticeable evolutionary effects are expected. High redshift objects under study in the MRSP are distant clusters of galaxies (Horstmann *et al.* 1989) and quasars

(Gericke 1988). A first step to analyse such objects is the preparation of a catalogue of candidates, selected according to well-defined criteria.

On the direct plates of the survey fields clusters of galaxies with $z < 0.6$ are detected. The redshifts are estimated from the magnitudes of the expected cluster galaxies and from the apparent diameters of the clusters using the isopleths at half mean galaxy surface density.

In order to detect significant numbers of clusters at redshifts $z > 0.6$, magnitudes near $m_J = 25^m$ must be reached for samples covering large areas. This is possible by superposition of several digitized Schmidt plates. Tests with 12 GPO Astrograph plates show a gain of about 2 magnitudes below the magnitude limit of individual plates, resulting in a factor 8 in the number of galaxies. Redshift estimates show that with this method clusters at $z = 0.7$ are detectable. The redshifts will be measured using the superposition of several digitized red objective prism plates of the same field. It is expected to obtain spectra for objects with $m_J = 22^m$. The observational programme has been started.

In order to separate stars and quasars, the spectra of all star-like objects are classified. The normal stellar spectra are employed for a study of the distribution of stars in the Galaxy and for the derivation of the stellar luminosity function. The remaining spectra are classified as quasars from their characteristic emission lines. Samples of automatically classified stellar and quasar spectra are shown in **Fig.5** and in **Fig.6**, respectively. For subgroups, *e.g.*

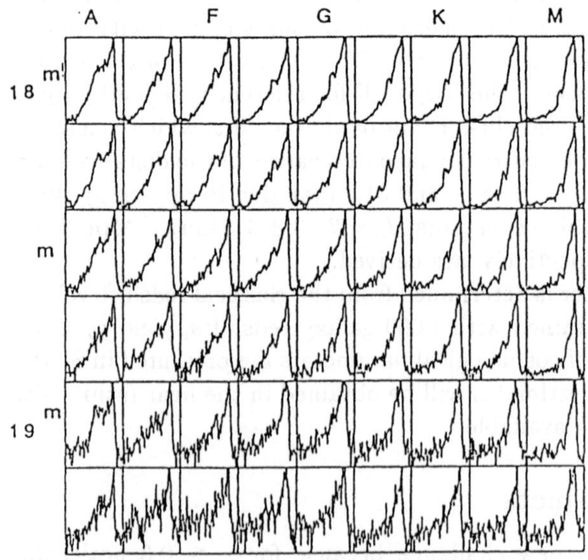

Fig. 5. Sample of automatically classified stellar spectra of different types in the magnitude range $m_J = 18^m - 19^m$.

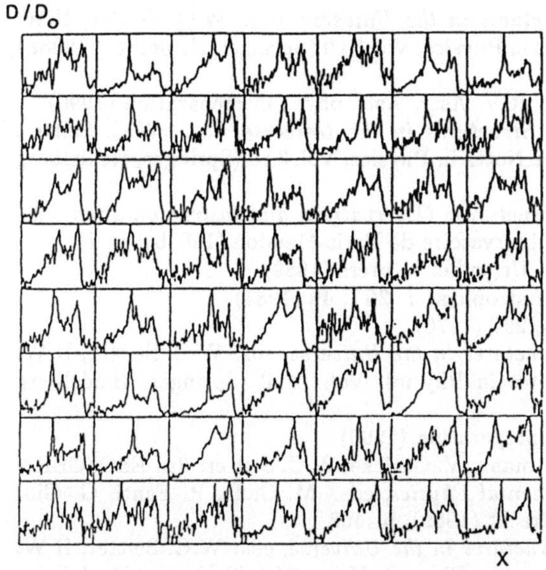

Fig. 6. Sample of automatically detected quasars up to $m_J = 19^m.3$.

Lyman-alpha quasars, redshifts are determined from the objective prism spectra (Gericke 1988).

Preliminary data on quasar clustering show that their b-values do not differ significantly from zero (Poisson distribution) near $z = 2$. This together with the observed galaxy clustering (see above) and with the fluctuations (or upper limits) found for equivalent masses in the various backgrounds, especially in the MWB, indicates fast evolution towards hierarchical clustering (Schuecker *et al.* 1989).

8 Acknowledgments

We cordially thank the UKSTU for supplying film copies of the UK objective prism plates and the Deutsche Forschungsgemeinschaft for supporting the project.

References

1. B. Binggeli, A. Sandage, G.A. Tammann: Astr. J. **90**, 1681 (1985)
2. A. Dressler: Astrophys. J. **236**, 351 (1980)
3. G. Efstathiou, R.S. Ellis, B.A. Peterson: Mon. Not. R. astr. Soc. **232**, 431 (1988)
4. S.M. Faber, J.S. Gallagher: Ann. Rev. Astr. Astrophys. **17**, 135 (1979)
5. J.E. Felten: Comments Astrophys. **11**, 53 (1985)

6. V. Gericke: In *Large-Scale Structures in the Universe,* eds. W.C. Seitter, H.W. Duerbeck, M. Tacke, Lecture Notes in Physics, Vol. **310** (Springer, Heidelberg 1988a), p. 235
7. R. Giovanelli, M.P. Haynes: Ann. Rev. Astr. Astrophys., in preparation (1990)
8. H. Horstmann: In *Large-Scale Structures in the Universe,* eds. W.C. Seitter, H.W. Duerbeck, M. Tacke, Lecture Notes in Physics, Vol. **310** (Springer, Heidelberg 1988a), p. 111
9. H. Horstmann: In *Large-Scale Structures: Observations and Instrumentation,* eds. C. Balkowski, S. Gordon, Publ. Observatoire de Paris-Meudon (1988b), p. 71
10. H. Horstmann, M. Naumann, R. Ungruhe: in press (1989)
11. W.C. Saslaw, A.J.S. Hamilton: Astrophys. J. **297**, 49 (1984)
12. P. Schechter: Astrophys. J. **203**, 297 (1976)
13. P. Schuecker: In *Large-Scale Structures in the Universe,* eds. W.C. Seitter, H.W. Duerbeck, M. Tacke, Lecture Notes in Physics, Vol. **310** (Springer, Heidelberg 1988), p. 142
14. P. Schuecker, H. Horstmann: in preparation (1990)
15. P. Schuecker, H.-A. Ott, H. Horstmann, V. Gericke, W.C. Seitter: 3rd ESO/CERN *Conf. on Cosmology and Fundamental Physics*, eds. M. Caffo, R. Fanti, G. Giacomelli, A. Renzini, Kluwer, Dordrecht (1989), p. 468
16. W.C. Seitter: In *Large-Scale Structures in the Universe,* eds. W.C. Seitter, H.W. Duerbeck, M. Tacke, Lecture Notes in Physics, Vol. **310** (Springer, Heidelberg 1988), p. 9
17. W.C. Seitter, H. Horstmann, H.-A. Ott, P. Schuecker, H. Boehringer, G. Hartner, S. Schindler, R.A. Treumann: Proposal for ROSAT Observations (1989a)
18. W.C. Seitter, H.-A. Ott, R. Duemmler, P. Schuecker, H. Horstmann: In *Morphological Cosmology,* eds. P. Flin, H.W. Duerbeck, Lecture Notes in Physics, Vol. **332** (Springer, Heidelberg 1989b), p. 3
19. S. van den Bergh: In *The Extragalactic Distance Scale,* eds. S. van den Bergh and C.J. Pritchet (Astronomical Society of the Pacific Conf. Ser. Vol. **4**, San Francisco 1988), p. 375

Galaxies in the Galactic Plane

R.C. Kraan-Korteweg

Astronomisches Institut der Universität Basel,
Venusstr. 7, CH-4102 Binningen, Switzerland

The anisotropy of the microwave background radiation (MWB) is assumed to be caused by a local motion that is presumably induced by the irregular mass distribution of galaxies in our closer or wider neighbourhood. Assuming that the mass to light ratio is constant, the overall extragalactic light distribution should reveal the direction of any excess gravitational force on the Local Group. A galaxy search to faint magnitudes was started in the southern galactic plane ($|b| \leq 10°$) to gain a more complete picture of the galaxy distribution close to the hardly explored dust equator of the Milky Way. So far, 15½ fields of the ESO/SERC survey near Hydra (i.e. the vicinity of the dipole) were searched. The distribution of more than 2200 detected galaxies with $D \geq 0.2$ arcmin suggests an extension of the Hydra and Antlia clusters across the galactic plane. If this large structure of over 35° on the celestial sphere can be confirmed it will contribute significantly to the peculiar velocity of the Local Group and hence in explaining the observed MWB dipole.

I. Introduction

The Milky Way obscures a large part of the extragalactic sky. Even the partially transparent regions have generally been avoided by extragalactic astronomers because the determination of the intrinsic properties of galaxies has proven very difficult due to the locally variable front absorption. Now the demand for additional information on galaxies behind the galactic plane has increased for the following two reasons: (1) the anisotropy of the 2.7°K MWB radiation [1, 2] should be the result of the all-sky distribution of galaxies [cf. 3, 4], and (2) from the irregular distribution of nearby galaxies (v_o<5000 km sec^{-1}) a complete picture should emerge as to their distribution in clusters, filaments and voids (cf. Fig. 6 in [5], the CFA and southern redshift slices [6,7], and the Perseus-Pisces area [8]).

The anisotropy of the MWB radiation is interpreted as a motion of the Local Group (LG) with respect to a cosmic MWB frame. This motion of v=630 km sec^{-1} towards l=268°, b=27° (see [9] for a review) is superposed on the

Reviews in Modern Astronomy 2 Editor: G. Klare

quiet Hubble expansion and shared, except for expected shear motions, by all nearby galaxies. Its origin stems from the irregular distribution of mass. Assuming (1) that the mass to light ratio of galaxies is approximately constant and (2) that there is no biasing in the sense that the distribution of extragalactic light follows the true mass distribution, the total gravitational acceleration can be deduced either from the known distribution of *galaxies, galaxy clusters and voids*, or else from the total extragalactic light flux [4, 10]. Both methods require whole sky coverage. Due to the front absorption of the Galaxy, a large part of the extragalactic sky remains hidden (>15%). Therefore, most determinations of the total gravitational field are based on the assumptions that the zone of avoidance is uniformly filled with galaxies with the same average density and selection function as the sample under consideration. However, the galaxy distribution of complete regions demonstrates that this distribution is nowhere uniform but highly clumped [3-8].

Part of the peculiar velocity of the LG can be attributed to the gravitational force of the nearest galaxy cluster, the Virgo cluster [5, 11]. Subtracting this component leaves about 500 km sec^{-1} towards $l=274°$, $b=12°$ to be explained [12, 13]. This direction points towards the Hydra and Antlia clusters [14] (which alone hardly suffice to induce a large acceleration on the LG) *and* accidentally to the dust equator; i.e. interesting galaxy structures might indeed be obscured in this little known area.

The knowledge about density structures behind the galactic plane can be considerably improved with deeper searches on available optical surveys. This was demonstrated by Weinberger [15] who searched for galaxies within ±2° of the northern galactic plane, using the POSS prints. Even though large areas disclose no galaxies, a relative excess was found exactly at the location where the Perseus-Pisces supercluster (PPS) disappears into the plane and the cluster A569 is found at the opposite side [16]. Later redshift measurements proved these highly absorbed galaxies at the same distance as the PPS [17].

A similar search with a broader latitude range was begun in the South using the ESO/SERC survey. As a starting point, the apex of the MWB corrected for the Virgocentric component was chosen on account of the above mentioned reasons. Moreover, the distribution of all southern galaxies (Fig. 2 in [18]), as well as the distribution of extragalactic optical light (Fig. 2 in [10]), and that of the IRAS galaxies (Fig. 2e, f in [19]) all seem to suggest that the Hydra and Antlia clusters are part of a larger filament that fades into the galactic plane.

II. The Data

The search is carried out by a systematic visual inspection of the IIIaJ film copies of the ESO/SERC survey, using a viewer with a magnification of f=50. The area searched so far encompasses 15½ fields (F91–F93, F125–F129, F165–F170, F211–F213). The 400□° roughly span a range in galactic coordinates of $l \approx 266°$ to $296°$ and $b \approx -10°$ to $+8°$ ($\alpha \approx 8^h30^m$ to 12^h, $\delta \approx -47.°5$ to $-67.°5$).

A minimal diameter limit of 0.2 arcmin was imposed to avoid confusion with faint stars, stellar blends and to warrant consistency in the identification of the galaxies. Depending on the surface brightness, the smallest galaxies have magnitudes in the order of B_J=18–19.5 mag. 1939 galaxies were discovered as well as 334 candidates. Only 85 galaxies were previously recorded in the ESO/Uppsala Survey of the ESO(B) Atlas [18]. Therefore less than 4% have diameters over 1 arcmin (but they are heavily absorbed!).

Diameters were determined, average surface brightness estimated, and the morphological types are described. The diameters, in combination with the average surface brightness give a magnitude estimate.

The positions were accurately determined with the OPTRONICs at ESO in Garching. A comparison of the positions of galaxies from overlapping borders of the fields suggests that the precision lies well below 1 arcsec.

An assessment of the reliability of the diameter and magnitude estimates was made possible by MacGillivray, who extracted the images determined by the automated measuring machine COSMOS for the visually detected galaxies of field 213. The correlation between the diameters is satisfactory with decreasing scatter towards smaller diameters. On average the present diameters are 10% smaller; they correspond to an isophote of about 24.5 mag arcsec^{-2}. This is confirmed independently by the comparison of the 85 galaxies in common with Lauberts. The relation between the estimated magnitudes and the COSMOS values is surprisingly good, with no deviation from linearity up to the faintest galaxies and a scatter of only $\sigma=0.^m47$.

Due to the heavy front absorption, the outer parts of the galaxy images – foremost the spiral structure – are heavily dimmed and often only the bulges remain distinct. Because of the local variability of obscuration, a homogenous type classification could not be achieved. Still, for about 2/3 of the galaxies, at least a rough morphological classification is made. The majority of the galaxies are spirals (≈75%).

A catalogue for the first 10 fields (cf. [20]), listing the above mentioned properties of the galaxies is in preparation [21].

III. The Distribution of the Detected Galaxies in the Galactic Plane

The distribution of the detected galaxies in galactic coordinates is illustrated in Fig. 1. The outlined area encloses the searched fields. The apex of the MWB dipole, corrected for Virgocentric infall, is marked with an arrow (b=12°) and coincides with the direction to the Antlia (b=18°) and Hydra (b=25°) clusters. The dots signify the certain galaxies (N=1939); the crosses mark the possible galaxies (N=334).

It is quite obvious from the distribution that the actual dust equator at these longitudes is offset to the south by about 1.5°-2°. This was already established for longitudes between l=200°-330° from the hydrogen column densities [22]. The detection of galaxies very close to the *true galactic equator* (|b|≤3°) is still not very successful.

The more transparent regions demonstrate that the number density of galaxies in the search area not only depends on galactic latitude but also on galactic longitude. Below the plane the number density distinctly increases from l=290° to around l=272°, whereas on the other hand the maximum density *at constant galactic latitude* is reached around l=280° (cf. at b=5°). The few extremely faint galaxies across the plane seem to connect the relative overdensities.

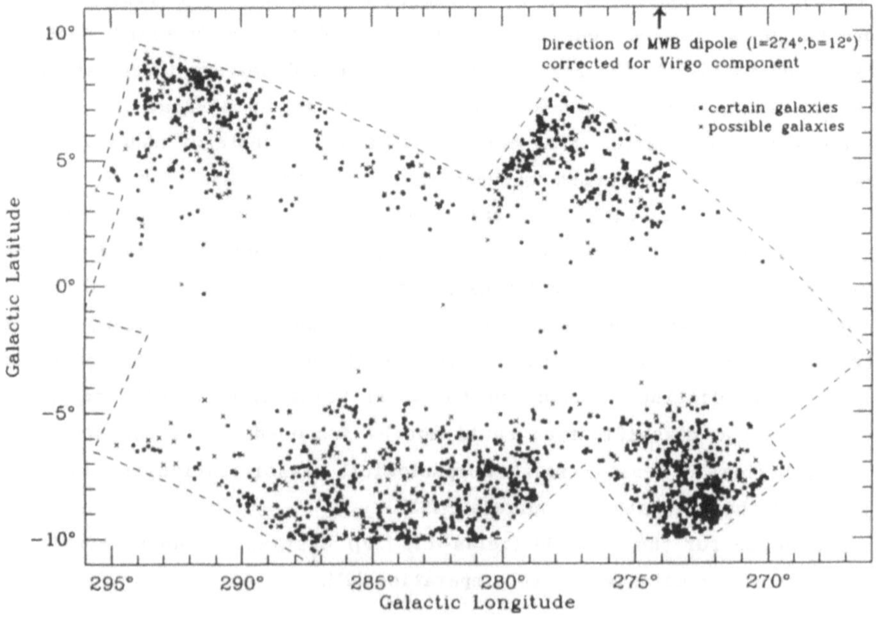

Fig. 1: Distribution of the optically detected 2273 galaxies with D≥0.2 arcmin. The outlined area demarcates the searched 15½ ESO/SERC fields.

The number counts of galaxies are, of course, no objective measure of densities since they are strongly influenced by the patchiness of the galactic dust clouds. However, an overlay of the galaxy distribution with the dark clouds and their different opacities (as analyzed in [23]) clearly exhibits that *the denser areas are not due to particularly transparent regions.* This is independently confirmed by the numerous extinction values in the galactic plane [24] and at $|b|=10°$ [25]. On the contrary, the extinction values and dark clouds - taken on average - show an increasing gradient towards $l=270°$. This is supported by the appearance of the galaxies themselves: below the plane, the galaxies around $l=285°$ are well resolved and the disks of the spirals distinct, whereas in the densest region ($l \approx 272°$) merely the bulges of the galaxies remain visible. Thus the evident overdense regions would actually be more pronounced *when corrected for extinction.*

IV. Properties of the Detected Galaxies

IV.1. The Effect of Absorption on the Observed Properties

The galaxies appear on average quite small ($<D> \approx 0.4$ arcmin) and faint ($<B_J> \approx 18.^m0$). However, even the galaxies around $|b|=10°$ suffer an absorption of about $A_B=1.^m0-1.^m5$ (A_B is assumed to be equal to A_J). The absorption increases for decreasing latitudes. The effect of the extinction on the observed parameters is reflected clearly in Fig. 2. Here the observed magnitude is plotted as a function of cosec(b): while at the highest

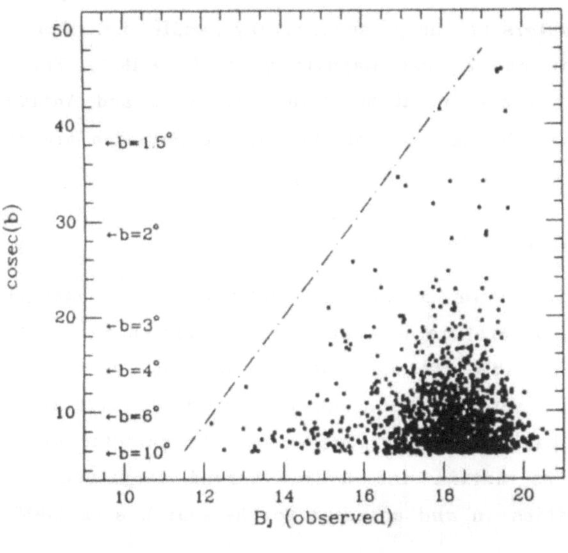

Fig. 2: Illustration of the effect of extinction on the observed magnitude. Even though the dust clouds are distributed very irregularly in the plane, an extreme systematic cut-off of apparently bright galaxies is evident as a function of cosec(b). The filled triangle would be even sharper defined if the galactic equator were replaced by the true dust equator.

latitudes galaxies with a broad range in magnitude are found, a systematic cut-off discriminates more and more against *apparently* large and bright galaxies for decreasing latitudes. Disregarding the patchiness of the dust cloud distribution and the asymmetry about the galactic equator, one might be tempted to read off the amount of absorption from Fig. 2 as a simple function of $|b|$. However, the cut-off which is marked by the dashed line in Fig.2 is defined by intrinsically bright, nearby galaxies with particularly low absorption. The average absorption of the present sample manifests a lower limit. Because the sample is optically selected the true absorption in the searched area must still be higher. Similar considerations can be made when the galaxy magnitudes are replaced by their observed diameters.

The extinction not only dims the isophotal magnitudes by the amount A_B, *but because the observed diameters are also considerably reduced*, the magnitudes will be subjected to an additional dimming Δ (i.e. $B_J^o=B_J-A_B-\Delta$). This effect has been studied in detail by Cameron [26], who artificially absorbed the intensity profiles of various galaxies in Virgo. He deduced analytical functions which satisfactorily describe the additive effects of absorption on magnitudes and the diameter reductions. The functions are distinct for spirals as compared to E and SO galaxies. For typical absorption values in the plane the effects are significant: e.g with $A_B=2.^m5$, the absorption-corrected diameter would be $D^o=2.9 \cdot D_{obs}$ and the additional dimming on top of the $2.^m5$ absorption is $\Delta=0.^m7$ for spiral galaxies (and $D^o=2.6 \cdot D_{obs}$ and $\Delta=0.^m4$ for E and SO galaxies). These values increase dramatically for increasing A_B.

These relations and an idealized absorption law (neglecting the patchiness in extinction) yield a preliminary correction for the absorption effects on the *average parameters* of the present galaxy sample, i.e. a mean diameter of $\langle D^o_{25}\rangle=0.8$ arcmin and a mean magnitude of $\langle B_J^o\rangle=16.^m3$. These values correspond roughly to the expectation values of Hydra and Antlia cluster members, implying that the majority of the sampled galaxies are at a similar distance range.

IV.2. Distances to the Detected Galaxies

Since most of the galaxies were previously unknown, the redshift information available in the literature is sparse and restricted mainly to higher latitudes ($|b|\geq 8°$). No reliable distance estimate for the overdensity across the galactic plane is yet possible. But the available velocities favour the notion that the high density bridge across the galactic plane is at a similar velocity as Antlia and Hydra: around longitudes of $l\approx 280°\pm 5°$ the measured velocities in *and* adjacent to the search area ($N\approx 60$)

show values around $v_o \approx 2800$ km sec^{-1} with very small scatter ($\Delta v \lesssim \pm 200$ km sec^{-1}). That holds on *both* sides of the obscured plane.

IV.3. Identifications with IRAS Sources and their Distribution

Part of the galaxies were rediscovered in the IRAS Point Source Catalogue (IRAS PSC). A cross identification is in principle viable even for the most absorbed galaxies, because the infrared fluxes hardly suffer from extinction. Yet the increasing level in the background radiation and the high source counts at these low latitudes also impede the qualification as reliable point sources.

Still, of the 2273 galaxies 111 were identified in the IRAS PSC. Judging from their colours they very likely are galaxies (i.e. $f_{60}>f_{25}$ or f_{12}, $f_{100}>f_{60}$; f_{60} stands for the flux measured at 60μ, etc.). In addition, the Reject Catalogue (RejC) was searched and another 61 identifications became plausible. The data from the RejC generally have only 1-hour-confirmed sightings and therefore low flux quality. Yet the existence of an optical image at the exact IRAS position enhances the probability of the reality of the source. At a later stage the co-added fluxes will enable a confirmation of the tentative identifications.

The clustering of IRAS galaxies is less pronounced than of optically selected samples, because early type galaxies are hardly sampled in the IRAS bands. In spite of the weaker tendency of IRAS galaxies to cluster, the supplementation of colour-selected galaxies from the IRAS PSC without an optical counterpart (as defined in [19]) reveal a connection across the plane, apparently uniting the optically detected high density regions (cf. Fig. 2 in [27]). This connection is at slightly smaller longitudes when compared to the tentative, optically detected bridge across the plane.

V. Conclusions

- The distribution of the optically detected 2273 galaxies in the galactic plane with D≥0.2 arcmin reveal a relative overdensity around $l \approx 275°$, reaching all the way across the plane. This feature does not seem to be an artifact of local variations of the opacity in the galactic plane.
- Corrections of the diameters and magnitudes for absorption make the majority of the galaxies comparable to Hydra and Antlia cluster members.
- The few available velocity measurements in and adjacent to the detected overdensity support the idea that the overdensity is due to galaxies at similar distances as the Hydra and Antlia clusters.

- The complementation of the optically selected galaxies with colour-selected galaxies from the IRAS PSC discloses a bridge connecting the higher density regions.

The available data so far suggest that the detected overdensity forms one large structure with Hydra and Antlia, hence extending over at least 35°. It is the only "nearby" large structure which does not lie in the super-galactic plane (SGP). It is an elongated feature 30° to the south of the SGP and more or less parallel to it. If its existence can be confirmed it will disturb the Hubble flow at the location of the LG with a significant component away from the SGP, towards the direction of the MWB dipole.

Indirect evidence that the inferred extension of the Hydra and Antlia clusters across the plane might play an important role in bringing the total gravitational acceleration of the LG - as derived from the mass distribution - in agreement with the apex of the MWB dipole, is found in the calculations by Lynden-Bell [10]. Taking advantage of the fact, that the light flux as well as the gravitational force decrease as r^{-2}, he determined the total gravitational field from the distribution of the extragalactic light, assuming that the mass to light ratio is constant. He used two different methods: (1) he filled the zone uniformly with the average light of the well known part of the two hemispheres and (2) he applied a cloning method in which the structure of the neighbouring area (b=15°-30°) is copied into the zone of avoidance. With this second method he artificially reproduces an extension of the Hydra and Antlia clusters into the galactic plane. This second method brings the direction of the gravitational acceleration on the LG closer to the observed apex of the MWB than any other previous determinations (cf. Fig. 3 in [10]). In addition, these calculations imply that the irregular distribution of the galaxies within $v_o \approx 4000$ km sec^{-1} suffices to explain the observed dipole direction of the MWB with no other sources needed.

VI. Future Plans

A verification of the suspected supercluster should be possible with the planned follow-up projects: The search will be extended at least to $-10° \lesssim b \lesssim +8°$ for the longitude range in Fig. 1.

It is planned to do CCD photometry in the R band for the majority of the galaxies. This will allow a better determination of the intrinsic properties of the absorbed galaxies. The addition of some B photometry will lead to estimates of extinction at specific locations in the plane.

Optical redshift measurements for some of the denser regions, using the multifiber system OPTOPUS of ESO will lead to an estimate of the 3-dimensional distribution. These measurements can be supplemented with HI observations for the strongly absorbed galaxies closer to the dust equator. In [28] it was shown that the detection rate of galaxies in the galactic plane in Puppis ($l \approx 245°$) is comparable to high-latitude regions and that the optically dimmed galaxies are not biased with respect to their global properties.

Acknowledgements

The careful reading of the manuscript by G.A. Tammann has proven very valuable. I would like to thank H. MacGillivray, who derived the diameters and magnitudes from COSMOS scans for the visually detected galaxies of field F213, E. Meurs for a list of the colour-selected IRAS galaxies in the relevant area, and B. Madore for the IRAS cross identifications at IPAC. Without the financial support of the Swiss National Science Foundation this project could not have been pursued.

References

1 E.K. Conklin: Nature 333, 46 (1969)
2 G.F. Smoot, P.M. Lubin: Astrophys. J. Lett. 234, 183 (1979)
3 O. Lahav, M. Rowan-Robinson, D. Lynden-Bell, Mon. Not. R. Astron. Soc. 234, 677 (1988)
4 D. Lynden-Bell, O. Lahav, D.Burstein: Mon. Not. R. Astron. Soc., in press (1989)
5 R.C. Kraan-Korteweg: Astron. Astrophys. Suppl. 66, 255 (1986).
6 M.J. Geller, J.P. Huchra: In Large Scale Motions in the Universe: A Vatican Study Week, eds. V.C. Rubin, G.V. Coyne (Princeton University Press, Princeton NJ, 1988), p. 3
7 L.N. da Costa, P.S. Pellegrini, C. Willmer, D.W. Latham: Astrophys. J. 344, 20 (1989)
8 M.P. Haynes, R. Giovanelli: In Large Scale Motions in the Universe: A Vatican Study Week, eds. V.C. Rubin, G.V. Coyne (Princeton University Press, Princeton NJ, 1988), p. 31
9 D.T. Wilkinson: In IAU Symposium 130, Large Scale Structures of the Universe, eds. J. Audouze, M.-C. Pelletan, A. Szalay, (Dordrecht: Reidel, 1988), p.7
10 D. Lynden-Bell: In Large-Scale Structures and Peculiar Motions in the Universe, eds. D.W. Latham, L.M. da Costa, (ASP Conference Series, 1989), in press
11 G.A. Tammann, A. Sandage: Astrophys. J. 294, 81 (1985)
12 A. Sandage, G.A. Tammann: In Large Scale Structure of the Universe, Cosmology and Fundamental Physics, First ESO-CERN Symp., eds. G. Setti, L. van Hove (Garching: ESO, 1984), p. 127
13 E.J. Shaya: Astrophys. J. 280, 470 (1984)
14 U. Hopp, J. Materne: Astron. Astrophys. Suppl. 61, 93 (1985)
15 R. Weinberger: Astron. Astrophys. Suppl. 40, 123 (1980)
16 M. Hauschildt: Astron. Astrophys. 184, 45 (1987)
17 P. Focardi, B. Marano, G. Vettolani: Astron. Astrophys. 136, 178 (1984)

18 A. Lauberts: In The ESO/Uppsala Survey of the ESO(B) Atlas (Garching: ESO, 1982)
19 E.J.A. Meurs, R.T. Harmon: Astron. Astrophys. 206, 53 (1988)
20 R.C. Kraan-Korteweg: In Large Scale Structure: Observations and Instrumentation, eds. C. Balkowski, S. Gordon (Meudon: Observatoire de Paris press, 1988), p.155
21 R.C. Kraan-Korteweg: in preparation
22 F.J. Kerr, G. Westerhout: In Galactic Structure (Chicago: University of Chicago Press, 1965), p. 186
23 J.V. Feitzinger, J.A. Stüwe: Astron. Astrophys. Suppl. 58, 365 (1984)
24 Th. Neckel, G. Klare: Astron. Astrophys. Suppl. 42, 251 (1980)
25 D. Burstein, C. Heiles: Astron. J. 87, 1165 (1982)
26 L.M. Cameron: Astron. Astrophys. (in press) (1989)
27 R.C. Kraan-Korteweg: In Large-Scale Structures and Peculiar Motions in the Universe, eds. D.W. Latham, L.M. da Costa, (ASP Conference Series, 1989), in press
28 W.K. Huchtmeier, R.C. Kraan-Korteweg: in preparation

Synchrotron Light from Extragalactic Radio Jets and Hot Spots

K. Meisenheimer

Max-Planck-Institut für Astronomie, Königstuhl 17,
D-6900 Heidelberg, Fed. Rep. of Germany

Summary: The present status of the detection of optical synchrotron radiation from extended extragalactic radio sources is reviewed. There is now firm evidence for optical synchrotron emission from the jets of M 87 and 3C 273 as well as from three radio hot spots (3C 20 west, 3C 33 south and Pictor A west). Another five objects are very good candidates for emitting synchrotron light. The relevance of optical synchrotron emission as tracer of *in situ* particle acceleration is outlined. Energy dependent propagation effects should lead to significant differences in polarization structure and morphology on radio and optical maps. I present the overall synchrotron spectra of radio jets and hot spots which are based on measurements at radio, millimetre, near-infrared and optical wavelengths. Detailed maps of the optical spectral index in the jets of M 87 and 3C 273 do not show the pronounced variations predicted by standard acceleration models (*e.g.* shock acceleration).

1. Introduction

From an observational point of view, optical synchrotron emission plays a minor role in investigations of extragalactic radio sources. Linked radio interferometers – e.g. the Very Large Array, VLA – beat the angular resolution of groundbased optical telescopes by a factor of 10 or more. With respect to sensitivity radio telescopes reach 1000 times deeper for typical radio-optical synchrotron spectra which are essentially flat in detected power νS_ν. Nevertheless, the detection of synchrotron light contributes a significant piece of evidence to our theoretical picture of these sources since it is emitted by freshly accelerated, ultra-relativistic electrons. Thus optical synchrotron radiation traces the place of particle acceleration in the source.

The morphology of an archetypical double radio source – for example Cygnus A – is characterized by four major constituents (Fig. 1): A bright extended double lobe structure symmetrically surrounds a very compact core, located in the center of an elliptical galaxy. The existence of hot spots with extreme surface brightness near the outer boundaries of the source requires a continuous energy supply which is provided by a twin jet (Scheuer 1974, Blandford & Rees 1974). Optical synchrotron emission is observed from all regions of high surface brightness, that is the cores (e.g. NGC 1275 and many highly polarized quasars), as well as several jets and hot spots (see table 1).

The most spectacular example of optical synchrotron emission from a hot spot has been found in the western lobe of Pictor A (Röser & Meisenheimer 1987, see Fig. 2). In classical double radio sources like Cygnus A or Pictor A the radio jets are weak and there is little hope to detect them in the optical. Brighter radio jets are common in a different class of radio sources which is characterized by the fact that their extended radio emission peaks within one half of the maximum source extension. The radio galaxy M 87 (= Virgo A) with its prominent radio jet is a good example of this class

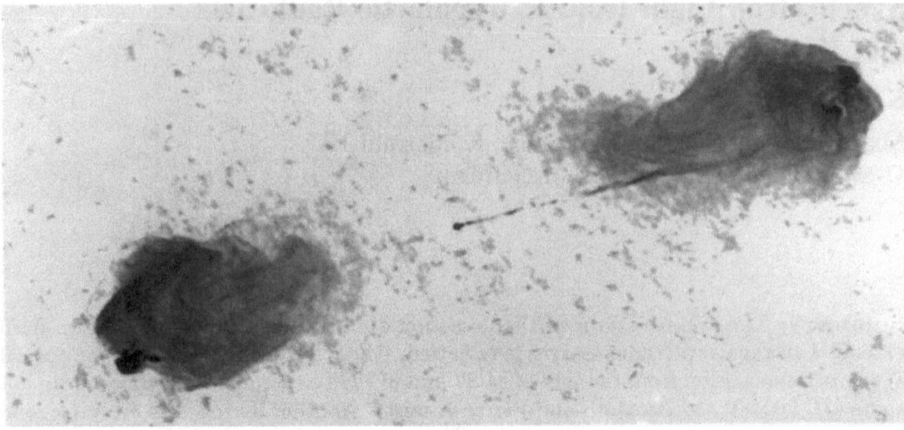

Fig. 1: The classical double radio source Cygnus A. This VLA map at 6 cm (resolution: 0$''$.4 FWHM) shows the jets on ***both sides*** of the core and the filamentary structure of the lobes (for further details refer to Carilli *et al.* 1989).

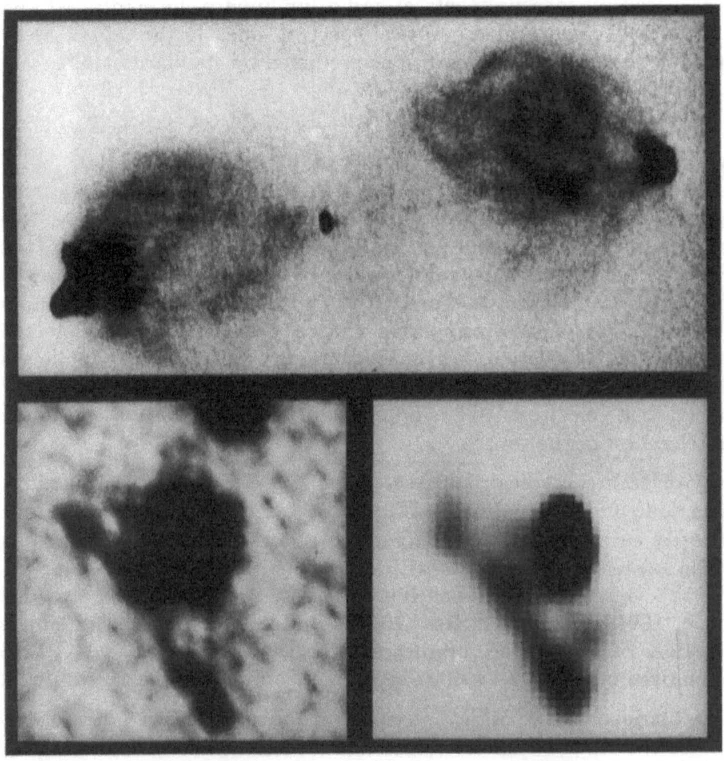

Fig. 2: The double radio source Pictor A with its bright optical hot spot in the western lobe (Röser *et al.* 1987). The **upper panel** shows the entire radio source. The **lower right panel** displays a VLA map of the western hot spot (6 cm, 1$''$.4 × 2$''$.8 *Beam*). The **lower left panel** shows the same region as seen on a red CCD image taken with the ESO 3.6m telescope. Besides the galaxy cutted in half at the upper edge of the image, the morphology of the hot spot and the adjacent filament are virtually identical.

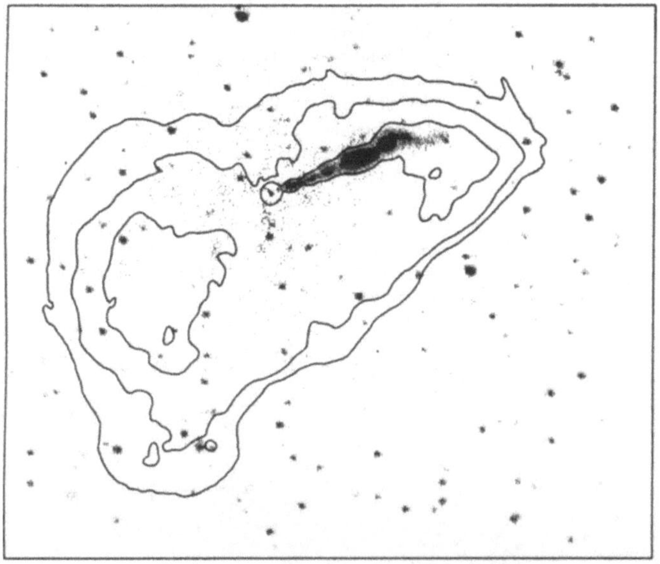

Fig. 3: The extended radio source around M 87 (contour lines) superimposed on an optical image (greyscale, starlight of M 87 removed). Many globular clusters are seen on the optical image. The radio map at 20 cm are has been provided by Frazer Owen (Owen *et al.* 1989b).

although the obvious asymmetry of the source (Fig. 3) is not typical for its class. Due to its relatively small distance (D = 16 Mpc) and its high surface brightness the radio jet of M 87 can be studied in great detail (with a linear resolution of $\lesssim$ 10 pc, see Owen *et al.* 1989a). At variance with the majority of extragalactic jets (see chapter 2, below) the M 87 jet has first been detected on optical photographs (Curtis 1918). The second jet readily seen at optical wavelengths (Schmidt 1963) is the jet of 3C 273 - the brightest quasar. This jet is a favourite of our group in Heidelberg, because its remarkable differences with the M 87 jet open up the possibility to separate *specific* and *general* properties of radio jets in a comparative study.

2. Observational evidence for optical synchrotron radiation

Optically thin synchrotron emission is characterized by high linear polarization — up to 100% in a homogeneous magnetic field. In fact, both compact and extended synchrotron sources occassionally show linear polarization values beyond p = 40%.[1] So optical polarimetry is certainly the best way to verify optical synchrotron emission. If the optical polarization coincides with a proper extrapolation from the radio polarization the synchrotron origin is established beyond any doubts (Faraday rotation and the dependence of linear polarization on the local spectral index should be taken into account). In those cases for which we have polarization maps of good signal to noise ratio and with a matched beam at both radio and optical wavelengths, we find almost a 1:1 correspondence between optical and radio polarization structure (see examples in Fig. 4 and 5).

[1] An optical polarization p = 47% has been found in the quasar 3C 279 on 1986 Aug 6th (Mead 1988). The western hot spot of Pictor A shows p = 46% (Röser 1989).

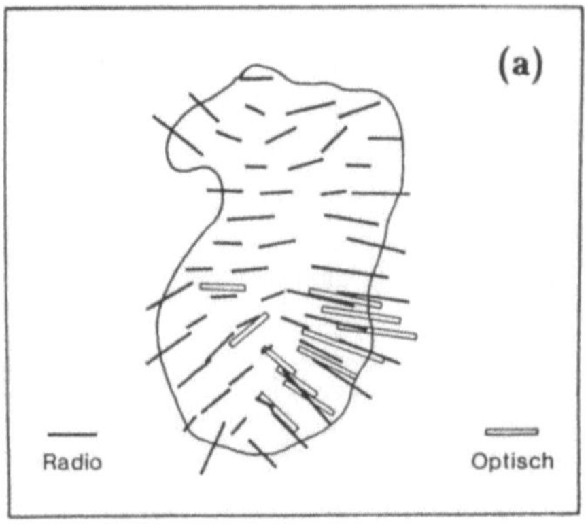

Fig. 4: Comparison between optical and radio polarization.
(a) Radio polarization map of the southern lobe of 3C 33 (Rudnick *et al.* 1981) with the optical polarization vectors from Meisenheimer & Röser (1986) overlayed. **E**–vectors are shown.
(b) Radio and optical polarization in the M 87 jet. **B**–vectors from Owen *et al.* (VLA, 2cm smoothed to the optical resolution) are shifted by 0.″1 towards NW with respect to the optical vectors from Schlötelburg *et al.* (1988).

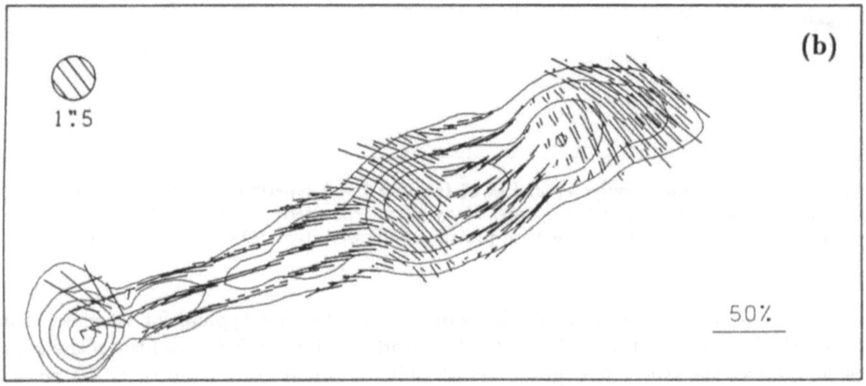

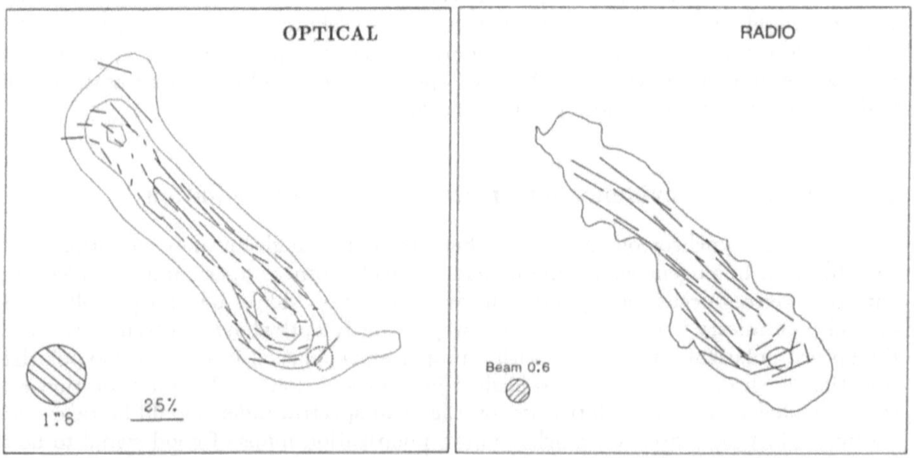

Fig. 5: Comparison of optical and radio polarization in the jet of 3C 273. **B** vectors are shown. The **left panel** displays our new optical polarimetry with an effective beam of 1.″6 FWHM. The dotted contour represents the radio peak at the hot spot (as drawn in the right panel). Note the turning of the polarization vectors between the optical peak and the hot spot position. The **right panel** gives the radio polarization at 6 cm (based on Perley 1984). In general the amount of polarization seems to be significantly higher at radio frequencies.

Besides the firm evidence of a radio-optical agreement in the polarization structure, an optical morphology matching that observed on radio maps can be taken as good evidence for optical synchrotron emission. The filament in the western lobe of Pictor A (Fig. 2) is a good example. For unresolved point sources a positional coincidence within $\leq 1"$ might be regarded as the zeroth order of this morphological criterion. On the other hand, from the well established cases and some counter examples (Röser et al. 1990) we know that obvious morphological differences combined with reasonable positional offsets certainly are incorrect identifications (e.g. a fuzzy optical object 2" away from an unresolved radio hot spot). A third, even less certain argument for optical synchrotron emission may be based on the fact that a standard synchrotron spectrum is able to connect the radio and optical flux measurements (e.g. 3C 111 east, see Meisenheimer et al. 1989a, or 3C 303 west, see Röser 1989).

In the past, it has been a common argument in favour of optical synchrotron emission that a "non-thermal spectrum" is observed in the optical. This criterium may work if relatively bright ($R \leq 19$ *mag*), compact objects are considered. For faint objects (optical counterparts of jets or hotspots have typically $R > 22$ *mag*) this criterion is not reliable: a normal galaxy without emission lines is often a much more likely interpretation for a noisy "featureless continuum".

An overview over extended extragalactic radio sources which show considerable evidence for the emission of synchrotron light is given in table 1. At present, there are five well established optical synchrotron sources, another five very good candidates and > 3 possible candidates. The classification as "established" or "very good" is based on the property encircled in table 1. Optical synchrotron emission on a level above

Table 1: Optical synchrotron emission from extended extragalactic radio sources.

Three criteria for synchrotron light are listed:

P: optical polarization || radio polarization detected, i.e. synchrotron light established,

M: optical morphology ≈ radio morhology,

SS: *overall* spectrum points to a common synchrotron origin of the radiation.

Encircled criterion led to the classification as "established", "very good" or "possible" candidate.

Object	P	M	SS	Reference
Established optical synchrotron light:				
M 87 Jet	✓⃝	•	•	Baade 1956
3C 273 Jet	✓⃝	•	•	Röser & Meisenheimer 1986
3C 33 south Hotspot	✓⃝	•	•	Meisenheimer & Röser 1986
Pictor A west Hotspot	✓⃝	•	•	Röser & Meisenheimer 1987
3C 20 west Hotspot	✓⃝	•	•	Hiltner *et al.* 1989
Very good candidates for optical synchrotron light:				
3C 303 west Hotspot		(?)	⊙	Kronberg *et al.* 1977
3C 66B Jet		⊙	?	Miley *et al.* 1981
Pks 0521-36 Jet		⊙	?	Keel 1986
3C 111 east Hotspot		⊙	•	Meisenheimer *et al.* 1989
Pictor A west Filament		⊙	•	Röser *et al.* 1987
Possible candidates:				
3C 351 east Hotspot(s)		•		*see* Röser 1989
3C 349 south Hotspot		•		new identification
Coma A Knots	?	•	- !	Miley *et al.* 1981, *see* Röser 1989
some additional candidates are discussed in Röser 1989				

$\sigma(\text{R}) = 22.mag/arcsec^2$ seems to be very rare: Our group has looked into almost all candidate identifications of radio hot spots proposed in the literature, plus another 25 objects (total: $N \gtrsim 50$). We confirmed the emission of synchrotron light from 3C 33 south and found 5 new candidates: 3C 20, 3C 111, 3C 349, 3C 351, Pictor A, two of which are now established by polarimetry. Many candidate identifications from the list of Crane *et al.* (1983) are not included here since they either show large radio-to-optical offsets or the identification has been disproved by polarimetric observations (Röser *et al.* 1990). Out of the hundreds of known radio jets only 4 are either established or good candidates, although deep optical images have been taken in many more cases. Nevertheless, it should be noted that the detection rate of radio jets may be much higher than that of radio hot spots if one refers to a limiting ratio of radio-to-optical flux rather than a limiting optical brightness. For instance, in the optically faint 3C 273 jet the radio-to-optical flux ratio is $\text{R} = S_\nu(6cm)/S_\nu(600nm) \simeq 10^5$, while the limit for the hot spot D in Cygnus A is $\text{R} \geq 3 \times 10^7$.

3. The astrophysical relevance of synchrotron light

Typical magnetic field strengths in hot spots and in the brightest knots of radio jets are $\text{B} \simeq 30\text{nT}$ (e.g. Meisenheimer *et al.* 1989a, Owen *et al.* 1989a). So the emission of synchrotron light requires electron energies $\gamma = E/m_e c^2$ in excess of 3×10^5. Since synchrotron losses depend quadratic on energy,

$$\frac{d\gamma}{dt} = 5 \times 10^{-9}\ \gamma^2 \left(\frac{B}{30nT}\right)^2\ years^{-1},$$

these high energy electrons lose their energy very quickly — i.e. half of the energy within a few hundred years. Even if the ultra-relativistic electrons could freely escape the acceleration region with the speed of light they could not travel further than about 100 pc in a field of 30 nT. This corresponds to a projected length of $\leq 0\farcs03$ at the distance of 3C 273 and $\leq 1''$ for the nearest source — M 87. That is, images in synchrotron light essentially map the regions of local particle (re-)acceleration.[2] On the other hand, electron radiating at 6 cm wavelength are 300× less energetic and accordingly could reach much greater distances from their place of acceleration. In a very simple source geometry where the particle acceleration occurs at the strong shock (Mach-disk) in the working surface of a supersonic jet and the accelerated electrons are advected downstream with the magnetized plasma, such energy dependent propagation effects would result in a thin layer of optical emission ($< 10\ pc$ near the Mach disk) while the radio emission would originate from a cylinder extending $> 1\ kpc$ downstream (Meisenheimer & Heavens 1986). Theoretical synchrotron spectra based on this model show a spectral break from a powerlaw index $\alpha = -0.5$ (for $S_\nu \sim \nu^\alpha$) observed at low frequencies to $\alpha = -1.0$ in the intermediate regions (Heavens & Meisenheimer 1987). Recently we found that this simple consideration of the propagation effects does provide a excellent fit to the overall synchrotron spectra of radio hot spots (see Fig. 6, details are given in Meisenheimer *et al.* 1989a).

The simple model has been developed for radio hotspots which are only marginally resolved even on the best VLA maps with about 0.1 arcsec resolution. Applying a similar naive model to clearly resolved sources like the jets of M 87 and 3C 273 one

---

[2] There is one way out of this *in situ acceleration* picture if the electrons propagate through regions void of any magnetic fields.

would expect that the optical image should show seperated spots of high acceleration efficiency while the radio image is much smoother due to the large range of electrons with lower energy. The visual inspection of the morphology and polarization structure (Fig. 4 and 5) shows that nature fails to fulfil our simple-minded expectations. I will give a more quantitative analysis of this apparent "lack of propagation effects" in section 5.3, below.

4. The overall synchrotron spectra of extended radio sources

The ultimate goal of the investigations of radio sources at frequencies beyond the radio domain ($\lambda < 1$ cm) aims to establish the overall synchrotron spectra of these sources and their spatial variations. The latter are expected due to the above described propagation effects. Optical observations play a key role in this business since they are sensitive enough to allow the polarimetric confirmation of synchrotron emission. Having established that synchrotron radiation dominates both the radio and the optical regime it does not seem too brave to interpret flux measurements at intermediate frequencies in terms of pure synchrotron radiation, provided a smooth radio-to-optical spectrum can be drawn through all flux measurements. Since extended radio sources are faint at $\lambda < 1$ cm we have to rely entirely on groundbased observations with the largest telescopes, in order to detect hot spots or jets in the region between 1 cm and 1 μm.

4.1. The synchrotron spectra of radio hot spots

In Fig. 6 I display the overall synchrotron spectra between 10^9 and 10^{15} Hz for five strong radio hot spots in classical double sources. The radio and optical data have been supplemented by flux measurements at 1.3 mm (at the 30 m telescope on Pico Veleta) and at 2.2 μm (at the ESO 3.6m and at UKIRT on Mauna Kea). A detailed

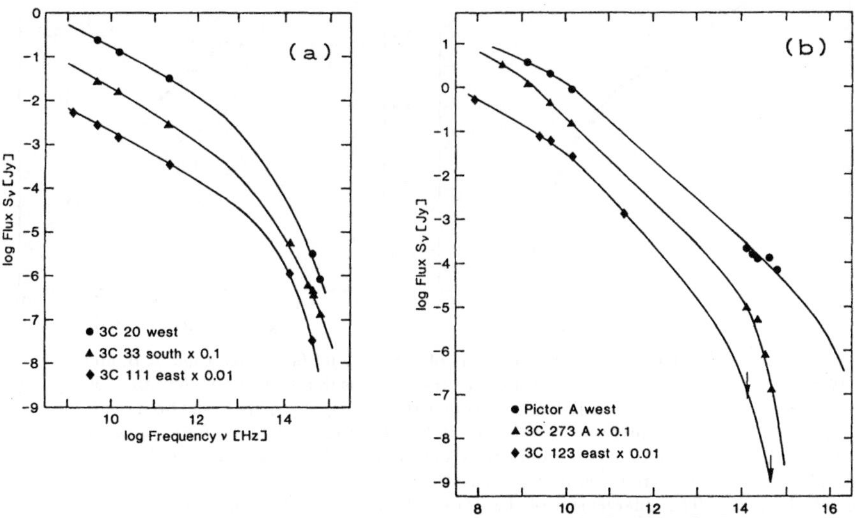

Fig. 6: The overall synchrotron spectra of radio hot spots. Panel **(a)** contains spectra without spectral break, while the spectra in panel **(b)** all show a break by $\Delta\alpha = -0.5$ at radio frequencies.

description of these data and their interpretation is given in Meisenheimer *et al.* (1989). The results of our study may be summarized as follows:

- With only one exception – Pictor A west – all spectra cut off around $\nu_c = 10^{14}$ Hz. Including the majority of hot spots which are undetected in the optical, we find that $\geq 90\%$ of all hot spot spectra cut off below 10^{15} Hz.
- All radio spectra are flat: $\alpha \simeq -0.5$. This is in perfect agreement with the prediction of diffusive shock acceleration at a strong shock (first-order Fermi acceleration).
- With respect to the spectral index at intermediate frequencies, 10^{10} to 10^{14} Hz, two types of radio hot spots may be distinguished: (i) hot spots which show a $\Delta\alpha = -0.5$ break at radio frequencies (Fig. 6b). This break is caused by synchrotron losses within the extended hot spot emission region itself (ii) Low loss hot spots show straight powerlaw spectra all the way from the radio band up to the cutoff.

4.2. The synchrotron spectra of radio jets

The overall spectrum of the brightest part of the M 87 jet (Knots $A + B + C$) as compiled from various authors is shown in Fig. 7a. The model fit through the data ignores the UV point at 2×10^{15} Hz (large systematic errors) and the X-ray point (synchrotron origin not established). The enlargement of the infrared-to-optical part shows that after ten years of controversies, there is now a perfect agreement between a model synchrotron spectrum with exponential cutoff and the near-infrared, optical and 240 nm data of five independent groups. I have no explanation why the infrared data of Smith et al. (1983) and Killeen et al. (1984) are > 35% below the mean spectrum although background subtraction should be more reliable than in the experiment done by Stocke et al. (1981).

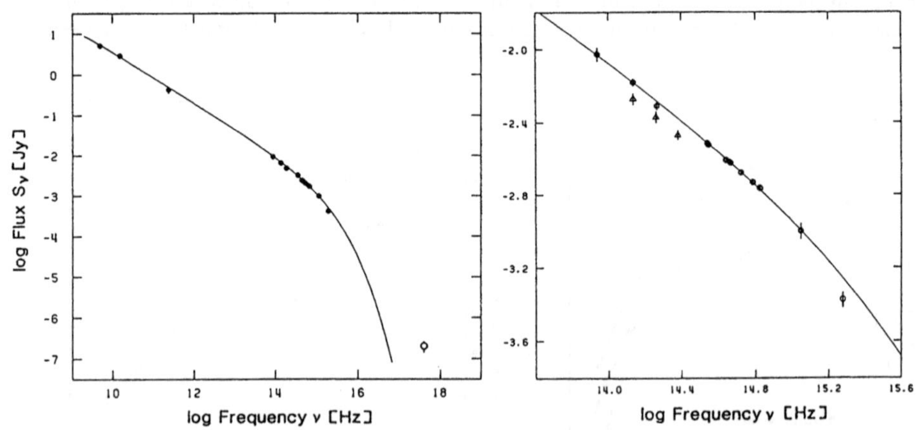

Fig. 7a: The overall spectrum of the brightest part of the M 87 jet (sum over knots A,B, and C, i.e. 10.″5 to 19.″0 from the core). In addition to our own measurements at 850, 650 and 450 nm, data from various sources are included: The radio data ($\lambda = 2$, 6 cm) are from Owen (priv. comm.), the point at $\lambda = 1.3$ mm is from Salter *et al.* (1989), infrared data are from Stocke *et al.* (1981), additional optical points are from Perez-Fournon *et al.* (1988) and Keel (1988). The UV data ($\lambda = 270$, 160 nm are taken from Perola & Tarenghi (1980), the X-ray point is from Schreier *et al.* (1982). The enlargement of the near-IR to optical region displayed in the *left panel*, shows that the photometry of Stocke *et al.*, the three independent optical measurement, and the UV point at 270 nm are in perfect agreement with a model synchrotron spectrum cutting off at 6×10^{15} Hz. There is no explanation for the discrepant near-IR measurements by Smith *et al.* and Killeen *et al.* the mean of which is shown by Δ's.

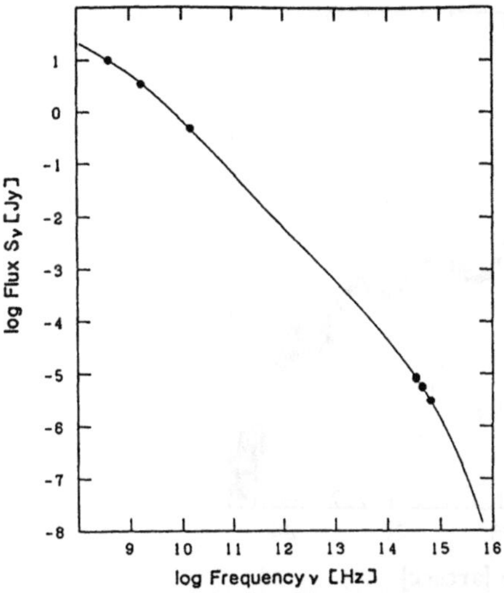

Fig. 7b. The optical spectrum of the brightest optical knot in the jet of 3C 273 (at 20" from the core). Radio data are from Conway (75 and 18 cm, *priv. comm.*) and Perley (2 cm, *priv. comm.*). The spectrum is relatively flat below 1 GHz and steepens to $\alpha_{RO} \simeq -1.0$ at about 3 GHz. In the optical region the spectral index is even steeper, $\alpha_{opt} = -1.6$. A model synchrotron spectrum provides a good fit to the data.

The synchrotron spectrum of the brightest optical knot in the 3C 273 jet (at $\simeq$20" from the QSO, Fig. 7b) presently can be based on measurements in radio and optical bands only, since observations at other wavelengths lack the resolution which would be required to isolate the various subcomponents with different spectra from each other and from the underlying host galaxy around 3C 273. Although the spectrum resembles that of the M 87 knots (steepening by $\Delta\alpha = 0.5$ at $\nu_b = 2$ GHz, cutoff at $\nu_c = 10^{15}$ Hz) the spectral index in the intermediate region (10^{10} to 10^{13} Hz) is significantly steeper ($\alpha_{RO} \simeq -1.0$) than that found in the M 87 jet.

4.3. Spectral variations along the jets of M 87 and 3C 273

The optical spectra of the radio jets in M 87 and 3C 273 show spectral indices which are steeper than the two-point indices inferred from straight radio-to-optical interpolations (Fig. 7). Thus a spectral *cutoff* or *break* has to occur at around 10^{14} Hz. Spatial variations of this cutoff or break frequency ν_c should result in pronounced variations of the optical spectral index α_{opt} at frequencies *above* ν_c. The first map of the optical spectral index of the M 87 jet (see Fig. 8) is based on 7 exposures in B, R, and I, taken at the 2.2 m telescopes II on La Silla under < 1" seeing (Schlötelburg 1989). A very careful positional alignment[3] and the adjustment of slightly different seeing profiles to a common effective PSF of 1.″5 resulted in a typical error of the spectral index $\sigma \lesssim 0.03$, even in steep intensity gradients.

No significant spectral index variations across the jet were found. Surprisingly also the run of the optical spectral index along the jet (Fig. 8) is very smooth: outside the inner 4" (where high background leads to large uncertainties) and inside 24" from the core (beyond which the jet is very faint) the spectral index ranges between $\alpha_{opt} = -0.75$ (at knot *A*, 12.″5 from the core) and $\alpha_{opt} = -1.5$. Interpreting the observed variations of α_{opt} in terms of a constant spectral shape (Fig. 7b) with changing critical frequency

[3] The 7 exposures have to be aligned with an accuracy of < 0.″03. At this level even the differential refraction in the B filter between globular clusters which served as reference and the blue jet have to be taken into account (0.″07).

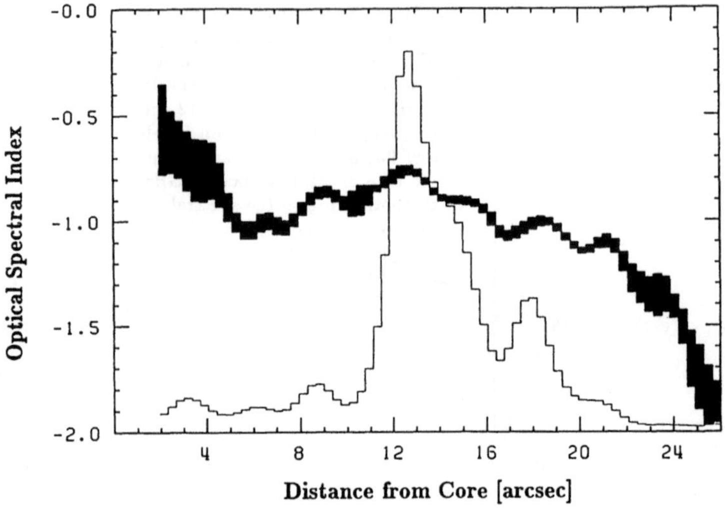

Fig. 8: Optical spectral index α_{opt} along the M 87 jet. The black band gives $\alpha_{opt} \pm 1\sigma$, a thin line represents the total flux. The effective beam is $1\overset{''}{.}5$ FWHM.

ν_c (Meisenheimer *et al.* 1989b), I find that ν_c does not deviate by more than a factor of 2 from the mean anywhere between 4" and 24" from the core.

On the other hand, considering the synchrotron losses (see Chap. 3, above) one concludes that $\nu_c \equiv 42\ (B_\perp/nT)\ \gamma_c^2$ Hz (where $B_\perp$ is the field $\perp$ the line-of-sight) should decrease by a factor > 4 if the electrons have left their place of acceleration by 1" projected distance. Since the electrons presumably propagate much slower than light, one gets that nowhere in the jet optically radiating electrons could be $> 0\overset{''}{.}5$ away from their place of acceleration, even if one allows for "beam smearing" effects. So the knots (typical separation: $2\overset{''}{.}5$) are much to sparse to be the sole acceleration centers. Even if one regards every knot as a series of internal shocks (cf. Nieto & Leliévre 1982) one is not able to explain the obvious lack of synchrotron losses in the inter-knot regions.

The smoothness of α_{opt} along the jet favours a quite opposite model - namely that the critical energy γ_c is more or less constant along the entire jet: Moderate variations of the transverse field $B_\perp$ (due to compression or mere projection effects) and some variations of the density of ultra- relativistic electrons n_{rel} are sufficient to explain both the apparent variations in ν_c and the peaks and dips in the intensity profile (because $S_\nu \sim n_{rel}\ B_\perp^{1-\alpha}$, where $\alpha \equiv dlog\ S_\nu/dlog\nu$ is the local spectral index.) Then the observed synchrotron losses would have to be balanced by a continuous re-acceleration. My feeling is that Fig. 8 tells us something very important which we do not yet comprehend because we are preoccupied with our present-day models. Note that the good coincidence of the radio and optical polarization (Fig. 4b) also argues for propagation effects playing a minor role.

Very recently we got a similar accurate map of the optical spectral index along the 3C 273 jet (Fig. 9). Presumably due to the inferior linear resolution (3C 273 is 50$\times$ more distant than M 87) the variation of α_{opt} seems even smoother than in M 87. Therefore the general trend - steepening of α_{opt} with distance from the core - is very clear. The most important result of our new photometry of the 3C 273 jet, however, concerns the outermost parts: here we now have firm polarimetric evidence for the

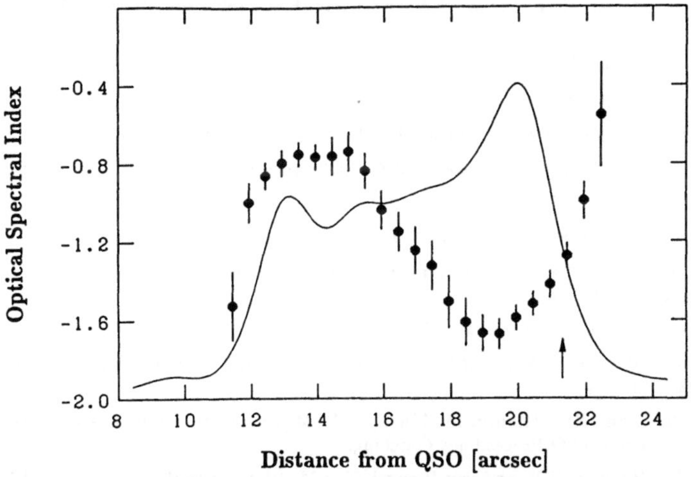

Fig. 9: Optical spectral index along the jet of 3C 273. A thin line shows the intensity profile at 650 nm (effective beam: 1.″6 FWHM). The position of the radio hot spot is marked by an arrow.

emission of optical synchrotron radiation from the bright radio hot spot (see Fig. 5): we find that the optical spectrum flattens beyond the optical peak at 20" from the core. This is in obvious conflict with our original claim of an extremely steep optical hot spot spectrum (Röser & Meisenheimer 1986) which was supported in subsequent papers by Fraix-Burnett and Nieto (1988) and Keel (1988). Our new data are so much better than the old ones that undoubtedly our old result is wrong. From the flat optical spectral index $\alpha_{opt} = -1.2$ I conclude that the relatively high radio-optical flux ratio $R = 2.5 \times 10^7$ is due to a steep radio-to-optical powerlaw rather than due to a spectral cutoff at around $\nu_c = 10^{14}$ Hz.

Acknowledgement: Since 1982 H.-J. Röser and myself are working very closely together in investigating synchrotron light from extragalactic radio sources. Members of our group in Heidelberg are M. Schlötelburg who did his thesis about the M 87 jet, and P. Hiltner who is involved in the survey for optical counterparts of radio hot spots. Measurements above $\lambda = 1\ \mu m$ have been carried out together with M.G. Yates, R. Chini, and R. Perley. I thank R. Conway, R. Laing, and F. Owen who kindly provided their beautiful radio maps on digital tapes to enable a direct comparison of radio and optical measurements.

Baade, W. : 1956, *Astrophys. J.* **123**, 550

Blandford, R.D. and Rees, M.J. 1974, *Mon. Not. R. astr. Soc.* **169**, 395

Carilli, C.L., Dreher, J.W. and Perley, R.A. 1989, in K. Meisenheimer and H.-J. Röser (eds.): *Hot Spots in Extragalactic Radio Sources.* Lecture Notes in Physics Vol. **327**. Springer Heidelberg, p. 51.

Crane, P., Tyson, J.A. and Saslaw, W.C. 1983, *Astrophys. J.* **265**, 681

Curtis, H.D. 1918, *Lick Obs. Publ.* **13**, 11

Fraix-Burnet and J., Nieto, J.-L. 1988, *Astron. Astrophys.* **198**, 87

Heavens, A.F., Meisenheimer, K. 1987, *Mon. Not. R. astr. Soc.* **225**, 335

Hiltner, P., Meisenheimer, K., Röser, H.-J., Perley, R.A. and Laing, R.A., 1989, submitted to Astron. Astrophys.

Keel, W.C. 1986, *Astrophys. J.* **302**, 296

Keel, W.C. 1988, *Astrophys. J.* **329**, 532

Killeen, N.E.B., Bicknell, G.V., Hyland, A.R. and Jones, T.J. 1984, *Astrophys. J.* **280**, 126

Kronberg, P.P., Burbidge, E.M., Smith, H.E. and Strom, R.G. 1977, *Astrophys. J.* **218**, 8

Mead, A.R.G., 1988, *Ph.D. Thesis*, Edinburgh

Meisenheimer, K.and Heavens, A.F. 1986, *Nature* **323**, 419

Meisenheimer, K. and Röser, H.-J. 1984, in W. Brinkmann (ed.): *X-ray and UV Emission of Active Galactic Nuclei*, proceedings, MPE München.

Meisenheimer, K. and Röser, H.-J. 1986,*Nature* **319**, 459

Meisenheimer, K., Röser, H-J., Hiltner, P., Yates, M.G., Longair, M.S., Chini, R. and Perley, R.A. 1989a, *Astron. Astrophys.* **219**, 63

Meisenheimer, K., Röser, H.-J. and Schlötelburg, M. 1989b, in E. Meurs and R. Fosbury (eds.): *Extranuclear Activity in Galaxies*, ESO Proceedings Garching.

Miley, G.K., Heckman, T.M., Butcher, H.R. and van Breugel, W.J.M. 1981, *Astrophys. J.(Lett.)* **247**, L5

Nieto, J.-L. 1983, in A. Ferrari and A.G. Pacholczyk (eds.): *Astrophysical Jets*, Proceedings Vol. 103, Reidel Dordrecht, p.113.

Nieto, J.-L. and Lelièvre, G. 1982, *Astron. Astrophys.* **109**, 95

Owen, F.N., Hardee, P.E. and Cornwell, T.J. 1989a, ApJ 340 698

Owen, F.N., Hardee, P.E., Cornwell, T.J., Hines, D.C. and Eilek, J.A. 1989b, in K. Meisenheimer and H.-J. Röser (eds.): *Hot Spots in Extragalactic Radio Sources.* Lecture Notes in Physics Vol. **327**. Springer Heidelberg, p. 77.

Perez-Fournon,I., Colina, L., Gonzalez-Serrano, J.I. and Biermann, P.L. 1988, *Astrophys. J.(Lett.)* **329**, L81

Perley, R.A. 1984, in R. Fanti, K. Kellermann, and G. Setti, (eds.): *VLBI and Compact Radio Sources*, Reidel Dordrecht, p. 153

Perola G.C. and Tarenghi, M. 1980, *Astrophys. J.* **240**, 447

Röser, H-J. 1989, in K. Meisenheimer and H.-J. Röser (eds.): *Hot Spots in Extragalactic Radio Sources.* Lecture Notes in Physics Vol. **327**. Springer Heidelberg, p. 91.

Röser, H.-J., Hiltner, P. and Meisenheimer, K. 1990, *in prep.*

Röser, H.-J. and Meisenheimer, K. 1986, *Astron. Astrophys.* **154**, 15

Röser, H.-J. and Meisenheimer, K. 1987, *Astrophys. J.* **314**, 70

Röser, H.-J., Perley, R.A. and Meisenheimer K. 1987, in E. Assio and D. Grissilon (eds.): *Magnetic Fields in Extragalactic Objects.* Cargese, p. 361.

Rudnick, L., Saslaw, W.C., Crane, P. and Tyson, J.A. 1981, *Astrophys. J.* **246**, 647

Salter, C.J., Chini, R., Haslam, C.G.T., Junor, W., Kreysa, E., Mezger, P.G., Spencer, R.E., Wink, J.E. and Zylka, R. 1989, *Astron. Astrophys.* **220**, 42

Scheuer, P.A.G. 1974, *Mon. Not. R. astr. Soc.* **166**, 513

Schlötelburg, M. 1989.*Ph.D. thesis*, Heidelberg

Schlötelburg, M., Meisenheimer, K. and Röser, H.-J. 1988, *Astron. Astrophys. (Lett.)* **202**, L23

Schmidt, M. 1963, *Nature* **197**, 1040

Schreier,E.J., Gorenstein, P. and Feigelson, E.D. 1982, *Astrophys. J.* **261**, 42

Smith, R. M., Bicknell, G.V., Hyland, A.R. and Jones, T.J. 1983, *Astrophys. J.* **266**, 69

Stocke, J.T., Rieke, G.H. and Lebofsky, M.J. 1981, *Nature* **294**, 319

Very High Energy X-Rays from Supernova 1987A

R. Staubert

Astronomisches Institut der Universität Tübingen,
Waldhäuserstr. 64, D-7400 Tübingen, Fed. Rep. of Germany

Supernova 1987A in the Large Magellanic Cloud was not only the brightest since Keplers supernova in 1604, but also the very first supernova in history which could be studied in all ranges of the electromagnetic spectrum with sensitive instruments based on ground and in space. In addition a flash of neutrinos was observed, confirming observationally the concept of production of neutrinos in the collapse of a star. Of equal importance were detailed theoretical model calculations which had become possible through the new generation of powerful supercomputers. So, SN 1987A is a test case for existing theories and has provided a wealth of new unexpected features. Next to the observation of neutrinos, especially remarkable features of SN 1987A are the bolometric light curve, the detection of infrared and gamma-ray line radiation and the detection of the very high energy X-ray continuum, all connected to the existence and spatial distribution of 0.075 $M_\odot$ of radioactive ^{56}Co in the debris of this stellar collaps.

Very high energy X-rays (>20 keV) were measured by the High Energy X-Ray Experiment (*HEXE*) onboard the soviet space station *Mir* and a few balloon instruments. Here we concentrate on results from the *Mir* observations. *HEXE* discovered hard X-ray emission from SN 1987A in August 1987. The continuous monitoring of SN 1987A by *HEXE* resulted in a high energy X-ray light curve (20-200 keV) showing a slow rise to a maximum in March 1988 and a nearly exponential fall off thereafter. This is in general agreement with an origin due to Compton scattered gamma-ray line photons from the decay of ^{56}Co, produced by thermonuclear fusion. The data require a substantial mixing of ^{56}Co into the supernova envelope. The existence of a pulsar is not needed to explain the observed high energy radiation.

I. Observations and early results

The High Energy X-Ray Experiment (*HEXE*) (Reppin et al. 1983) is a low background phoswich detector system with a geometric area of 800 cm^2 sensitive for photon energies of 15-200 keV. *HEXE* was launched as part of the *Kvant* Röntgen observatory on March 31, 1987 and docked onto the soviet space station *Mir* on April 12, 1987. After the discovery of hard X-rays from SN 1987A on August 10, 1987 (Sunyaev et al. 1987) this source had become the prime target and was regularly monitored with a total observing time of 148 hours until mid 1989. Operational details about the *Kvant* Röntgen observatory and the *HEXE* detector performance and first results of observations on SN 1987A are given by Sunyaev et al. 1987, Trümper et al. 1987, Reppin et al. 1987 and Sunyaev et al. 1988.

Since *HEXE* is a non-imaging instrument it was crucial for the interpretation of the observations of SN 1987A to establish from which point in the sky the high energy X-rays were received. This was done by performing a series of pointings around the source at various angular separations and position angles. The final analysis confirms and strengthens the earlier conclusion: there is no doubt that SN 1987A, lying well within the 1 standard deviation contour, is the source. The two nearest known other sources, PSR 0540-693 and LMC X-1, are well outside the 5 std. dev. and 7 std. dev. contours, respectively (Staubert et al. 1989). Although both of these sources are inside the field of view of the *HEXE* collimator (1.6^0 FWHM) when pointing towards SN 1987A, there is no sign for any contamination in the energy range of 15-200 keV, very likely because of their soft spectra.

Reviews in Modern Astronomy 2 Editor: G. Klare

The appearance of high energy X-rays as well as line radiation at 847 keV and 1238 keV from the decay of ^{56}Co (Matz et al. 1988) as early as August 1987 was quite unexpected. According to model calculations (e.g. Woosley et al. 1988) such radiation should have appeared only about two hundred days later. It had also been speculated (Bandiera et al. 1988) that hard X-rays, including those in the 2-15 keV range as observed by *Ginga* (Tanaka et al. 1988), may be produced by a pulsar which might have been formed in this supernova event. It was not clear, however, how a pulsar could be the cause of a high flux of hard X-rays at the surface of the still very dense envelope of about 11 $M_\odot$ surrounding any central object without significantly affecting the bolometric light curve (even under the assumption of a clumpy structure of the shell). In addition, the fast rise of the hard X-ray intensity expected from a pulsar origin was not observed. It now seems quite clear that 0.075 $M_\odot$ of radioactive ^{56}Co are responsible for the observed high energy X-ray continuum. The early appearance as well as the shape of the overall light curve can be explained by substantial mixing of this material into the outer parts of the expanding envelope.

II. Evidence for ^{56}Cobalt

There is convincing evidence for the existence of radioactive ^{56}Co from various observational data.

The optical light curve

Both, the visible light curve (Burki et al. 1989, see Fig. 1) and the bolometric (U to M) light curve (e.g. Catchpole et al. 1989) follow an exponential decrease between day 150 and about day 350, in excellent agreement with the decay time for ^{56}Co of 111.3 days. This suggests that the optical supernova is entirely powered by energy stored and released by ^{56}Co. The decay energy (on average 3.3 MeV per nucleus) is mainly released into gamma-rays of 847 keV and 1238 keV and thermalized by successive Compton scatterings and final photo absorption in the envelope. The amount of originally produced ^{56}Ni (decaying into ^{56}Co with a time constant of 8.8 days) is 0.075 $M_\odot$. From the shape of the overall light curve it can be concluded that ^{56}Ni must be mixed into the outer parts of the envelope (Kumagai et al. 1989).

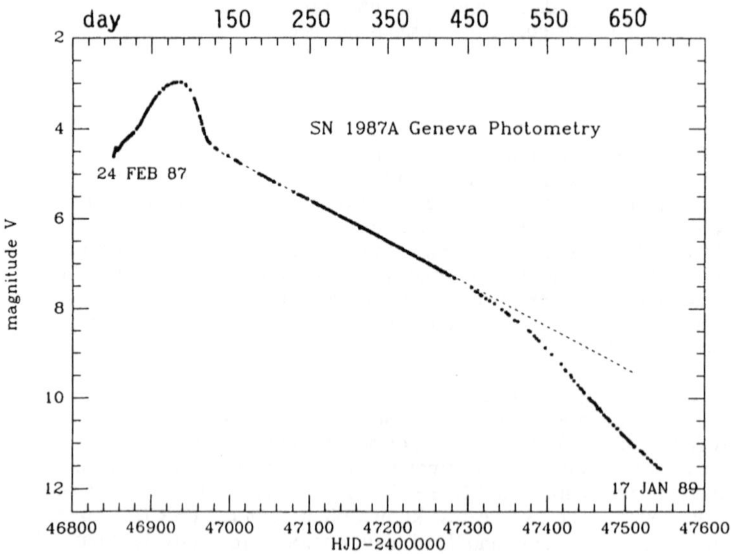

Fig. 1 The visible light curve of SN 1987A (Burki et al. 1989). The dashed line represents the decay time of ^{56}Co.

Gamma-ray lines

A confirmation of the suspected radioactive source of energy was achieved when gamma-ray line radiation at 847 keV and 1238 keV was detected directly, first by instruments onbord the SMM satellite (Matz et al. 1988) and then by a number of balloon payloads (see the review by Gehrels et al. 1987 and references therein). The observed intensity and its development with time are quite well reproduced by 'mixed' models like model 10HMM of Pinto and Woosley (1988b) or model 11E1 by Kumagai et al. 1988.

IR lines

Further direct evidence for the existence of Cobalt in the supernova shell was the detection of an IR line at 10.5 μm (Rank et al. 1988), which was identified as due to CoII.

The high energy X-ray continuum

The appearance of a high energy X-ray continuum is an unavoidable consequence of the release of gamma-rays within a scattering medium. While only a few of the original decay gamma-rays can escape without interaction most of them suffer multiple Compton scatterings, thus producing a very hard high energy X-ray spectrum. The spectrum originally measured by the *HEXE* and *PULSAR X-1* instruments (Sunyaev et al. 1987) is reproduced in Fig. 2. With a power law photon index of -1.4 it was the hardest spectrum ever observed from a celestial source in this energy range. The subsequent analysis of all *HEXE* data using improved attitude information, a revised energy calibration and optimum background reduction procedures has reproduced the early results (Englhauser et al. 1989). A few balloon experiments have measured spectra which are in good agreement with the results from *Kvant* (see the review by Gehrels et al. 1987 and references therein).

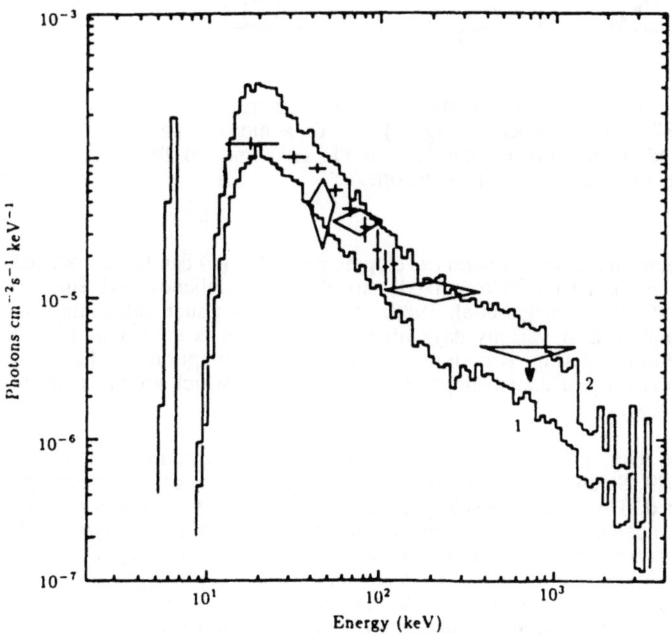

Fig. 2 High energy X-ray spectrum of SN 1987A as observed by *HEXE* (crosses) and *PULSAR X-1* (diamonds) in Aug./Sep. 1987. The histograms show model predictions by Grebenev and Sunyaev 1988 for day 180 (1) and day 240 (2).

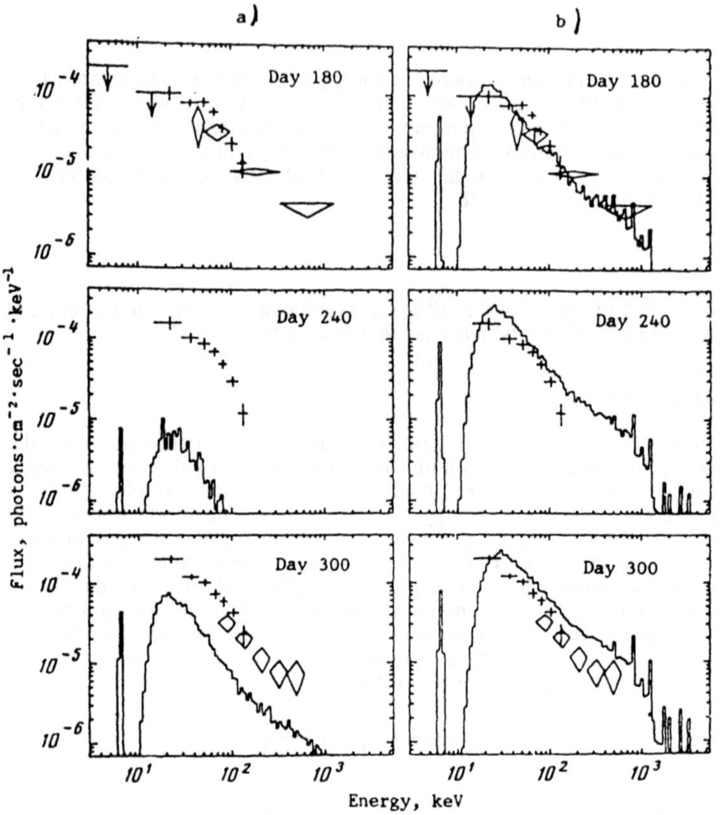

Fig. 3 Comparison of observed and predicted high energy X-ray spectra of SN 1987A (symbols as in Fig. 2). a) for a model where ^{56}Co is concentrated at the inner boundary of the envelope, b) for substantial mixing of cobalt into the envelope.

The high energy X-ray spectrum and its temporal development (see below) can be reproduced by Monte Carlo calculations based on ^{56}Co 'mixed' models (e.g. Grebenev and Sunyaev 1988, Pinto and Woosley 1988b, Kumagai et al. 1988). An example for such calculations in comparison to observational data at various days after the explosion is shown in Fig. 3 (Grebenev and Sunyaev 1988). 'Mixing' is definitely required, the assumption of ^{56}Co being concentrated at the inner boundary of the envelope leads to predictions which are inconsistent with the observations (panel a).

Comparing spectra of Sept./Oct. 1988 with those from January 1988 a spectral hardening is evident (Sunyaev et al. 1989, Staubert et al. 1989). It appears, that the flux at the low energy end was particularly reduced. This kind of development is consistent with Compton scattering models. It is qualitativly understandable, if one recalls that a gamma-ray line photon needs several interactions to be down-scattered to an energy of 20 keV. An expanding supernova shell, however, allows for an increasing number of decay photons to escape unscattered (thus keeping the number of observable line photons up despite the rapid decay of ^{56}Co nuclei) or scattered only a few times, such that the lower energies will be reached less efficiently.

III. The high energy X-ray light curve

The temporal development of the X-ray intensity is determined by the increasing transparency of the expanding shell and by the rapid decay of ^{56}Co. The overall high energy X-ray light curve of SN 1987A in the energy range 45-105 keV, as observed by *HEXE*, is shown in Fig. 4. The light curves for 15-45 keV and 105-200 keV are very similar in shape (see Sunyaev et al. 1989). Their characteristics are a slow rise to a maximum intensity in March 1988 (~50% over half a year), followed by a faster fall off, the time constant of which is close to that of the decay of ^{56}Co.

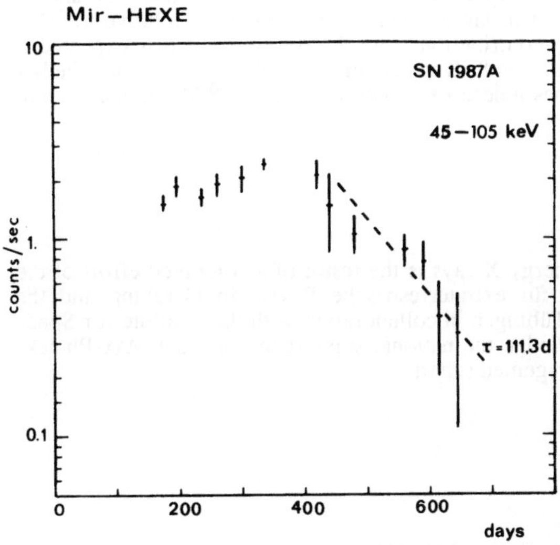

Fig. 4

The high energy X-ray light curve in the 45-105 keV range as measured by *HEXE*.

Mixing of ^{56}Co into the outer envelope can explain both, the early appearance of high energy X-rays and the rather slow increase of the intensity with time. Extreme models, either the concentration of all ^{56}Co in a thin shell close to the center or a uniform mixing into e.g. the inner six $M_\odot$ of the envelope (Grebenev and Sunyaev 1988) can be ruled out. But also models with a more moderate mixing, like the '10HM model' by Pinto and Woosley (1988 a,b) or the 'HY model' by Kumagai et al. 1988 seem to predict much more peaked light curves than required by the data. Sunyaev et al. 1989 have made a first attempt to actually use the observed light curve to derive the radial distribution of ^{56}Co. The conclusion is that a model with 80% of ^{56}Co concentrated in the center and 20% mixed almost uniformly into a 12 $M_\odot$ envelope seems to reproduce the observed intensities reasonably well.

IV. Does SN 1987A contain a pulsar ?

Supernovae of Type II, as SN 1987A, are thought to be the result of the gravitational collaps of the iron core of a massive star at the end of its development, and are the suspected birth places of neutron stars or pulsars. The observed burst of neutrinos in SN 1987A has particularly enhanced the expectation to find a pulsar in this supernova. The intensive search for any pulsed flux in all wavelength ranges had finally resulted in a report about the detection of optical pulsations with a period of 0.51 ms (Kristian et al. 1989). However, besides the fact that the assumed pulsar seemed to have some unconventional properties, this observation has never been confirmed, nor has any other periodic signal been found.

As described above, the observed high energy X-ray continuum and its temporal characteristics can be understood in the frame work of 'mixed' ^{56}Co models. There is no need for the existence of a pulsar.

The same statement can probably also be made with respect to the bolometric light curve and the overall energy budget of the remnant. The bolometric light curve had started (at about day 300, see Fig. 1) to decay faster than ^{56}Co, which can only in part be explained by the increasing fraction of energy escaping as gamma-rays and hard X-rays. In fact, on day 560 the combined luminosity in gamma-ray lines and the high energy X-ray continuum is less than 10^{39} erg/s, which is less than one third of the difference between the extrapolated cobalt luminosity and the observed bolometric (U-M) luminosity. So, no additional energy source is required and a possibly existing pulsar cannot significantly contribute to the overall energetics at this time. It was then noticed that the optical light curve (compare Fig. 1) started to flatten again, which seemd to indicate the emergence of some additional energy source. This could be a pulsar. There are, however, competing explanations: together with ^{56}Co also ^{57}Co and ^{44}Ti should have been produced, possibly even in larger than solar abundances relative to ^{56}Co. With its decay time of 271 d the energy contribution from the decay of ^{57}Co is expected to become significant at this time (Pinto et al. 1988, Kumagai et al 1989). The question whether a pulsar (or an accreting neutron star) does indeed exist within the SN 1987A remnant is still open.

ACKNOWLEDGEMENT

The observation of SN 1987A in high energy X-rays is the result of a combined effort of the *HEXE* team at the Max-Planck-Institut für extraterrestrische Physik in Garching and the Astronomisches Institut der Universität Tübingen in collaboration with the Institute for Space Research (IKI) in Moskow. We acknowledge the national support through the Max-Planck-Gesellschaft and the Deutsche Forschungsgemeinschaft.

REFERENCES

Bandiera, R., Pacini, F. and Salvati 1988, *Nature* **332**, 418

Burki, G., Cramer, N., Burnet, M., Rufener, F., Pernier, B. and Richard, C. 1989, *ESO Messenger* No. 55, p.51

Catchpole, R.M., Whitelock, P.A., Menzies, J.W., Feast, M.W., Marang, F., Sekiguchi, K., Wyk, F.van, Roberts, G., Balona, L.A., Egan, J.M., Cartere, B.S., Laney, C.D., Laing, J.D., Spencer Jones, J.H., Glass, I.S., Winkler, H., Fairall, A.P., Lloyd Evans, T.H.H., Cropper, m.S., Shenton, M., Hill, P.W., Payne, P., Jones, K.N., Wargau, W., Mason, K.O., Jeffery, C.S., Hellier, C., Parker, Q.A., Chini, R., James, P.A., Doyle, J.G., Butler, C.J. and Bromage, G. 1989, *Mon.Not.R.astr.Soc.* **237**, 55p

Englhauser, J. et al. 1988, to be published in the Proc. of the 23rd ESLAB-Symp. on X-Ray Astronomy, Bologna, Sept. 1989

Gehrels, N., Leventhal, M., MacCallum, C.J. 1987, AIP Conf. Proc. No.170, in *Nuclear Spectroscopy of Astrophysical Sources*, Washington D.C., (Eds. Gehrels, N. and Share, G.H.), p87

Grebenev, S.A. and Sunyaev, R.A. 1988, *Soviet Astron. Lett.* **14**, 675 (288 in english translation)

Kristian, J., Pennypacker, C.R., Middleditch, J., Hamuy, M.A., Imamural, J.N., Kunkel, W.E., Lucinio, R., Morris, D.E., Muller, R.A., Perlmutter, S., Rawlings, S.J., Sasseen, T.P., Shelton, I.K., Steinman-Cameron, T.Y. and Tuohy, I.R. 1989, *Nature* **338**, 234

Kumagai, S., Itoh, M., Shigeyama, T., Nomoto, K. and Nishimura, J. 1988, *Astron. Astroph.* **197**, L7

Kumagai, S., Shigeyama, T., Nomoto, K. Itoh, M., Nishimura, J. and Tsuruta, S. 1989, to be publ. in *Ap.J.* **345**

Matz, S.M., Share, G.H., Leising, M.D., Chupp, E.L., Vestrand, W.T., Purcell, W.R., Strickmann, M.S. and Reppin, C., 1988, *Nature* **331**, 416

Pinto, P.A. and Woosley, S.E. 1988a, *Ap.J.* **329**, 820

Pinto, P.A. and Woosley, S.E. 1988b, *Nature* **333**, 534
Pinto, P.A., Woosley, S.E. and Ensman, L.M 1988, *Ap.J. (Letters)* **331**, L101
Rank, D.M., Pinto, P.A., Woosley, S.E., Bregman, J.D., Witteborn, F.C., Axelrod, T.S. and Cohen, M. 1988, *Nature* **331**, 505
Reppin, C. Pietsch, W., Trümper, J., Kendziorra, E., and Staubert, R. 1983, in *Non-thermal and Very High Temperature Phenomena in X-Ray Astronomy*, Rome, (Eds. Perola, G.C. & Salvati, M.), p. 279
Reppin, C., Englhauser, J., Hanson, C., Pietsch, W., Trümper, J., Voges, W., Kendziorra, E., Bezler, M., Staubert, R., Sunyaev, R., Kaniovskiy, A., Efremov, V., Grebenev, M., Kuznetzov, A., Melioranskiy, A., Stepanov, D. and Chulkov, I. 1987, AIP Conf. Proc. No.170, in *Nuclear Spectroscopy of Astrophysical Sources*, Washington D.C., (Eds. Gehrels, N. and Share, G.H.), p37
Staubert, R., Kendziorra, E., Maisack, M., Mony, B., Döbereiner, S., Englhauser, J., Pietsch, W., Reppin, C., Trümper, J., Sunyaev, R., Efremov, V., Kaniovsky, A., Kuznetzov, A., Melioransky, A. and Rodin, V. 1989, to appear in Proceed. of the *GRO Science Workshop*, GSFC Washington, Apr. 1989
Sunyaev, R., Kaniovsky, A., Efremov, V., Gilfanov, M., Churazo,E., Grebenev, S., Kuznetzov, A., Melioranskiy, A., Yamburenko, N., Yunin, S., Stepanov, D., Chulkov, I., Pappa, N., Boyarskiy, M., Gavrilova, E., Loznikov, V., Prudkoglyad, A., Reppin, C., Pietsch, W., Englhauser, J., Trümper, J., Voges, W., Kendziorra, E., Bezler, M., Staubert, R., Brinkman, A.C., Heise, J., Mels, W.A., Jager, R., Skinner, G.K., Al-Emam, O., Patterson, T.G., and Willmore, A.P.,1987, *Nature* **330**, 227
Sunyaev, R.A., Efremov, V.V., Kaniovskiy, A.S., Stepanov, D.K., Yunin, S.N., Kuznetzov, A.V., Loznikov, V.M., Melioranskiy, A.S., Rodin, V.G., Prudkoglyad, A.V., Grebenev, S.A., Reppin, C., Pietsch, W., Englhauser, J., Trümper, J., Voges, W., Kendziorra, E., Bezler, M., and Staubert, R. 1988, *Soviet Astronomy Letters* **14**, 579 (247 in english translation)
Sunyaev, R., Kaniovsky, A., Efremov, V., Grebenev, S., Kuznetzov, A., Loznikov, V., Englhauser, J., Döbereiner, S., Pietsch, W., Reppin, C., Trümper, J., Kendziorra, E., Maisack, M., Mony, B. and Staubert, R. 1989, *Soviet Astronomy Letters* **15**, 291
Tanaka et al. 1988, in *Physics of Neutron Stars and Black Holes* (Ed. Tanaka), Tokyo, p.431
Trümper, J., Reppin, C., Pietsch, W., Englhauser, J., Voges, W., Kendziorra, E., Bezler, M., Staubert, R., Sunyaev, R., Kaniovskiy, A., Efremov, V., Grebenev, M., Kuznetzov, A., Melioranskiy, A., Stepanov, D. and Chulkov, I. 1987, in *Proc. 4th George Mason Astrophysics Workshop*, Fairfax (Eds. Kafatos, M. and Michalitsianos, G., 1988), p.355
Woosley, S., Pinto, P.A. and Ensman, L. 1988, *Ap.J.* **324**, 466

Optical Spectrophotometry of the Supernova 1987A in the LMC

R.W. Hanuschik

Astronomisches Institut, Ruhr-Universität Bochum,
Postfach 102148, D-4630 Bochum 1, Fed. Rep. of Germany

> *"... directly overhead, a certain strange star was suddenly seen, flashing its light with a radiant gleam and it struck my eyes ... I began to doubt the faith of my own eyes, and so ... I asked [the servants] whether they too could see a certain extremely bright star ... They immediately replied ... that they saw it completely."*
>
> Tycho Brahe, *Astronomiae Instauratae Progymnasmata (1602)*

1. Introduction: The Guest Stars

In AD 1054, Chinese astronomer Yang Wei-De observed a guest star (*k'o-hsing*) in the constellation *T'ien-kuan* (near ζ Tauri), yellow in colour. He interpreted its brightness as a heavenly sign that there was "a person of great worth", a hypothesis willingly accepted by the Emperor. Yang Wei-De noted the brightness and colour changes of the guest-star which became what is now known as the Crab Nebula.

The guest star of 1054 was one of at least five galactic supernovae observed in the past 1000 years, namely SNe 1006 (the brightest one ever observed), 1054, 1181, 1572 (the one that amazed Brahe) and 1604 (observed by Johannes Kepler). While the ancients assumed a new star (= *nova stella*) to be born, theories elaborated in the following centuries assumed the sudden explosion of ashes from the surface of a star, or the collision of two celestial bodies. Modern theories assume that a supernova explosion marks the death of a star instead of its birth: In the case of a type-II explosion, the core of a massive (10 - 20 $M_{\odot}$) star runs through all fusion processes, until silicon burning synthesizes an iron core of $\approx 1.5\ M_{\odot}$ which finally collapses, turns into a neutron star by inverse β-decay and produces a shock front which overruns and expells the entire envelope.

Recording a supernova explosion with modern instruments can reveal a vast amount of information about three completely different physical regimes: (1) core collapse and the birth of a compact object; (2) explosive nucleosynthesis and the expulsion of the stellar envelope, revealing the inner structure of the progenitor star; and (3) the interaction of the expanding shell with the circumstellar and interstellar medium. The dimensional range spanned by these phenomena, i.e. between the neutron star radius (10^6 cm) and the circumstellar interaction region (a few parsecs), is more than 12 orders of magnitude.

Reviews in Modern Astronomy 2 Editor: G. Klare

Belonging to the most luminous objects in the sky (releasing $\approx 10^{49}$ erg of electromagnetic energy within ≈ 100 days), supernova events are quite rare. Selection effects, such as obscuration by interstellar dust and the geographical latitude of the ancient observers, in addition influence the observed rate of *galactic* supernovae. Due to their rareness, galactic supernovae define a unique class of astronomical objects: they have never been studied by telescope, only with the unaided eye.

Several dozens of *extragalactic* supernovae are observed each year; a total number of N > 500 is known. At relatively well-known distance and reddening, they offer, however, only few photons, and the steady struggle for observing time at the larger telescopes leaves the supernova investigator finally with a few noisy spectra spread over a couple of months.

The *optimum supernova*, i.e. a supernova with the largest impact on modern astronomy, therefore is a *nearby extragalactic* supernova, circumpolar for the large southern observatories. Its signals should approach Earth not before the 1980's (as otherwise neither neutrino detectors nor UV, X- and γ-ray satellites would have been available) and, of course, not too late after that date (as then other obervers instead of us would receive the privilege of studying it). This dream indeed became true when Supernova 1987A exploded in the Large Magellanic Cloud.

2. Supernova 1987A in the LMC: Star of the Century

Discovered 1987 February 24.2 UT by Canadian astronomer Ian Shelton (IAU circular No. 4316) at Las Campanas Observatory, SN 1987A must already have brightened shortly before Feb. 23.443 UT (McNaught, IAU circular No. 4389). Its hydrogen-line spectrum defined it immediately as a type-II event. Its precursor star was Sanduleak -69° 202, a *blue* supergiant of MK type B3 Ia with V = $12.^{m}24$ and B-V = $+0.^{m}04$ (Isserstedt, 1975). These parameters require $\log L/L_{\odot} \approx 5.0$, $T_{eff} \approx 14\,000$ K and $M \approx 15 - 20\ M_{\odot}$ (e.g., Woosley, 1988a). Although the brightest observed extragalactic supernova in historical times, it was, with $V_{max} = 2.95$ corresponding to $M_V \approx -16.1$, intrinsically rather faint in comparison to other type-II events.

The burst of 19 neutrinos measured by the KAMIOKANDE and IMB experiments (Hirata et al., 1987; Bionta et al., 1987) marks the epoch of core collapse: t_0 = 1987 Feb. 23.316 UT. The total energy emitted in all kinds of neutrinos amounts to 2...3 10^{53} erg within a few seconds (Burrows, 1988), dwarfing both the total kinetic energy of the expelled envelope (few 10^{51} erg) and the electromagnetic energy radiated from it ($\approx 10^{49}$ erg).

3. Optical Spectrophotometry: Record of the Afterglow

Despite its tiny contribution to the total *energy* budget, the electromagnetic radiation contains the bulk of the physical *information* about the envelope. At effective temperatures in the range of 5 - 10 000 Kelvin, the supernova envelope emits most of the electromagnetic energy in the optical and near-infrared range. Optical spectrophotometry therefore constitutes the most important kind of data from supernovae. A record of this energy provides key information about the physical conditions at the photosphere (temperature, radius, and density), as well as in the layers above it (geometry, composition, and ionization state from the line spectrum and from detailed line profiles). Thus, optical spectrophotometry provides an essential key for the understanding of the physical processes in the debris from the fatal explosion.

However, achieving spectrophotometric data, i.e. flux-calibrated spectroscopic information at photometric accuracy, is not an easy task, especially when a detailed record is desired. Therefore, only very few supernovae could be studied spectrophotometrically in some detail before 1987. A nearby supernova like SN 1987A offers the unique opportunity to gain such data at unprecedented detail and time coverage, provided the following is available: (1) the optimum telescope and detector, (2) a well-calibrated system of standard stars, and (3) last but not least: lots of observing time.

Spectroscopic work on SN 1987A was undertaken at virtually any larger southern observatory (for a list see Hearnshaw & McIntyre, 1988). Four of these campaigns collected spectrophotometric data (see Table 1), namely the campaigns at the Bochum telescope, at CTIO, at ESO, and at the SAAO. Truly spectrophotometric work, i.e. direct calibration by spectrophotometric standards instead of by broad-band photometry, has been performed by and published from only two telescopes: the 0.61m Bochum telescope on La Silla and the 1m telescope of CTIO.

Table 1: List of optical spectrophotometric measurements of SN 1987A

Observatory	wavelength range (Å)	resolution (Å)	calibration	atlas (A), other references (O)
Bochum 61cm telescope	3200 - 8700	10-15/3	absolute, flux standards	A: Hanuschik & Dachs, 1988; Hanuschik et al., 1989; O: Hanuschik, 1988
Cerro Tololo 1m telescope	3700 - 7100	5	absolute, flux standards, δ Doradus (rel.)	A: Phillips et al., 1988; O: Phillips, 1988
ESO La Silla 1.5m telescope	3500 - 9400	≈10	relative, flux standards	A: –; O: Danziger et al., 1987
South African 1.9m telescope	3400 - 7600	7/3.5	relative, flux standards, BV photometry	A: Menzies et al., 1987; Catchpole et al., 1987, 1988, 1989; Whitelock et al., 1988, 1989; O: Feast, 1988

4. The Bochum Telescope Campaign: The Right Telescope at the Right Time

The 0.61m telescope of the Ruhr University of Bochum is located at the ESO site at La Silla. It is equipped with a rapid spectrum scanner (Haupt et al., 1976) and a single-channel photomultiplier, thus combining low efficiency (rather essential for a naked-eye object like SN 1987A) with high linearity. The entrance slit can be opened as wide as 20 - 30", essential for an accurate flux record. A system of bright southern standard stars, connected to α Lyrae and the equatorial standard stars of Hayes (1970), has been constructed by Tüg (1980 a,b). For SN 1987A, the Bochum telescope thus was the right telescope at the right time.

The scanner spectrograph was used until day 314 (1988 Jan. 3); we further continued observations with a different detector (OMA, a multi-channel detector kindly provided by E.H. Geyer from Hoher List Observatory), until on day 450 the signal from the supernova finally faded below a reasonable level at our telescope. At present, data for the first 314 days are fully reduced. They cover the wavelength range 3200 – 8700 Å (at some dates the range starts at 3900 Å) at 10 Å. After day 200, the resolution is 15 Å. A number of wavelength bands enclosing interesting line features were measured at higher (3 – 5 Å) resolution and flux-calibrated. Absolute calibration was performed by comparison to several flux standards from the Tüg list. Accuracy of the flux calibration is ≈3 %, and slightly worse in the UV and near-IR range. The average signal-to-noise ratio is 100 – 200. For days 2 - 20, we had to employ Walraven and Cousins photometry to achieve absolute flux calibration as the entrance slit at that epoch was chosen too narrow.

The Bochum spectrophotometric flux atlas of SN 1987A thus combines homogeneity, high-precision calibration, full resolution, excellent signal-to-noise ratio and high time coverage. Parts I and II, covering days 2 - 157, are published (Hanuschik & Dachs, 1988; Hanuschik et al., 1989); part III presenting data from days 158 - 314 is in preparation.

5. The Continuum Radiation

5.1 The Evolution

Three different evolutionary stages of the expanding envelope of SN 1987A are visible in Fig. 1: Our first measurement on day 1.8 shows a spectrum which resembles closely a black body, with extremely broad P Cygni lines of hydrogen and helium (λ5876). On

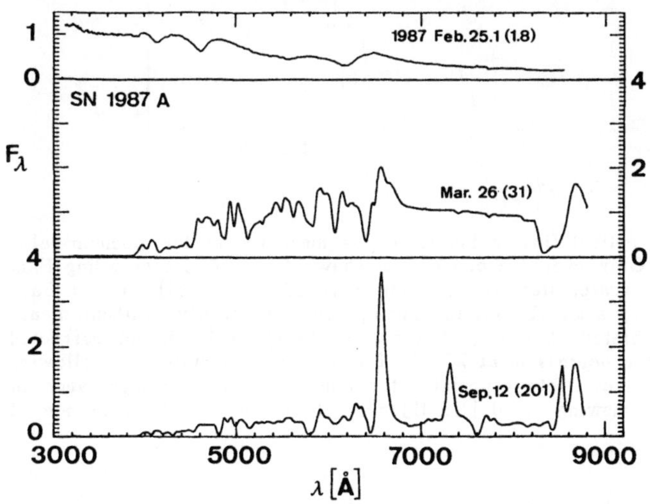

Figure 1: Fluxes F_λ (in 10^{-10} erg s^{-1} cm^{-2} $Å^{-1}$) for SN 1987A in three different evolutionary phases. Days since 1987 Feb. 23.316 UT are given in brackets

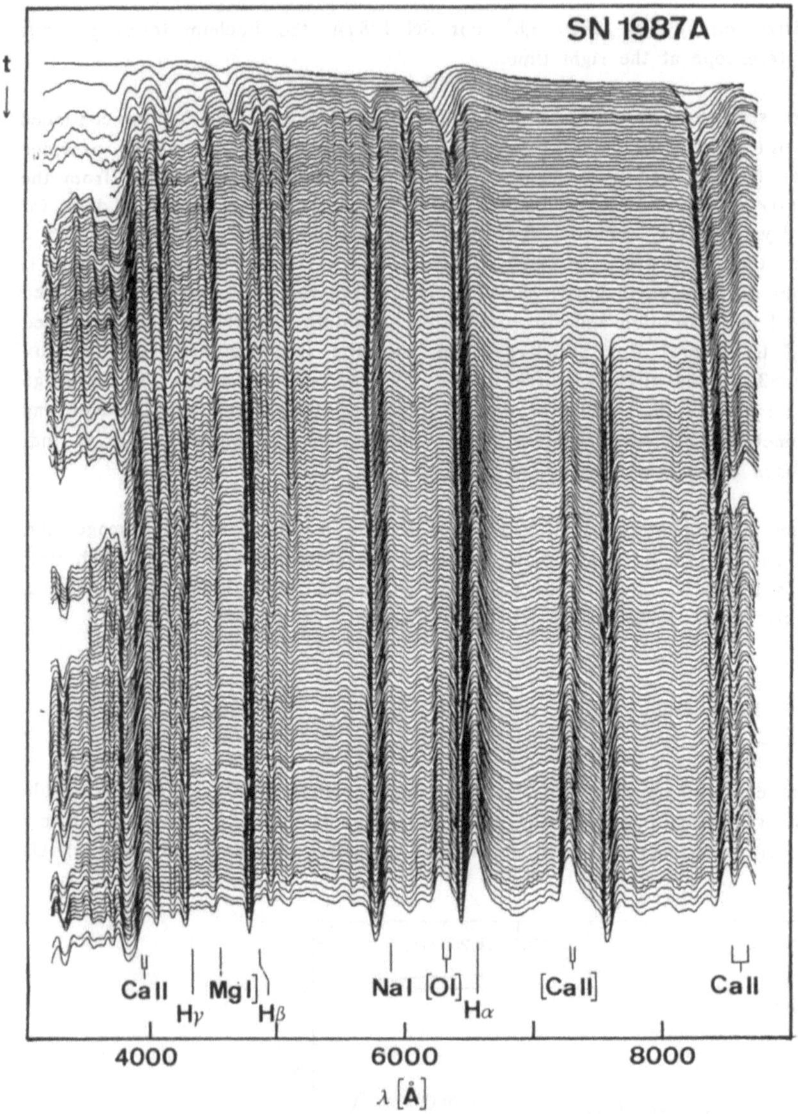

Figure 2: Evolution of the optical flux in SN 1987A as measured at the Bochum telescope, for days 2 - 314. Only every second day is shown to avoid overcrowding. Flux is shown on a logarithmic scale, starting with day 2 (= 1987 Feb. 25) on top. Each spectrum has been shifted by a small constant amount. Spectra for days without measurements have been interpolated. A strong atmospheric absorption band, not corrected for after day 111, causes the depression at λ7500. This plot is best suited for following the *evolution of spectral lines*. Note especially the appearance and disappearance of spectral features, e.g. the showing up of [Ca II] and [O I] and the disappearance of Ba II λ6142

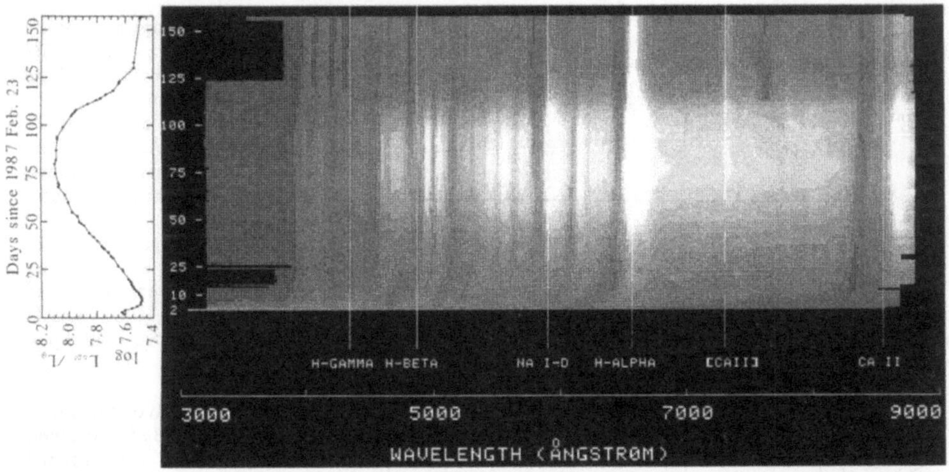

Figure 3: Intensity-coded fluxes of SN 1987A for days 2 - 157. Flux scale is *linear*, dark-grey denoting minimum, and white maximum, flux. Ordinate is in days since 1987 Feb. 23. The optical lightcurve is shown for comparison. Some prominent lines are marked. This plot clearly shows the brightness evolution, with the maximum of optical flux occurring around day 82

day 31, the line spectrum is well developed. The total brightness has increased as a result of the expanding photosphere. On day 201, the photosphere has shrunk, while the emission-line-forming atmosphere has further increased in size: the result is a nebular spectrum, with a large amount of radiation emitted in permitted and forbidden lines.

An overview of the *spectral evolution* during the first 314 days, emphasizing the appearance and disappearance of spectral features, is shown in Fig. 2. The *brightness evolution* is more clearly visible in Fig. 3 (here only the first 157 days are shown).

The evolution of the continuum and the line spectrum of SN 1987A occurred initially very fast, subsequently much slower. This behaviour is unusual for type-II supernovae. In the optical UV region (at 3500 — 4000 Å), flux decreased by almost a factor of 100 within the first 10 days, while the flux longwards of 5000 Å continuously increased towards the rather flat brightness maximum which occurred on day 82 ± 2 (day 87 at shorter wavelengths) after core collapse.

Integrating the optical fluxes yields the *optical* lightcurve, L_{opt} (Fig. 4a), which differs from the total *bolometric* lightcurve mainly by neglecting the IR and UV flux. Effective temperatures, determined from our data by black-body fitting for the first 5 days and by determination of the approximate continuum maximum thereafter, are depicted in Fig. 4c. Photospheric radii, r_{ph}, have been calculated from these temperatures and the SAAO (Catchpole et al., 1987) bolometric lightcurve (Fig. 4b).

The optical lightcurve demonstrates that the flux evolution during the first ≈300 days can be empirically divided into four phases:

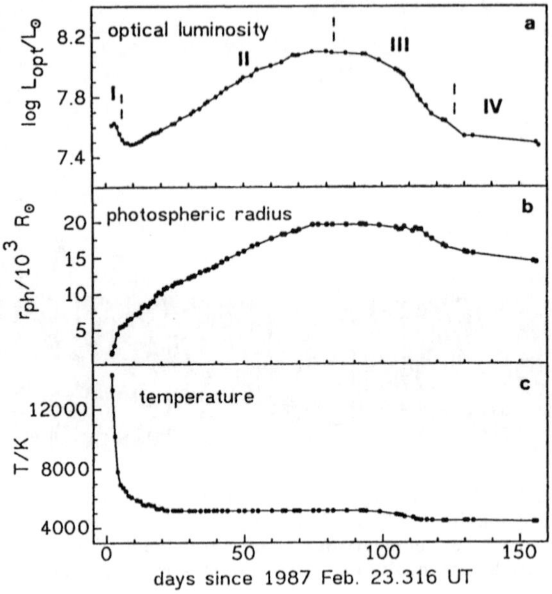

Figure 4a: The optical bolometric lightcurve for SN 1987A derived from the Bochum fluxes. Integration of fluxes has been carried out between 3200 Å and 8600 Å. A reddening of 0.16 mag and a distance of 50 kpc have been assumed.

Figure 4b: Photospheric radius in units of 10^3 $R_{\odot}$ as calulated from the temperatures in Figure 4c and the SAAO bolometric lightcurve (Catchpole et al., 1987)

Figure 4c: Effective temperatures for SN 1987A, derived from the Bochum flux record by black-body fitting (days 2 - 5) and by estimation of the Wien flux maximum thereafter

(1) The phase of *rapid cooling* (lasting from day 0 to day 5) is characterized by the adiabatic cooling of the envelope, from $T \approx 15000$ K on day 2 to 5500 K on day 10. Models predict an initial temperature of $T \geq 10^5$ K for the time of shock breakthrough at the photosphere and a resulting short-lived UV flash (e.g., Arnett, 1988). The UV radiation, as a result, falls off rapidly. The spectrum shows hydrogen and helium lines, Ca II λ8500 starts to develop.

(2) Phase II, the *brightening phase*, comprises days 5 — 82 and exhibits a slow and steady brightness increase until maximum. The optical (λλ3200 - 8700 Å) luminosity then is $L_{opt} = 5.1\ 10^{41}$ erg s^{-1} = $1.3\ 10^8\ L_{\odot}$. The temperature stays at an almost constant value, $T \approx 5100$ K, favourable for recombination of hydrogen, thus releasing the energy originally stored by the shock wave. A rich absorption and P Cygni-type spectrum develops. Forbidden emission of [Ca II] starts around day 36.

(3) The *maximum phase*, phase III (days 82 — 126), exhibits a steep continuum decline at an average rate of 0.045 mag per day, a slightly decreasing temperature and the transition into a nebular-type spectrum.

(4) Finally, phase IV as the *exponential decay phase* shows a less steep, exponential flux decline. It extends beyond day 314 and is best visible in Fig. 5, an extended version of Fig. 4a. The exponential decline rate of the *optical* lightcurve is 0.008 mag per day between days 126 and 314, yielding a mean half-life of 96 days. The SAAO *bolometric* lightcurve declines faster, its half-life is ≈76 days (Catchpole et al., 1988). This exponential decline strongly supports the model in which 0.07 $M_{\odot}$ of ^{56}Ni are explosively synthesized and decay with a half-life of 6.1 days into ^{56}Co and finally with $\tau_{1/2}$ = 77 days into ^{56}Fe, thus powering the observed lightcurve which otherwise would have dropped steeply after about day 40 (Woosley, 1988b). A well-

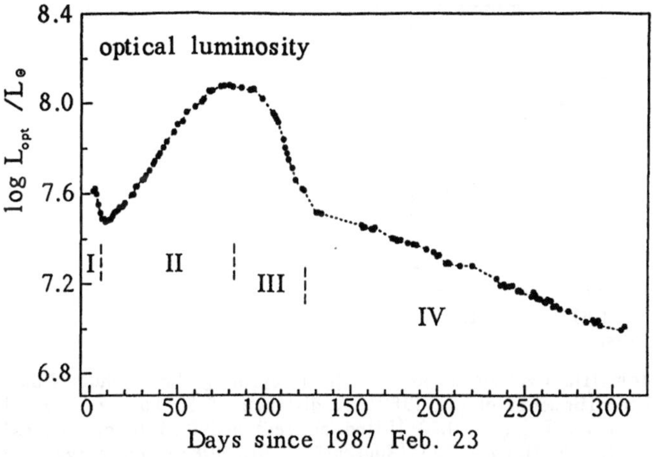

Figure 5: Optical luminosity (3200 - 8600 Å) of SN 1987A for days 2 - 314. Distance and reddening are the same as in Fig 4a

developed nebular spectrum shows up with strong allowed and forbidden emission lines (among them [O I] λλ6300/6363) which emit increasing fractions of the total flux and are important, by their cooling effect, for the energy balance in the envelope. Envelope opacity drops as indicated by fading absorption troughs and by the appearance of emission lines from iron-group elements in the infrared (Aitken, 1988): the envelope becomes more and more transparent, and also fragmentates as implied by the unexpectedly early emergence of γ and X-rays in August 1987 (Dotani et al., 1987; Matz et al., 1988). Deeper going analyses of the physical implications of the lightcurve slope are reviewed, e.g., by Hillebrandt & Höflich (1989), Arnett et al. (1989) and McCray (1989).

5.2 The Bolometric Lightcurve

The bolometric lightcurve of a supernova, as a record of its total emitted electromagnetic energy, is an especially valuable input parameter for envelope modelling. Supernova 1987A is the first supernova from which a bolometric lightcurve could be derived from the observations without crude assumptions. However, all presently existing bolometric lightcurves for SN 1987A (Catchpole et al., 1987; Suntzeff et al., 1988; Dopita et al., 1988) are derived from fits to optical and infrared *photometry*, thus suffering from two biasing effects: (1) the notorious uncertainties in the transformation of broad-band into monochromatic fluxes, and (2) numerical errors introduced by the method how the lightcurve is integrated from these flux points, namely by black-body or by spline fitting. The photometric transformations always assume some "average" spectrum type which is satisfied for normal stars but is hazardous for a supernova spectrum with its broad P Cygni-type and emission lines.

The optimum solution avoiding both problems is the integration of optical and infrared (and, in the first few days, UV) *spectrophotometry*. As a first step towards this final goal, the Bochum fluxes were combined with ESO infrared photometry (Bouchet

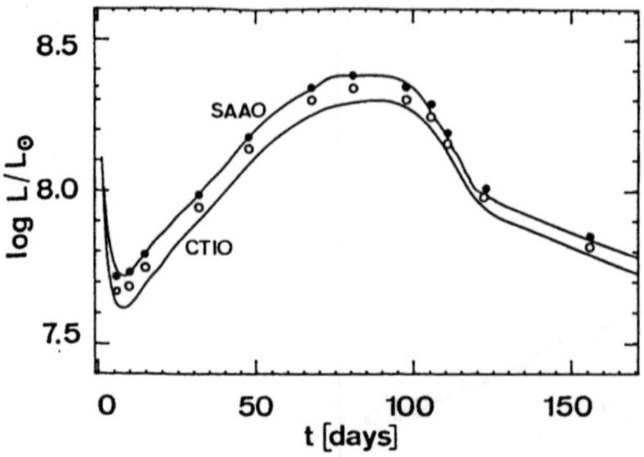

Figure 6: The Bochum bolometric lightcurve, obtained by integrating the Bochum fluxes and IR photometry from ESO (Bouchet et al., 1989). A distance of 50 kpc is assumed; calculations were carried out for E_{B-V} = 0.20 (filled circles) and 0.15 (open circles). The SAAO (Catchpole et al., 1987) and CTIO (Suntzeff et al., 1988) lightcurves are shown as solid lines

et al., 1989) for a number of selected days, thus covering the wavelength range 0.32 μ – 20 μ.

Figure 6 shows our results, calculated with two values of the interstellar reddening: with $E_{B-V} = 0.^{m}20$, the value chosen by the SAAO observers, and with $0.^{m}15$, corresponding to the CTIO choice (a true distance modulus of 18.5, or D = 50 kpc, is adopted in either case). The agreement of the SAAO and the Bochum lightcurve is excellent until maximum light. After day 100 when the relative strength of emission lines increases dramatically, photometric lightcurves become more clearly inadequate. The CTIO lightcurve seems to underestimate the total flux from SN 1987A: only half of the downshift towards the SAAO and the Bochum lightcurves is due to the lower E_{B-V}. A more thorough interpretation of Fig. 6 will be undertaken when a final Bochum lightcurve has been finished.

6. The Line Spectrum

6.1 Velocities and Radii

A rich line spectrum developed very quickly, one of the many surprising facts in SN 1987A. Usual type-II spectra evolve more slowly. Line identifications can be found, e.g., in Williams (1987) and Hanuschik & Dachs (1988). A very helpful tool for following and identifying individual features is the calculation of ΔF/F, essentially equivalent to the (normalized) first spectral derivative of the fluxes F(λ,t) (see Hanuschik et al., 1989); Fig. 2 can also be used very efficiently for line identification.

The first spectrum on day 1.8 merely shows hydrogen and helium lines. The maximum observed outflow velocity is -31 000 km s^{-1}, observed in the blue Hα wing (Hanuschik & Dachs, 1987a). On day 4, Ca II λ8500 as well as its H+K lines at λ3900 appear, and Na I-D around day 7. By day 10, most of the spectral lines are present, among them several lines of Fe II, Sc II and, somewhat later, the s-process elements Ba II and Sr II. They remain essentially unchanged afterwards apart from velocity shifts. Most of the weaker lines disappear again by day ≈160. However, the Balmer

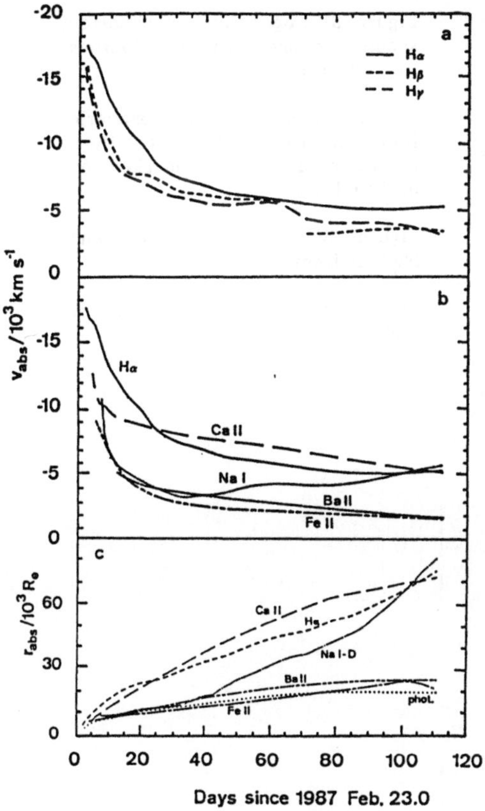

Figure 7: Trough velocities and radii of maximum absorption in SN 1987A for days 2 - 111.

Figure 7a: Trough velocities v_{abs} for the Balmer lines Hα, Hβ and Hγ.

Figure 7b: A comparison of trough velocities for Hα, Ca II λ8542, Na I-D, Ba II λ6141 and Fe II λ5018.

Figure 7c: Effective radii of maximum absorption, r_{abs}, calculated from (1), for the lines in Fig. 7b. The *photospheric radius* (from Fig. 4b) is shown as a dotted line

lines, as well as Ca II H+K and λ8600, Fe II λ5018 and Na I-D, remain visible throughout the first 300 days of SN 1987A. The rather strong lines of s-process elements (e.g. Ba II λ6141) have been interpreted as a sign for enhancement, e.g. by a factor of 10 times solar for barium (cf. Williams, 1987; Hillebrandt & Höflich, 1989).

Absorption trough velocities, as documented by Hanuschik & Dachs (1987b), Blanco et al. (1987), Phillips et al. (1988), and Hanuschik & Schmidt-Kaler (1989a), show a rather complex evolution (Fig. 7a,b), characterized by: (1) a steep decrease within the first ≈20 days and a rather flat fall-off thereafter, and (2) a stratification indicated by the fact that the velocities of the Balmer lines, as well as of Ca II λ8600, are larger by a factor of two or more than those of "weaker" lines, e.g. of Ba II and Fe II. The Na I-D line is unusual in that its trough velocity *increases* after about day 30 (as does Hα after day ≈85). Finally, fine-structure in some of the lines (see next chapter) makes the definition of the trough minimum rather doubtful between days ≈20 and 70.

The appearance of the optical line spectrum is influenced mainly by three factors: abundance stratification, geometrical dilution, and ionization state, the latter two being a function of time. The picture of line opacity decreasing by geometrical dilution and thus revealing *peu à peu* layers of smaller radial velocity is a good first-order approxim-

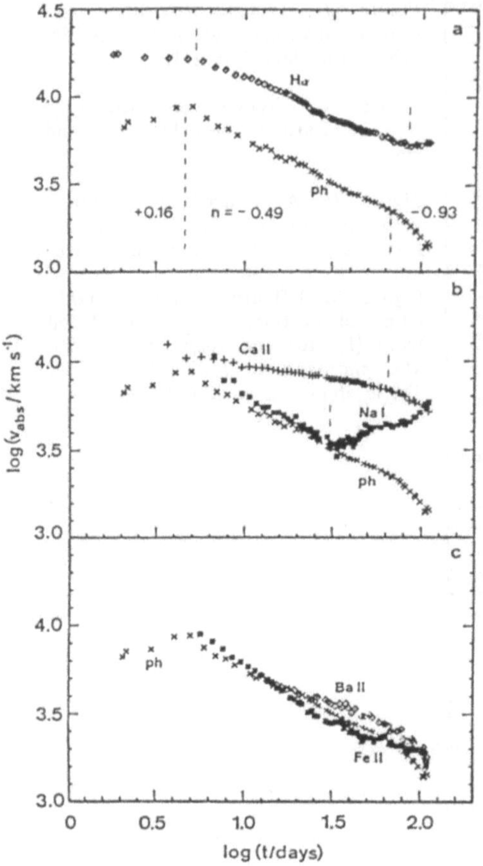

Figure 8: Logarithms of spectroscopic and photospheric velocities vs. logarithm of time.

Figure 8a: Hα and photospheric (ph) velocities. Phase transitions in the evolution are marked by broken lines.

Figure 8b: Ca II λ8542 and Na I-D lines.

Figure 8c: Ba II λ6141 and Fe II λ5018

ation. However, as detailed spectral line information is available for the first time from SN 1987A, this simple picture needs further refinement e.g. by NLTE calculations of the ionization equilibrium (e.g., Höflich, 1988).

Assuming homologous expansion ($v(r,t) = r/(t-t_0)$), observed trough velocities v_{abs} can be related to effective radii of maximum absorption (Kirshner & Kwan, 1974; Hanuschik & Schmidt-Kaler, 1989a), r_{abs}, letting

$$r_{abs} = v_{abs} \cdot (t - t_0) \qquad (1)$$

with t_0 = 1987 Feb. 23.316 UT (see Fig. 7c). Around day 100, maximum absorption in the stronger lines (Ca II, Hα, Na I-D) occurs at radii ≈2.5 times larger than in the weaker lines, which arise close to the photosphere.

An interesting fact is that both line and photospheric velocities [as calculated from the photospheric radii in Fig. 4b inverting (1)] show distinct phases of evolution. Fig. 8 demonstrates that the photospheric velocity v_{ph} shows *three distinct phases* within the first 111 days. Assuming a power law, $v_{ph} \sim t^n$, the power law exponents *n* are +0.16, -0.49 ± 0.02 and -0.93 ± 0.07 (Fig. 8a). As opacity in supernova photospheres is electron-scattering dominated, these powers are related to the electron density at the photosphere by the requirement that the electron-scattering optical depth be unity at r_{ph}:

$$\tau_e = 1 = \int_{\infty}^{r_{ph}} N_e(r)\, dr \; . \tag{2}$$

If the initial density distribution can be described by a power law, i.e. $N_e \sim r^{-\alpha}$, the slope of v_{ph} in Fig. 8 and the density law α are related by

$$v_{ph} \sim t^n = t^{2/(1-\alpha)} \quad , \tag{3}$$

(derived from Branch, 1987). Consequently, $\alpha = 5.1 \pm 0.2$ before day 73, and $\alpha = 3.2 \pm 0.2$ thereafter. Before day 5, the density law must have been very steep (Branch, 1987; Dopita et al., 1988).

Interestingly enough, some of the *absorption lines* also show distinct velocity phases, i.e. straight lines in the log v — log t plot of Fig. 8. This is especially remarkable because relating trough velocities to geometric radii and physical conditions there (namely density and ionization state) is to some degree model-dependent. These conditions are obviously constant over longer periods and then change fairly quickly (cf. Hα in Fig. 8a, Ca II and Na I-D in Fig. 8b). This supports the suggestion that the trough velocity evolution is mainly governed by the local density law (because temperature remains almost constant, as Fig. 4c shows).

6.2 Line fluxes

Emission line fluxes f_{emi} and absorption trough equivalent widths W_{abs} can be measured when a local continuum can be defined in a sufficiently reliable manner. The Ca II lines at λ8542 (permitted) and λ7300 (forbidden) show a correlated behaviour (Fig. 9).

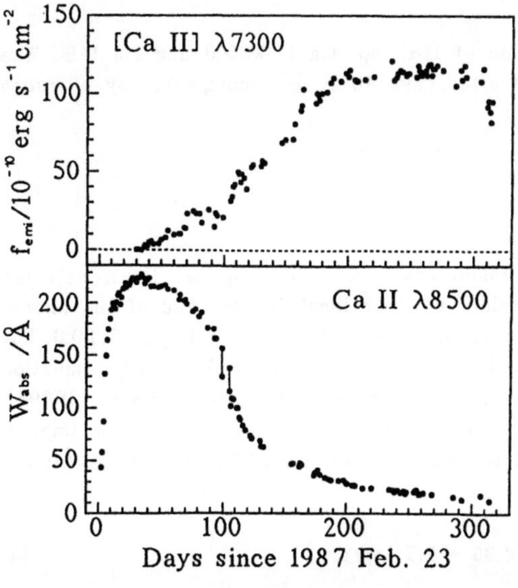

Figure 9: Integrated line flux f_{emi} of [Ca II] λ7300 (top) and trough equivalent width W_{abs} of Ca II λ8542 (bottom). As the two main components of the latter line deblend with time, the definition of W_{abs} changes: before day 100, only a single trough is visible, which splits up afterwards. The impact of this change of definition can be estimated from the data points of day 100 and 105 when values for both definitions are connected by a line

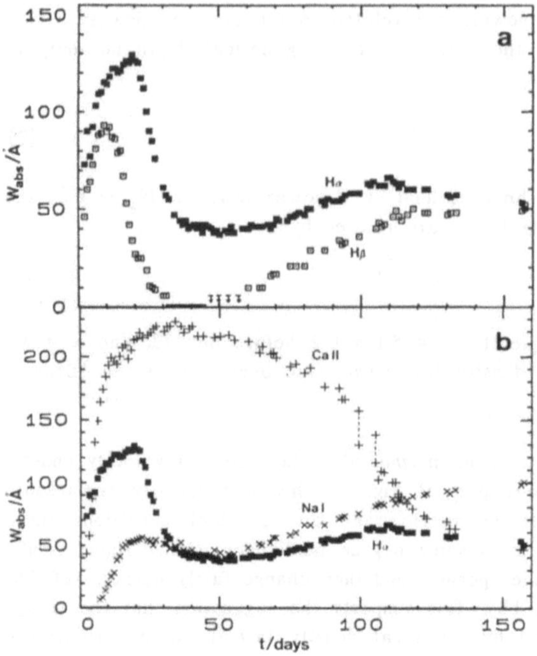

Figure 10: Trough equivalent widths.

Figure 10a: A comparison of Hα and Hβ.

Figure 10b: Ca II λ8542, Na I-D and Hα

The λ8542 trough equivalent width reaches a saturation value of 225 Å on day 30 and starts decreasing again when the [Ca II] emission starts (around day 36). This is caused by the fact that the lower metastable level of the λ8542 transition, 3d, is more and more de-populated by the λ7300 forbidden line (Hanuschik et al., 1988) as suggested earlier by Kirshner & Kwan (1975). The slope of W_{abs} follows the optical lightcurve after day ≈120, while f_{emi} remains constant between day 200 and 300, despite a fading of the local continuum by a factor 2.7.

Fig. 10 demonstrates the flux evolution of Hα, Hβ, Ca II λ8500 and Na I-D. The *opacity increase* in Hα, Hβ and Na I-D after days 40-50 is accompanied by a *trough velocity increase* (cf. Fig. 7).

6.3 The distance to SN 1987A

A comparison of effective radii of maximum absorption for 9 optical lines, selected for their weakness and for the high certainty of their identification, and of the photospheric radius (cf. Fig. 4b) yields a Baade-Wesselink distance of D_{BW} = 50.0 kpc (distance modulus 18.5) for SN 1987A, if spherical symmetry, black-body radiation field and a reddening of $0.^m16$ are assumed (Hanuschik & Schmidt-Kaler, 1989b). Correcting for ellipticity with a true axis ratio of μ = 0.7 on day 20 which is implied by combining speckle and polarimetric data (Karovska et al., 1989; Schwarz & Mundt, 1987), yields

$$D_{ell} = 46.9\ \varepsilon^{1/2} \pm 1.3\ \text{kpc},\ (m-M)_0 = 18.36 + 2.5 \log \varepsilon. \qquad (4)$$

Here ε is the model-dependent ratio of the true emergent flux, πF, and the black-body flux at temperature T, σT^4, and is expected to be close to unity (for a discussion of ε see, e.g., Branch [1988]). The good agreement of the SN 1987A distance in (4) and the LMC distance derived from other methods, viz. 49.4 ± 3.4 kpc (Feast & Walker, 1987), gives confidence that the use of the Baade-Wesselink method for supernova distances is justified.

7. Fine-Structure in Spectral Lines: the Bochum Event

One of the most interesting spectroscopic surprises during the early days of SN 1987A was the detection of spectroscopic fine-structure in Hα and other spectral lines (Hanuschik & Dachs, 1987a,b; Blanco et al., 1987; Hanuschik et al., 1988; Phillips & Heathcote, 1989; Hanuschik & Thimm, 1989) dubbed the Bochum event. It was followed up by the Bochum observers in virtually any night at full resolution (3 Å) and high signal-to-noise ratio (> 100). Fig. 11 shows detailed plots of the Hα, Hβ and Na I-D lines all showing different aspects of this event (discussed in detail by Hanuschik et al., 1988, Phillips & Heathcote, 1989, and Hanuschik & Thimm, 1989).

The fine-structure shows up in the absorption trough, *as well as in the emission hump* when the latter is visible (in Hα and Na I-D). Its typical radial velocity is declining from ±5000 to less than ±3000 km s^{-1}, close to the photospheric value. Another characteristic property is its sudden appearance as a peak-like trough structure (on day 19 in Hα and on day 17 in Hβ (Hanuschik & Thimm, 1989) and its gradual fading until it disappears by days 70 — 100. The phenomenon is obviously not restricted to the optical (Hanuschik et al., 1988; Hanuschik & Thimm, 1989) and IR (Larson et al., 1987; Phillips & Heathcote, 1989) hydrogen lines, but is also visible in Na I-D and, very faintly, in Ca II λ8600. Thus no doubt is left that we are dealing with an *intrinsic line phenomenon*, rather than with a transient line blend.

A number of possible explanations for the origin of the Bochum event have been discussed extensively (Larson et al., 1987; Hanuschik et al., 1988; Lucy, 1988; Höflich, 1988; Phillips & Heathcote, 1989; Hanuschik & Thimm, 1989). The width of the peak-like trough fine-structure, as well as its radial velocity, time evolution and symmetry, rule out the suggestion that these satellite features arise in the circumstellar environment of SN 1987A. An origin within the expanding envelope, namely closely above the photosphere, seems much more likely: the layers causing the fine-structure then would be hidden inside the photosphere before day 17 and suddenly become visible around that date, thus providing an attractive explanation for the observed temporal behaviour (Lucy, 1988).

The net (= emitted minus absorbed) flux in Hα is observed to vanish on day 20, but is larger than zero before and afterwards (Thimm et al., 1989; Phillips & Heathcote, 1989). We are thus observing the onset of a new Hα heating source around day 20, after the shock-deposited energy is released by hydrogen recombination. This energy source presumably is heating by clumps or fingers of synthesized, mixed-up and now decaying ^{56}Co. The coincidence of the Bochum event and the sudden change in the Hα energy budget then would imply that this event is the earliest evidence for ^{56}Co mixing in SN 1987A.

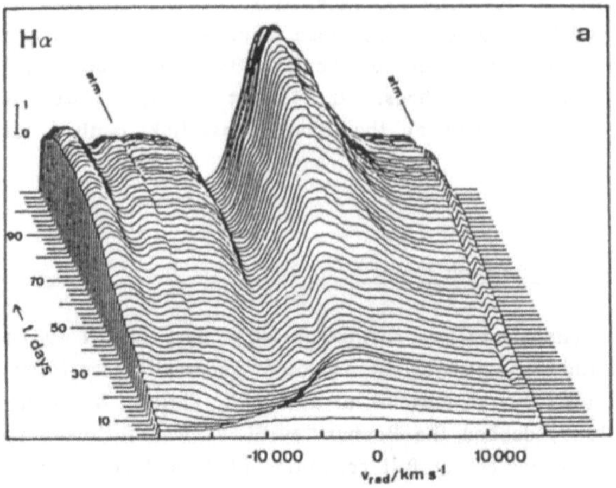

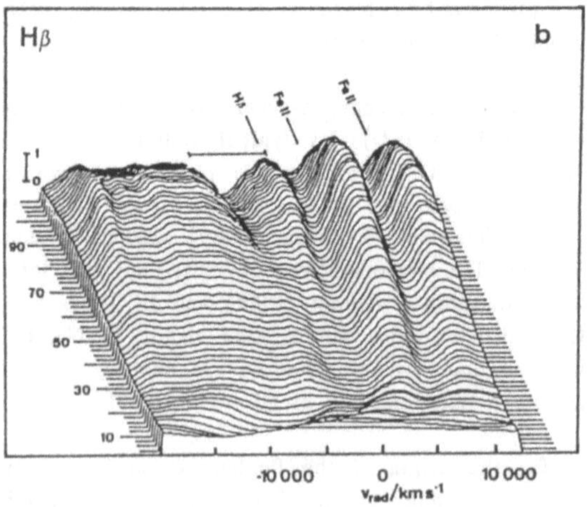

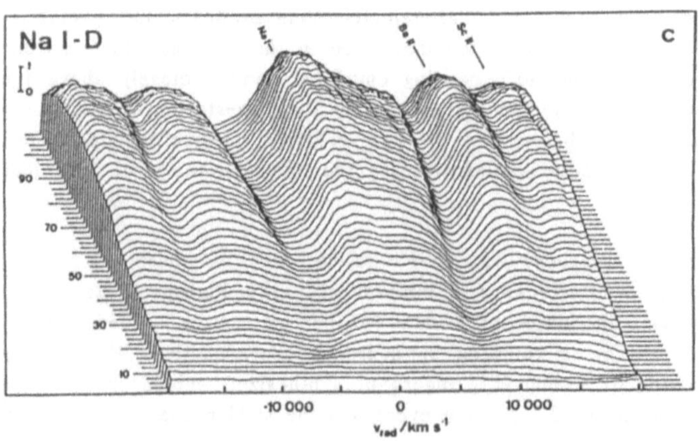

Figure 11: Detailed flux records at 3 Å resolution. Scans were averaged for every two days; the few observational gaps were filled by interpolation. The first scan is from day 2. Elapsed time in days is indicated. The radial velocity scale has been corrected for the LMC system velocity of +280 km s^{-1}. The vertical bar indicates 10^{-10} erg s^{-1} cm^{-2} $Å^{-1}$.

Figure 11a: The Hα line. Fine-structure is visible as blueshifted, peak-like flux excess in the trough and as redshifted flux deficit in the emission hump.

Figure 11b: The Hβ line. Only the trough is visible, redshifted Hβ photons are suppressed by the Fe II lines at λλ4924/5018. Fine-structure shows up as blueshifted flux excess.

Figure 11c: Na I-D λλ5890/5896. The depression at v ≈ 0 km s^{-1} is caused by the interstellar galactic and LMC components. Fine-structure appears as a depression in the red side and as a weak flux excess in the trough

Type-II SNe 1985L (Filippenko & Sargent, 1986) and 1985P (Chalabaev & Cristiani, 1987) show some fine-structure in their Hα profiles which resembles the Bochum event in SN 1987A. This provides some evidence that mixing is not unique to SN 1987A, but instead might be of general relevance in type-II supernovae.

8. Glories in the Sky: The Light Echoes

The largest-scale phenomena hitherto observed in connection with SN 1987A are light echoes (Gouiffes et al. 1988; Suntzeff et al., 1988; AAO Newsletters No. 47+49). Detected by A. Crotts in March 1988 (IAU circular No. 4561), they show up in CCD images obtained as early as August 1987. Their characteristics in February 1988 were (cf. Fig. 12a): two complete and nearly concentric rings of radii 32" and 52", with a typical half-width of 5" and an average blue surface brightness of 22 mag per square arcsec. Recently, a third inner ring has been detected (IAU circular No. 4791). The outermost ring is slightly distorted or eccentric. Recent frames show an increasing degree of complexity (cf. AAO Newsletter No. 49). Their spectra unambiguously show supernova light (Gouiffes et al., 1988; Suntzeff et al., 1988).

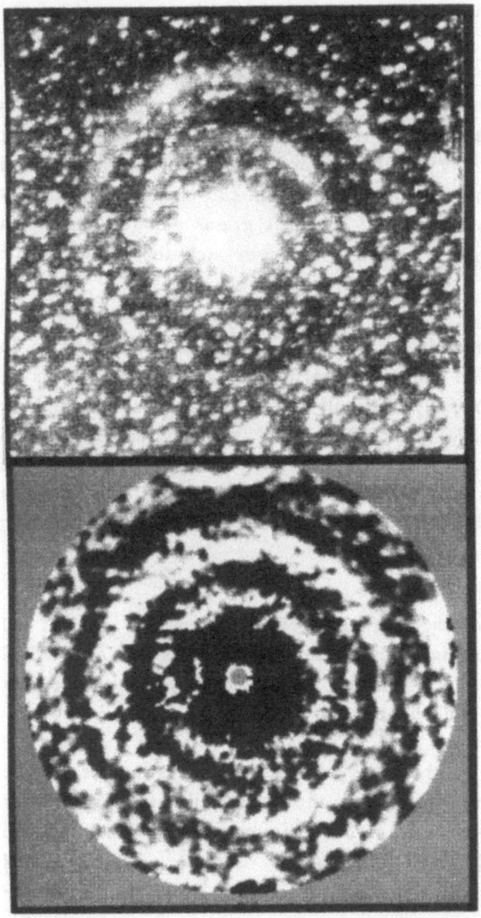

Figure 12: Light echoes around SN 1987A.

Figure 12a: CCD image from ESO 3.6m telescope + EFOSC at 4700 Å, 60 Å width, obtained on February 13, 1988 (Gouiffes et al., 1988).

Figure 12b: Contrast-enhanced and filtered version of Fig. 12a. First, an especially efficient filter routine (written by S. Kimeswenger at Bochum Institute) was run twice across the CCD frame. Then, the resulting picture was transformed into polar coordinates centered on SN 1987A, differentiated in radial direction (with $\Delta r \approx 3"$), and transformed back into cartesian space. Thus the transition from a black ring (= positive intensity gradient) to a white ring (= negative intensity gradient) in Fig. 12b marks the contrast-enhanced intensity maximum of a true light echo ring in Fig. 12a

(The CCD frame in Fig. 12a is courtesy of Christian Gouiffes at ESO)

Light echoes are not a SN 1987A-first, but have been observed earlier, the most famous example being the complex features around Nova Persei 1901 detected by Max Wolf (cf. e.g. Felten, 1988; Bührke & Hessmann, 1989). They originate from supernova light reflected by interstellar dust, and require both a pulse-like (instead of a continuous) light source and a layer-like (instead of spread-out) geometry of the dust. Otherwise they would appear as a normal, i.e. smeared-out and steady reflection nebula. Light echoes are related to the atmospheric halo phenomenon, with the difference that in a halo the continuity of the light source is compensated by the discontinuity of the refraction angles in ice crystals.

Following-up light echo geometry and time evolution can yield the detailed *three-dimensional* structure of the interstellar dust. Given the observed angular radius, α, the distance to the illuminating light source, D, and the time difference Δt between the emission and the observation of the reflected light are known, the true distance R of the reflectors from the supernova is given by

$$R \approx (\alpha D)^2 / 2\, c\Delta t \quad , \tag{5}$$

and its inclination as seen from the light source against the line of sight, Θ, by

$$\sin \Theta = \alpha D / R = 2\, c\Delta t / \alpha D \quad . \tag{6}$$

The two outer rings around SN 1987A correspond to two sheets of material at R_1 = 110 pc and R_2 = 320 pc *in front* of the supernova. The recently detected innermost ring belongs to material at R_3 = 2 pc *behind* it. The surface density of the dust across the ring defined by R and Θ can in principle be calculated from the observed brightness profile of the lightecho, if the reflection properties of the dust grains and the supernova lightcurve are known. A record of the echoes then yields step by step the three-dimensional structure of the cloud.

At present, Joachim Stüwe from Bochum University and myself are undertaking an effort to derive the spatial structure of the clouds. Determination of the brightness of the lightechoes around SN 1987A is extremely difficult as the region is crowded by stars and has a complex HII background. Star filtering as well as comparison to pictures of the same region at other epochs will help derive the exact shape of the light echoes and construct the dust cloud geometry. As a first result, Fig. 12b shows a filtered version of Fig. 12a, additionally differentiated in radial direction. The improvement of contrast is obvious.

Availability. The already published parts of the Bochum spectrophotometric atlas of SN 1987A (= days 2 - 157; days 158 - 314 are in preparation) can be obtained as a FITS or ESO-IHAP formatted magnetic tape copy from the author. Please send an empty tape and specify format and density (800 - 6250 bpi).

Acknowledgements. This review is based on the work of many co-investigators at the Bochum Institute. Especially worthwile to mention are the dedicated and hard efforts of the observers at the Bochum telescope: J. Dachs, G. Thimm, K. Seidensticker, J. Gochermann, S. Kimeswenger, R. Poetzel, G.F.O. Schnur, U. Lemmer, J. Stüwe, and M. Werger. The personal engagement of our Institute Director, Prof. Th. Schmidt-

Kaler, who provided ideal and financial support for this project, is gratefully acknowledged. We are indebted to the former ESO Director General, Prof. L. Woltjer, who allotted scheduled ESO time at the Bochum telescope for our programme. M. Werger did an excellent job assisting with and performing a large part of the reductions. The Bochum telescope is due to grant Schm 160/9 of the Deutsche Forschungsgemeinschaft.

References

Aitken, D.K.: 1988, Proc. Astron. Soc. Austr. **7**, 462

Arnett, D.: 1988, in 4th *George Mason University Workshop in Astrophysics "Supernova 1987A in the LMC"*, ed. by M. Kafatos & A. Michalitsianos (Cambridge University Press), p. 301

Arnett, D., Bahcall, J.N., Kirshner, R.P. & Woosley, S.E.: 1989, Ann. Rev. Astron. Astrophys. **27**, 629

Bionta, R.M., et al.: 1987, Phys. Rev. Lett. **58**, 1494

Blanco, V.M. et al.: 1987, Astrophys. J. **320**, 589

Bouchet, P., Moneti, A., Slezak, E., Le Bertre, T. & Manfroid, J.: 1989, Astron. Astrophys., *in press*

Branch, D.: 1987, Astrophys. J. **320**, L23

Branch, D.: 1988, in *Proc. ASP 100 Ann. Symp. "The Extragalactic Distance Scale"* (ASP Conf. Ser., San Francisco), p. 146

Bührke, Th. & Hessman, F.: 1989, Sterne Weltraum **28**, 148

Burrows, A.: 1988, in 4th *George Mason University Workshop in Astrophysics "Supernova 1987A in the LMC"*, ed. by M. Kafatos & A. Michalitsianos (Cambridge University Press), p. 161

Catchpole, R.M. et al.: 1987, M.N.R.A.S. **229**, 15P

Catchpole, R.M. et al.: 1988, M.N.R.A.S. **231**, 75P

Catchpole, R.M. et al.: 1989, M.N.R.A.S. **237**, 55P

Chalabaev, A.A. & Cristiani, S.: 1987, in *ESO Workshop on the SN 1987A*, ed. by I.J. Danziger (ESO Garching), p. 655

Danziger, I.J., Fosbury, R.A.E., Alloin, D., Cristiani, S., Dachs, J., Gouiffes, C., Jarvis, B. & Sahu, K.C.: 1987, Astron. Astrophys. **177**, L13

Dopita, M.A. et al.: 1988, Astron. J. **95**, 1717

Dotani, T. et al.: 1987, Nature **330**, 230

Feast, M.W.: 1988, in 4th *George Mason University Workshop in Astrophysics "Supernova 1987A in the LMC"*, ed. by M. Kafatos & A. Michalitsianos (Cambridge University Press), p. 51

Feast, M.W. & Walker, A.R.: 1987, Ann. Rev. Astron. Astrophys. **25**, 345

Felten, J.E.: 1988, in 4th *George Mason University Workshop in Astrophysics "Supernova 1987A in the LMC"*, ed. by M. Kafatos & A. Michalitsianos (Cambridge University Press), p. 232

Filippenko, A. & Sargent, W.L.W.: 1986, Astron. J. **91**, 691

Gouiffes, C., Rosa, M., Melnick, J., Danziger, I.J., Remy, M., Santini, C., Sauvageot, J.L., Jakobsen, P., Ruiz, M.T.: 1988, Astron. Astrophys. **198**, L9

Hanuschik, R.W.: 1988, Proc. Astr. Soc. Austr. **7**, 446

Hanuschik, R.W. & Dachs, J.: 1987a, Astron. Astrophys. **182**, L29

Hanuschik, R.W. & Dachs, J.: 1987b, in *ESO Workshop on the SN 1987A*, ed. by I.J. Danziger (ESO, Garching), p. 153

Hanuschik, R.W. & Dachs, J.: 1988, Astron. Astrophys. **205**, 135

Hanuschik, R.W., Thimm, G.J. & Dachs, J.: 1988, M.N.R.A.S. **234**, 41P

Hanuschik, R.W., Thimm, G. & Seidensticker, K.J.: 1989, Astron. Astrophys. **220**, 153

Hanuschik, R.W. & Thimm, G.J.: 1989, Astron. Astrophys. *in press*

Hanuschik, R.W. & Schmidt-Kaler, Th.: 1989a, M.N.R.A.S. *in press*

Hanuschik, R.W. & Schmidt-Kaler, Th.: 1989b, *in preparation*

Haupt, W., Desjardins, R., Maitzen, H.M., Rudolph, R., Schlosser, W., Schmidt-Kaler, Th. & Tüg, H.: 1976, Astron. Astrophys. **50**, 85

Hayes, D.S.: 1970, Astrophys. J. **197**, 593

Hearnshaw, J.B. & McIntyre, V.J.: 1988, Proc. Astr. Soc. Austr. **7**, 424

Hillebrandt, W. & Höflich, P.: 1989, Progress in Physics *in press*
Hirata, K., et al.: 1987, Phys. Rev. Lett. **58**, 1490
Höflich, P.: 1988, Proc. Astron. Soc. Austr. **7**, 434
Isserstedt, J.: 1975, Astron. Astrophys. Suppl. **19**. 259
Karovska, M., Koechlin, L., Nisenson, P., Papaliolios, C. & Standley, C.: 1989, in Highlights in Astronomy **8**, ed. by W.C. Liller (Reidel, Dordrecht)
Kirshner, R.P. & Kwan, J.: 1974, Astrophys. J. **193**, 27
Kirshner, R.P. & Kwan, J.: 1975, Astrophys. J. **197**, 415
Larson, H.P., Drapatz, S., Mumma, M.J. & Weaver, H.A.: 1987, in *ESO Workshop on the SN 1987A*, ed. by I.J. Danziger (Garching), p. 147
Lucy, L.: 1988, in *4th George Mason University Workshop in Astrophysics "Supernova 1987A in the LMC"*, ed. by M. Kafatos & A. Michalitsianos (Cambridge University Press), p. 323
Matz, S.M., et al.: 1988, Nature **331**, 416
McCray, R.: 1989, in *Proc. Yellow Mountain Summer School "Structure and Evolution of Galaxies*, ed. by L.-Z. Fang (Singapore: World Science), p. 8
Menzies, J. et al.: 1987, M.N.R.A.S. **227**, 39P
Phillips, M.M.: 1988, in *4th George Mason University Workshop in Astrophysics "Supernova 1987A in the LMC"*, ed. by M. Kafatos & A. Michalitsianos (Cambridge University Press), p. 11
Phillips, M.M., Heathcote, S.R., Hamuy, M. & Navarette, M.: 1988, Astron. J. **95**, 1087
Phillips, M.M. & Heathcote, S.R.: 1989, Publ. Astr. Soc. Pacific **101**, 137
Schwarz, H.E. & Mundt, R.: 1987, Astron. Astrophys. **177**, L4
Suntzeff, N.B., Hamuy, M., Martin, G., Gomez, A. & Gonzalez, A.: 1988, Astron. J. **96**, 1864
Suntzeff, N.B., Heathcote, S., Weller, W., Caldwell, N., Huchra, J.P., Olowin, R.P., Chambers, K.C.: 1988, Nature **334**, 135
Thimm, G.J., Hanuschik, R.W. & Schmidt-Kaler, Th.: 1989, M.N.R.A.S. **238**, 15P
Tüg, H.: 1980a, Astron. Astrophys. Suppl. **39**, 67
Tüg, H.: 1980b, Astron. Astrophys. **82**, 195
Whitelock, P.A. et al.: 1988, M.N.R.A.S. **234**, 5P
Whitelock, P.A. et al.: 1989, M.N.R.A.S. *in press*
Williams, R.E.: 1987, Astrophys. J. **320**, L117
Woosley, S.E.: 1988a, in *4th George Mason University Workshop in Astrophysics "Supernova 1987A in the LMC"*, ed. by M. Kafatos & A. Michalitsianos (Cambridge University Press), p. 289
Woosley, S.E.: 1988b, Astrophys. J. **330**, 218

Planetary Nebulae in Late Evolutionary Stages

R. Weinberger

Institut für Astronomie der Universität Innsbruck,
Technikerstr. 25, A-6020 Innsbruck, Austria

This article is intended to highlight the importance of planetary nebulae (PN) as objectives of modern astronomical research. We do this by discussing an interesting subgroup of these sources, namely the aged ones. These old PN represent stages immediately before the nebulae become part of the general interstellar medium and before their central stars turn into white dwarfs.

1. Introduction

In the introduction to her review on CATALOGUES OF PLANETARY NEBULAE Acker (1989) wrote the beautiful sentence *Planetary nebulae are fascinating objects, varied, not easily classifiable, interesting in all spectral regions, each area bringing some specific information (dust in the infrared, molecules in the radio, the central star in the ultraviolet and the visible,...), from all of which the object called "planetary nebula" can be preceived and described.* Acker also reported that the object with the largest number of bibliographical references in the whole Strasbourg Stellar Data Center is a PN (NGC 7027).

To demonstrate the development of the research on PN within the last decade worldwide or in the Fed. Rep. of Germany + Switzerland + Austria, a comparison of the number of papers is shown in Fig. 1; it is evident that the growth of papers on PN worldwide conforms to the growth of the total of astronomical papers worldwide and that the interest in PN in the above three countries is distinctly higher in the eighties than before the year 1981.- Every 5 years since 1967 IAU-symposia on PN are held; in addition various meetings, like workshops, on special topics take place in irregular time intervals (e.g., Preite Martinez 1987, Kwok and Pottasch 1987), both demonstrating the rapid progress in this field.- Those that are looking for details of the physical background of PN may find these in the textbook of Pottasch (1984), and for those interested in more general informations reviews like the one by Kaler (1985) are of importance.

Reviews in Modern Astronomy 2 Editor: G. Klare

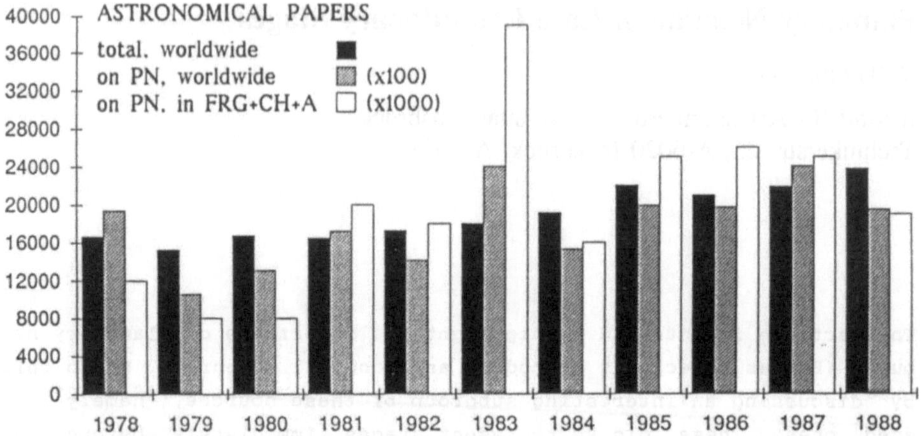

Fig. 1: Number of astronomical papers since 1978. Black: total numbers, worldwide; dotted: papers on PN, worldwide; white: papers on PN in the domain of the "Astronomische Gesellschaft", i.e., FRG + CH + A

The spread of physical parameters among PN is very high. For instance, electron densities n_e range from 10^6 to 10 cm^{-3}, ionized masses M_i from 10^{-9} to 1 $M_\odot$, central star (CS) temperatures from T_{eff} = 2.5 10^4 to 3.5 10^5 K and absolute magnitudes M_V of CS from -4 to +10. This heterogeneity is, on the one hand, the cause for the unsolved distance problem that is responsible for numerous uncertainties, but on the other hand a main source for the scientific fascination of PN.

The most widely known fact concerning PN is their important role in the late evolution of stars. Probably, all stars with about 1 - 8 $M_\odot$ have to pass through the PN-phase, which lasts for a few 10^4 years only; within this short period, a large portion of the HRD is traversed, connecting red giants with white dwarfs (WD). For a recently published diagram containing CS with wellknown distances and theoretical evolutionary tracks, see fig. 3 in Gathier and Pottasch (1989).

Besides their role in stellar evolution, PN are of importance, because they return ca. 1 - 3 $M_\odot$ per year to the interstellar medium. Moreover, this material is partly enriched in He, N etc.; that is, PN contribute to the chemical enrichment. Furthermore, the ionizing radiation of CS of optically thin nebulae at medium or high $|z|$ successfully competes with the radiation of O-stars. The brightest planetaries in other galaxies were recently recognized as promising standard candles to determine extragalactic distances. Finally, PN remain to serve as extremely valuable "astrophysical laboratories".

2. Late phases of PN = "old" PN

2.1 Difficulties in defining

How to tell an "old" PN from a "normal" one? A sharp separation is impossible. In practice, a separation against the domain of WD is easy - as soon as the nebula falls below the detection limit (as accessible with sensitive instrumentation). In short, old PN can hardly be "defined", but they can be "characterized" as follows:

the nebulae can be considered as old, if n_e is small (e.g., <500 cm^{-3}), the linear radius r large (e.g., >0.2 pc), the surface brightness, like $S(H\alpha+[NII])$, low (e.g., >20 mag/arcsec^2 after correction for extinction) etc.

the CS are old, a soon as their shell burning has ceased (i.e., they start to cool along the WD cooling tracks), the surface gravity g is high (e.g., $\log g/g_\odot > 6$), the luminosity L low (e.g., $\log L/L_\odot$ <3), the effective temperature T_{eff} high (e.g., >70.000 K) etc.

A few extreme values (Kaler et al., in prep.): $n_e = 2$ cm^{-3}, r = 1.03 pc (LoTr 5); $L/L_\odot$ = 4-12 (PW 1), CS-mass $M_c = 1.2$ $M_\odot$ (K 1-13).

By the way, the famous "old PN" NGC 7293 is largely atypical!

2.2 Significance of old PN

In the following, some highlights are presented in a condensed form:

As old PN are of low surface brightness, accurate measurements of CS are only possible in late evolutionary PN phases (e.g., apparent magnitudes, spectral details, variability etc.);

the low surface brightnesses allow the detection and investigation of compact nebulosities of newly ejected matter within old PN (these nebulosities are of considerable importance with respect to theoretical evolutionary models);

with the help of old PN it will be possible to largely close the gap in the HRD between PN and WD;

the disvovery of old PN means detection of mainly massive CS (since CS with small masses evolve slowly and may have lost their nebulae before being able to ionize them);

as massive CS stem from massive progenitors, where dredge-up processes had occurred, old nebulae are favourable to study enrichment;

the expansion characteristics of old PN are hardly known; it would be rewarding to study how the dilute nebulae become part of the ISM;

old PN are the main constituent of the local PN-population; a sound knowledge of the latter is of importance in many respects;

only few old PN are known in detail up to now; the detection of numerous unpredicted interesting peculiarities can be expected.

2.3 Discovery of old PN

In a pioneering work, Abell (1966) reported on the discovery and investigation of old planetary nebulae; he had detected altogether 86 PN candidates on Palomar Observatory Sky Survey (POSS) red-sensitive (E-) prints and wrote *"Those PN that lie north of −33°, that have surface brightnesses brighter than 24.5 (photored) mag/arcsec² [note: POSS E-limit = 25.0 according to Abell], and with angular diameters greater than 10" are probably nearly completely identified".* Although the above work was a big step forward, the conclusion cited proved to be wrong and possibly was responsible that i) for about a decade no systematical search for new PN was begun on the POSS and ii) the steadily increasing number of newly detected old PN (many of them found on the POSS) was practically neglected by the specialists in the field.- It is obvious today that Abell (1966) missed about 50% of the evolved planetaries detectable on the POSS.

In Innsbruck, members of the Institute für Astronomie are engaged in the detection and investigation of old PN since more than a decade (Dengel, Hartl, Melmer, Weinberger, Saurer, Weinberger, partly in cooperation with Ishida, Purgathofer, Sabbadin, and Tritton). Altogether, 82 PN-candidates could be discovered, among them 50 on the POSS, the rest on the ESO/SERC surveys. Most of the objects were found in scanning the POSS-prints by eye, since this is still the optimum method for finding such extremely faint nebulae.- Other groups also were successful in searching POSS prints for new PN (e.g., Ellis et al. 1984). All these PN will be contained in the forthcoming "Strasbourg-ESO Catalogue of Galactic Planetary Nebulae".

There are two procedures in use to discover new PN in late evolutionary stages: i) the conventional method, i.e., the search on existing deep material, preferentially on broadband sky-surveys like the POSS, the ESO/SERC surveys or on emission line surveys like the one published by Parker et al. (1979), and ii) a new method, where faint nebulae are looked for around hot WD or subdwarfs with the help of CCD's and line filters. With the second method, only few objects could be detected to date.

In all probability, a number of new evolved PN will be discovered in the near future, although we are approaching not only a sensitivity limit (at present, sky-surveys with a limiting magnitude deeper than the ESO/SERC survey hardly seem feasible) but also a limit, where the oldest nebulae have gas densities comparable to the interstellar medium. The majority of them will again be rather nearby PN.

2.4 The local PN-population

A precise knowlege of the local PN-population is of great importance in several respects. Local PN are a significant subset of the galactic PN, since i) they have considerable influence on conceptions on the late stellar evolution (e.g., via birthrates of PN and WD); ii) the total number of PN in the Galaxy (and consequently the effects on it like mass return etc.) is based on them; iii) they allow to pursue CS as far as possible on their way towards WD, because it is the local PN that contain several nebulae in the latest observable stages of evolution; iv) the true luminosity function of the nebulae and CS will depend on them (especially with regard to the faint end).

The following data on the local PN-population were calculated by Ishida and Weinberger (1987), who took also the nearby "post-Abell" - PN for the first time into account. For the determination of the distances mean values from up to 5 published distance scales were used and for the residual objects an ionized mass of 0.2 $M_\odot$ was assumed.

Within 500 pc there are 31 PN; 26 of them are "old" ($r \geq 0.2$ pc) and 40% of these local PN were found after 1966! 16 out of the 31 objects were originally detected on the POSS. 20 are located in the north, 11 in the south. The numbers below are already corrected for this disparity.

We found an average space density for the local PN of 326 kpc^{-3}; with V_{exp} = 20 km/sec and an r-interval of 0.8 pc a PN-birthrate of 8.3 10^{-3} $kpc^{-3}yr^{-1}$ was estimated. This birthrate is much higher than the WD-birthrate of 0.50-0.75 10^{-3} $kpc^{-3}yr^{-1}$ (Fleming et al. (1986).

A difference this large between both birthrates is entirely at variance with the current concepts on late stellar evolution; an increase of 1.9 times the PN distances used would bring them into agreement, but such an amount appears to be improbable. Apparently, there are also other explanations necessary.- The situation is impaired yet, since the local sample is incomplete even in the north, at least beyond ca. 400 pc; we present 2 out of several arguments:
i) the run of the PN-densities as projected on the galactic plane; these are 56, 60, 57, 60, 52, 46 (q = 0.0-0.25 up to 0.0-0.50 kpc);
ii) a diagram (fig. 2) which shows that the larger a nebula is, the more difficult is it to find it at larger distances.

An additional interesting information from fig.2: The ca. 2000 galactic PN known (we estimated a total of 140.000 in the Galaxy) obviously are highly unrepresentative. The "typical PN" is not of young or intermediate age but in a considerably late evolutionary stage!

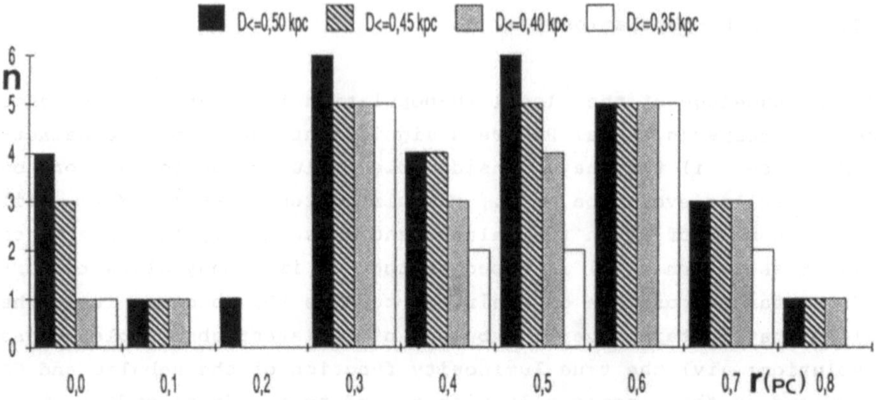

Fig. 2: Number of local PN vs. linear radii within 4 distance ranges

2.5 Morphology and expansion

Most of the old nebulae do not show a richness in morphology comparable to younger ones, because the pronounced differences in gas density within a young nebula had enough time to smear. Consequently, many of the old nebulae are of rather smooth appearance:the largest (S 216 and the 3rd largest (PW 1) of all PN (as to angular diameter) are typical examples (Fesen et al. 1981, Purgathofer and Weinberger 1980).

Over and above, in the case of extremely dilute nebulae it may for the first time be possible to see effects of the interstellar medium on the nebular material, influencing both the morphology and the expansion. There are different kinds of examples for these effects:

a) bow-shocks. Close to the CS of A 35 and the nebula $6^h 23^m + 71^o$, bowshock-like emission is visible (Jacoby 1981, Krautter et al. 1987)

b) off-center CS. In several PN, the CS is not centrally located (S 216, PHL 932, S 188, A 21...). In A 21, for example, a hitherto unrecognized extremely faint extension of the nebula reaches far out to the north-west (see POSS E-print); the filamentary main nebular body, the location of the CS, and this extension might well be the consequences of an interstellar wind blowing from the south-east;

c) decelerated expansion. Gieseking et al. (1986) and Hippelein and Weinberger (in press) argued that the low expansion velocity of several very old PN near the galactic plane might be due to a deceleration caused by the interstellar medium.- If true, this effect would lower the PN-birthrate (where constant V_{exp} is assumed).

The research on morphology and expansion of old PN is still in its infancy; intense observational and theoretical efforts are required.

2.6 From the radio range to the X-ray region

In this section a short account of the importance of the different wavelength ranges for old planetary nebulae is given.

Radio region: Old nebulae are weak emitters, often below the present detection limits. For objects large in angular extension there is a considerable risk for confusion with background sources. An instructive schematic diagram showing how the radio spectrum changes with increasing age is presented by Gathier (1986).

Infrared region: With exception of a few peculiar old nebulae that could be detected from the ground, the IRAS mission proved to be of pioneering importance, although only a fraction of the evolved PN could be detected. In several objects, the emission was shown to originate both in the internal dust and in IR line radiation from the gas.There are clear indications that the dust is continuously destroyed as the nebula ages and that the material eventually returned to the interstellar medium is heavily depleted in dust (Pottasch 1987).

Optical region: The nebular continuum of old nebulae is beyond the reach of the present instrumentation; usually, only few lines are visible and the physical interpretation therefore is limited, but n_e, T_e, and sometimes V_{rad} and V_{exp} were derived. Most CS can be well observed due to the low surface brightnesses of the nebulae, but several are of extremely low apparent (and absolute) brightness.

Ultraviolet region: The evolved nebulae cannot be detected with the IUE. For the CS, however, the UV is the prime region; several parameters like T_{eff}, wind velocities, interstellar extinction etc. were obtained.

X-ray region: Tarafdar and Apparao (1989) published an HRD diagram, where not only upper limits but also definite detections with the EINSTEIN satellite are shown. X-ray emission was observed from CS in the oldest nebulae only; probably, because the interstellar extinction is nearly negligible (i.e., they are at small distances) and due to the settling of heavy elements in the CS, which otherwise prevents escape of X-rays by providing opacity.

2.7 Some titbits among old PN

The late evolutionary stages of planetary nebulae permit the detection and investigation of various phenomena. In the following, we present a small selection thereof.

Compact nebulae at the center of PN: With the recent discovery of an unresolved nebular core component around the CS of the old planetary EGB 6 (Liebert et al., in press) altogether four such nebulae are known (A30, A58, A78, EGB 6). Whereas the components in A 30, A 58, and A 78 are extremely hydrogen deficient or show no hydrogen at all (result of He-shell flashes?), but are He-rich, the nebula within EGB 6 shows strong emission lines of [OIII], [NeIII], H, and He and might perhaps have another origin. If additional nebulae of this type could be found, one might ask whether this could be a part of the answer to the discrepancy in PN and WD-birthrates (the old and the new nebula could be entirely decoupled as to their time of formation). Then, however, the question arises: How do these compact nebulae reveal themselves after a few thousand years? Where are they?

Extremely nitrogen-rich nebulae: Possibly, we can answer the above questions. As is well known, a high He-abundance is coupled with a high nitrogen abundance. There are at least 2 extended nebulae with extraordinarily strong nitrogen lines. For both of them, a CS could not be identified and they might, consequently, be no planetaries. One is the object PL1547.3-5612 (Ruiz 1983), the other HT 10 (Hartl and Tritton 1985). In Fig. 3, we show a spectrum and a reproduction of the latter. The absence of a CS could be explained by the strong interstellar extinction as is obvious for HT 10, but if these objects indeed are nebulae of a second generation, the CS might already have cooled down and is of very low luminosity.

Old PN around non-post-AGB CS: The Asymptotic Giant Branch (AGB) appeared to be an inescapable track for the pre-WD surrounded by a planetary nebula. Recently, however, Mendez et al. (1988) showed that this need not be the case: for 2 CS (PN EGB 5 and PHL 932) the mass must be less than 0.5 $M_\odot$ and these stars could therefore not have been able to ascend the AGB; probably, they are components of a close binary and the nebulae are results of a common-envelope ejection.

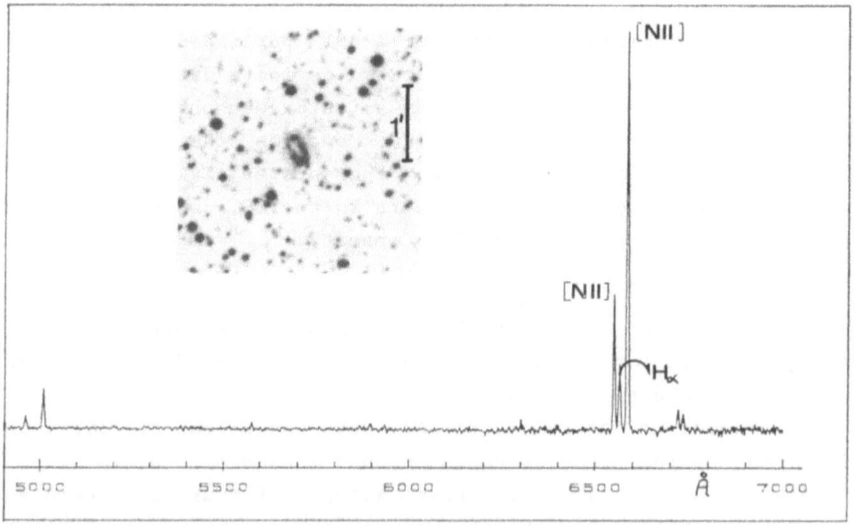

Fig. 3: HT 10, a nebula with extremely strong nitrogen lines

Variable central stars: To search for and monitor the variability of CS, low nebular surface brightnesses are necessary or at least favourable. It is thus not unexpected that the few variable CS known were predominantly found in old PN. At present, only 1 pulsating CS is known, 7 variables are known with a period <1 day (probably produced via common-envelope interaction), and a comparable number of CS with longer periods was registered (Bond 1989, Mendez 1989).

Molecules in old PN: The detection of molecules in NGC 6720, NGC 6853 and NGC 7293 came as a total surprise (Rodriguez 1989). Obviously, these evolved nebulae are only partly optically thin; it is not quite clear how the molecules could survive for such a long time.

3. Epilogue

The majority of the planetary nebulae in late evolutionary stages is not yet explored. In the following we present some general proposals:

1) Starting from shipment 12, the ESO/SERC R and J surveys should be examined for new PN (the film copies of the shipments 1-12 were already successfully searched by Melmer and Weinberger (in press)). The same should be done for the forthcoming POSS II.

2) Further examples of the effects of the interstellar medium on evolved nebulae should be looked for; a promising technique would be the use of CCD's + interference filters + focal reductors.

3) The centers of evolved nebulae should be examined for additional compact nebulae.

4) Spectroscopic measurements of hitherto unexplored extended PN should be done in order to find additional examples of nitrogen-rich or helium-rich nebulae.

5) The 3 old PN containing molecules might not be exceptional cases. A search for further objects of this kind is desirable.

6) Attempts to improve the distances of old PN must be undertaken. For the nearby objects, trigonometric parallaxes should be measured with the Hubble Space Telescope.

7) A large number of old PN are rather nearby; a search for visual companions to the CS might be fruitful and could, e.g., lead to accurate (spectroscopical) distances.

8) As only a fraction of the old PN are checked for variability of their CS, further searches and observations will prove to be useful.

Acknowledgements: Without financial support by the "Fonds zur Förderung der wissenschaftlichen Forschung", project Nr. 5708, this work would not have been possible.

References

Abell, G.O.: 1966, *Astrophys. J.* **144**, 259

Acker,A.: 1989, *Proceed. of the 131st IAU-Symp. on Planetary Nebulae,* Kluwer Academic Publishers, ed. S.Torres-Peimbert, p. 39

Bond, H.E.: 1989, *Proceed. of the 131st IAU-Symp. on Planetary Nebulae,* Kluwer Academic Publishers, ed. S.Torres-Peimbert, p.251

Ellis, G.L., Grayson, E.T., Bond, H.E.: 1984, *Publ. Astron. Soc. Pac.* 96, 283

Fesen, R.A., Blair, W.P., Gull, T.R.: 1981, *Astrophys. J.* **245**, 131

Fleming, T.A., Liebert, J., Green, R.F.: 1986, *Astrophys. J.* 308, 176

Gathier, R.: 1987, *Proceed. of the workshop on Late Stages of Stellar Evolution,* Astrophys. and Space Sci. Library, ed. S.Kwok and S.R. Pottasch, p. 371

Gathier, R., Pottasch, S.R.: 1989, *Astron. Astrophys.* 209, 369

Gieseking, F., Hippelein, H.H., Weinberger, R.: 1986, *Astron. Astrophys.* 156, 101

Hartl, H., Tritton, S.B.: 1985, *Astron. Astrophys.* 145, 41

Hippelein, H.H., Weinberger, R.: 1989, *Astron. Astrophys.* (submitted)

Ishida, K., Weinberger R.: 1987, *Astron. Astrophys.* 178, 227

Jacoby, G.H.: 1981, *Astrophys. J.* **244**, 903

Kaler, J.B.: 1985, *Annual Rev. of Astron. and Astrophys.,* p. 89

Kaler, J.B., Shaw, R.A., Kwitter, K.B.: 1989, in prep.

Krautter, J., Klaas, U., Radons, G.: 1987, *Astron. Astrophys.* 181, 373

Kwok, S., Pottasch, S.R.: 1987, *Proceed. of the workshop on Late Stages of Stellar Evolution,* Astrophys. and Space Sci. Library, Reidel Publishing Company

Liebert, J., Green, R., Bond, H.E., Holberg, J.B., Wesemael, F., Fleming, T.A., Kidder, K.: 1989, in press

Melmer, D., Weinberger, R.: 1989, *Monthly Not. Roy. Astron. Soc.* (in press)

Méndez, R.H.: 1989, *Proceed. of the 131st IAU-Symp. on Planetary Nebulae,* Kluwer Academic Publishers, ed. S.Torres-Peimbert, p. 261

Méndez, R.H., Groth, H.G., Husfeld, D., Kudritzki, R.P., Herrero, A.: 1988, *Astron. Astrophys.* 197, L25

Parker, R.A.R., Gull, T.R., Kirschner, R.P.: 1979, *An Emission-Line Survey of the Milky Way,* NASA SP-434

Pottasch, S.R.: 1984, *Planetary Nebulae,* Reidel Publishing Company

Pottasch, S.R.: 1987, *Proceed. of the workshop on Late Stages of Stellar Evolution,* Astrophys. and Space Sci. Library, ed. S.Kwok and S.R.Pottasch, p. 355

Preite Martinez, A.: 1987, *Proceed. of the workshop on Planetary and Protoplanetary Nebulae,* Astrophys. and Space Sci. Library, Reidel Publishing Company

Purgathofer, A., Weinberger, R.: 1980, *Astron. Astrophys.* 87, L5

Rodriguez, L.F..: 1989, *Proceed. of the 131st IAU-Symp. on Planetary Nebulae,* Kluwer Academic Publishers, ed. S.Torres-Peimbert, p.129

Ruiz, M.T.: 1983, *Astrophys. J. Lett.* 268, L103

Tarafdar, S.P., Apparao, K.M.V.: 1989,*Proceed. of the 131st IAU-Symp. on Planetary Nebulae,* Kluwer Academic Publishers, ed. S.Torres-Peimbert, p. 304

Structural Variations in the Quasar 2134+004

I.I.K. Pauliny-Toth[1], *A. Alberdi*[1], *J.A. Zensus*[2], *and M.H. Cohen*[3]

[1]Max-Planck-Institut für Radioastronomie, Auf dem Hügel 69, D-5300 Bonn 1, Fed. Rep. of Germany
[2]National Radio Astronomy Observatory, Socorro, NM 87801, USA
[3]California Institute of Technology, Pasadena, CA 91225, USA

Observations by means of Very Long Baseline Interferometry (VLBI) have revealed an increase in the extent and a change in the orientation of the parsec-scale radio structure of the quasar 2134+004. The changes occurred without any large radio outburst and are very different from those generally seen in superluminal radio sources.

The radio source 2134+004, one of the brightest extragalactic sources at centimetre wavelengths, is identified with a 18mag quasar at a redshift z=1.936. Its variations at radio wavelengths are relatively small, typically about 15 percent, and occur on a time-scale of several years (e.g. ref. 1). The radio spectrum is peaked and reaches a maximum flux density of about 8 Jy at present, at a frequency of 5GHz. There is little emission below a frequency of 1 GHz, which suggests the absence of extended radio structure. In both respects, the source differs from typical superluminal sources, which show large, rapid radio outbursts and have extended structure on scales of tens of kpc or greater.

Early VLBI measurements at frequencies of 10.7 and 5 GHz[2,3,4] showed that the structure on a scale of milli-arcsec (mas) was basically double, with components of roughly equal strength separated by about 1.8 mas (20 pc: we assume H_0=55 km s^{-1} Mpc^{-1}, q=0.05 throughout) in P.A. 65 deg. Further measurements made by us at epoch 1984.4 (unpublished data) gave the same morphology and orientation, but suggested that the separation of the components had decreased to about 1.4 mas (15 pc). Such a decrease need not imply a physical contraction of the source, but could be due to changes in the brightness distribution of one or both components, such as a fading of the outer, or a brightening of the inner parts.

Encouraged by these signs of changes in an apparently quiescent object, we carried out observations at 10.7 GHz with a global, 8-station VLBI array at epoch 1987.2. These revealed a dramatic change in the structure, as illustrated in Fig.1 which shows the structure at 10.7 GHz at epochs 1981.4 and 1987.2. At the earlier epoch, we see the original double structure,

Reviews in Modern Astronomy 2 Editor: G. Klare

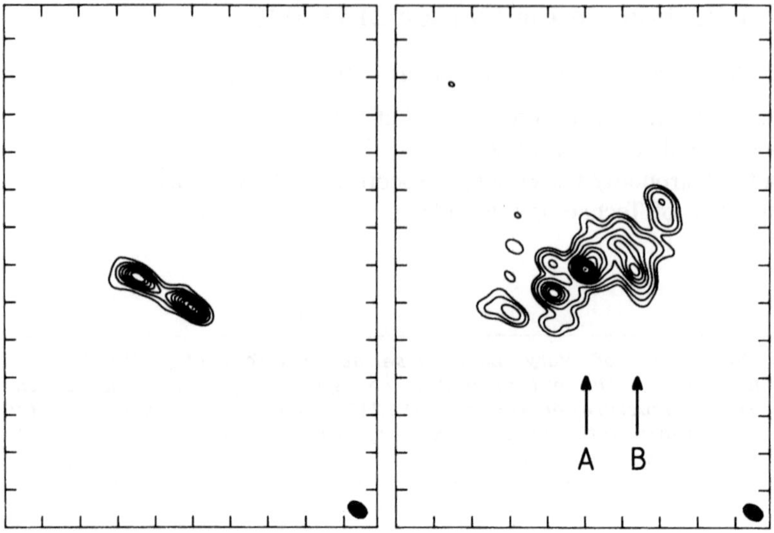

Fig.1: VLBI maps of the quasar 2134+004 at a frequency of 10.7 GHz. The epochs of observation are 1981.43 (left) and 1987.15 (right). The maps were made with a global array of 5 and 8 antennas respectively. Both maps were restored with the same beam (shown shaded) for ease of comparison; in fact, the earlier map lacked North-South baselines, and has a poorer resolution in the North-South direction. East is to the left and North to the top. Tick marks are at 1 mas (11 pc) intervals. Contours are 10,20,.....90 percent of the maximum brightness of each map. In the right-hand map, the 5 and 2.5 percent contours are plotted in addition. Features labelled as A and B are referred to in the text.

oriented in P.A. 65 deg, while the 1987.2 map shows a complex structure, containing at least 4 components. The total extent of the second map is about 5 mas (55 pc) and the structure is oriented predominantly along P.A. 120deg. Since the source was still double at epoch 1984.4 (see above), the change must have occurred in a time of 2.8 years or less.

There is no clear relation of the new to the old structure, that is, it is not obvious which (if either) of the components of the original double is the centre of activity and gave rise to the new structure. Nevertheless, the large change in extent along P.A. 120 deg implies an apparent velocity of (two-sided) expansion of at least 70c !

We note that there was no major radio outburst in the source in the 1980s: monitoring of the flux density (refs. 1,5 and Valtaoja, priv.comm.)only shows a slow increase beginning in 1981 which reached a peak (about 20 percent increase) in late 1984 at frequencies of 22 and 15 GHz. However, monitoring of the radio polarisation[1] did show a sharp rotation of the polarisation angle of about 90 deg during the period 1983 to 1985. This rotation was, however, different at different frequencies and therefore does not simply reflect a change in the orientation of the structure. Rather, it is probably due to changes in the opacity of different parts of the source, and/or to the appearance of new components.

We made further VLBI observations after 1987.2, to follow the subsequent behaviour of the source. The resulting maps are not shown here, but indicate that no further expansion occurred. Rather, the outer parts of the source faded, leaving only the brightest feature (labelled A in Fig.1) and the ridge-like feature (B) about 1.5 mas to the West. The ridge-like feature B became more prominent after 1987.2 relative to A.

The behaviour of the source is quite unlike that of "classical" superluminal sources, where components are successively ejected from a "core" following radio outbursts, and move away from the core monotonically along a roughly constant path (e.g. ref. 6). The behaviour of 2134+004 resembles that of the quasar 3C454.3, which has also shown a rapid increase in size and complexity of its core region, followed by a period in which the components show no relative motion, but gradually fade[7]. The mechanism proposed for 3C454.3 may also apply to 2134+004 : the "components" are stationary shocks (e.g. ref.8) in a relativistic beam or jet. However, the structure of 2134+004 is considerably more complex and more distorted than that of 3C454.3 and we have, as yet, been unable to identify the core. We hope that further VLBI observations will help us to understand the remarkable variations in the structure of this source.

References:

1 Aller, H.D., Aller, M.F., Latimer, G.E., and Hodge, P.E. 1985, Astrophys. J. Suppl. 59, 513

2 Schilizzi, R.T., Cohen, M.H., Romney, J.D., Shaffer, D.B., Kellermann, K.I., Swenson, G.W., Yen, J.L., and Rinehart, R. 1975, Astrophys.J. 201, 263

3 Pauliny-Toth, I.I.K., Preuss, E., Witzel, A., Graham, D., Kellermann, K.I., and Ronnang, B. 1981, Astron.J. 86, 371

4 Pauliny-Toth, I.I.K.,Porcas, R.W., Zensus, J.A., and Kellermann, K.I. 1983, IAU Symposium No 110,"VLBI and Compact Radio Sources", p.149, publ. D. Reidel, Dordrecht, eds. R.Fanti, K.Kellermann, G.Setti.

5 Salonen, E., Terasranta H., Urpo, S., Tiuri, M., Moiseev, I.G., Nesterov, N.S., Valtaoja, E., Haarala, S., Lehto, H., Valtaoja, L., Teerikorpi, P., and Valtonen, M. 1987, Astron. Astrophys. Suppl. 70, 409

6 Zensus, J.A., 1987, in "Superluminal Radio Sources", p.26, publ. Cambridge University Press, Cambridge, eds. J.A. Zensus, T.J. Pearson

7 Pauliny-Toth, I.I.K.,Porcas, R.W.,Zensus, J.A.,Kellermann, K.I., Wu, S.Y., Nicolson, G.D., and Mantovani, F., 1987, Nature 328,778

8 Lind, K.R., and Blandford, R.D. 1985, Astrophys. J. 232, 34.

Submillimeter Observations of Galactic and Extragalactic Objects

R. Chini

Max-Planck-Institut für Radioastronomie, Auf dem Hügel 69, D-5300 Bonn 1, Fed. Rep. of Germany

ABSTRACT. The present review is intended to give a brief introduction into technical and physical aspects of submm astronomy. After that, an overview on interesting topics and recent discoveries in galactic and extragalactic research is made, including submm continuum observations of pre-main sequence and main sequence stars, normal and starburst galaxies and radio-loud and radio-quiet quasars. In the case of thermal emission, the observation of submm radiation provides a promising way to estimate the temperature of the dust and the total mass of gas associated with the individual classes of objects. For quasars, where the emission is generally assumed to be non-thermal in origin, the submm regime turns out to be a powerful tool in order to detect separate spectral components and to discriminate between possibly thermal and non-thermal radiation.

1. INTRODUCTION

The IRAS "all-sky" survey revealed a huge amount of data in the wavelength range from 12 to 100μm and increased our knowledge about various classes of objects considerably. On the other hand, it was a surprise that the energy distributions of most galactic and extragalactic objects rise from optical wavelengths until 60μm - very often even to 100μm - indicating the presence of a large amount of cool dust with temperatures below 40K; this holds for stars with circumstellar matter, HII regions, galaxies and even quasars, though the origin of their FIR radiation is less clear. While there is little doubt that the FIR emission from dust embedded stars, HII regions and normal galaxies is due to stellar radiation re-processed by the interstellar material, the situation is less clear for "ultra-luminous infrared" galaxies. Having essentially similar spectral distributions, their total energy output comes close to that of quasars ($L \geq 10^{12} L_{\odot}$) and they appear to be sites of highest luminosity in the local universe. Finally, also the spectra of active galactic nuclei rise from X-ray to FIR wavelengths, commonly interpreted, however, as to origin from non-thermal Synchrotron components.

Despite the huge increase of information, due to the results of IRAS, there remain important open questions concerned with the continuum emission from various types of objects and even unexpected problems have been brought up by the new FIR data. The increasing spectra from optical to FIR wavelengths on one hand, and the (expected) decrease of several orders of magnitude in the range between 100μm and radio

Reviews in Modern Astronomy 2 Editor: G. Klare

wavelengths on the other hand make the submm regime a key tool for understanding major physical properties.

Just in time, several large submm telescopes have come into operation during the past three years and the sensitivity of the receivers could be improved considerably. This gives now the opportunity to fill the last gap in the astronomically important part of the electromagnetic spectrum for a variety of galactic and extragalactic sources.

2. TECHNICAL ASPECTS OF SUBMM OBSERVATIONS

There is no general adopted definition of the submm wavelength range but one can derive some limits caused by technical conditions. The FIR regime - explored by airborne (KAO, SOPHIA) and satellite (IRAS, ISO) observatories - extends to about 200μm and is not accessible from the ground. Radio telescopes and receivers at the long wave side are capable to observe radiation down to several mm. In the following, the range between 200μm and about 3mm is defined to be the domain of submm astronomy.

Unfortunately, the earth's atmosphere limits observations at submm wavelengths severely. Not only that the measurements are confined to a few spectral windows, as shown in Fig. 1, but also the total transmission in some of these windows may be decrease to zero, depending on the amount of water vapor. For that reason, dry sites of high altitude, the best of which is currently the 4200m volcano Mauna Kea, Hawaii are inevitable for obtaining data at e.g. 350μm. Another important factor is of course a suited telescope. Until recently, submm observers had to use optical or infrared telescopes, which - on one side - were perfect as concerns the surface accuracy of about 30μm rms, required for observations of e.g. 350μm, but which - on the other side - were too small in respect to collecting area and spatial resolution for most astrophysical applications.

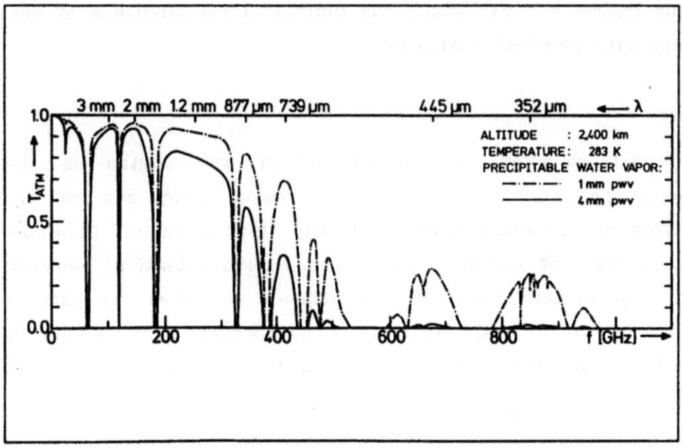

Fig. 1: The atmospheric transmission is shown as a function of wavelength and water vapour content for an altitude of 2400m.

In 1986, the IRAM 30m MRT (Baars et al. 1987) on Pico Veleta, Spain started its performance at 1300μm and has been used since then even for continuum observations at 870μm, making it the world's largest telescope for these wavelengths. On Mauna Kea, the 15m JCMT, built jointly by the United Kingdom and the Netherlands, works since two years in the entire submm range from 1300 to 350μm. Close to the JCMT, CalTech constructed the 10m CSO also designed to observe at all submm wavelengths. In the southern hemisphere, ESO and Sweden operate the 15m SEST (Booth et al. 1989) on La Silla which is supposed to work until a shortest wavelength of about 650μm. Currently, an array of three 15m antennas, each identical to SEST, is under construction by IRAM on Plateau de Bure, France and a 10m submillimeter telescope (SMT) is planned to be raised on Mt. Graham, Arizona in a joint collaboration of the MPI für Radioastronomie, Bonn and the University of Arizona.

The observations described in the following are concerned with the study of continuum emission. For this kind of measurement, ^{3}He cooled bolometers are now commonly used. The MPIfR system (Kreysa, 1985) consists of a neutron transmutation doped Ge-crystal operating at 0.27K.

The selection of the observing bandpass is done via interference mesh filters generally adapted to the atmospheric windows in order to obtain maximum bandwidth and sensitivity. The actual observing procedure is identical to that used in IR astronomy, i.e. chopping between object and sky by means of a focal plane chopper or a wobbling secondary mirror and beam-switching the telescope to eliminate background emission and sky fluctuations. The planets which are the brightest objects in the submm sky and whose temperatures are fairly known serve as calibration standards.

3. RADIATION MECHANISMS

At submm wavelengths three components may contribute to the observed flux density. Thus, for an interpretation of the measurements a careful separation of them is necessary. Due to the fact that thermal dust emission and non-thermal radiation overlap only in the submm regime one can study the individual contributions of both components and derive important physical quantities.

3.1. Thermal dust radiation

The most important thermal emission process at submm wavelengths is the re-radiation of high energy photons absorbed by dust grains. Many sources like star-forming regions or circumstellar disks remain optically thick even out to 100μm so that one sees only the surface of the emitting volume. Submm radiation, however, is optically thin for most objects (but there are exceptions!) and thus one samples the volume, i.e. the total mass of the emitting grains. The observed flux density S_ν correlates with the mass of dust M_d and its temperature T_d as

$$S_\nu = M_d \kappa_\nu B_\nu(T_d) D^{-2} \qquad (1)$$

where κ_ν is the dust opacity and D the distance to the source. The spectral slope in the Rayleigh-Jeans region is $\nu^{2+ß}$, where a is the frequency dependence of the dust opacity $\kappa_\nu \propto \nu^\beta$. In principle equation (1) then allows the determination of the total gas mass M_g by adopting a certain gas-to-dust ratio R. In practice, however, there remain many uncertainties, particularly with the determination of κ_ν and R (see e.g. Krügel et al. 1989).

3.2. Free-free emission

The second significant thermal process is the emission from ionized gas. While it dominates the millimeter spectrum of star-forming regions its contribution at submm wavelengths is generally masked by emission from dust. Likewise, it is not clear what fraction of free-free emission is present in the spectra of active galactic nuclei as there it is difficult to separate from possible non-thermal radiation. In any case, it is important to take into consideration a possible contamination before interpreting submm data in terms of dust or synchrotron radiation.

3.3. Synchrotron radiation

The radio spectrum of active galactic nuclei is dominated by synchrotron emission and it is likely that this radiation mechanism also contributes to the submm and FIR regime. The observations of steeper than canonical ($S_\nu \propto \nu^\alpha$, $\alpha > 2.5$) FIR turnovers in the spectra of several radio-quiet quasars, however, have led to a vivid discussion about a possibly thermal origin of the FIR radiation. Clearly, only submm data can provide a discrimination between thermal and non-thermal components in quasar spectra.

4. RESULTS OF RECENT SUBMM OBSERVATIONS

In the following, an attempt is made to give an overview about some recent outstanding results obtained from submm observations. This review is not intended to be complete but it will demonstrate the versatility of the new wavelength range. Most data discussed below were obtained with the MPIfR bolometer (Kreysa, 1985) attached to the IRAM 30m telescope - currently the most sensitive combination for continuum measurements at 870 and 1300μm.

4.1 Circumstellar disks around pre-main sequence stars

The solar system was born from a disk of gas and dust encircling the sun about 5 10^9 years ago and there is evidence that similar disks surround many young, solar-mass stars in the Galaxy today. Strom et al. (1989) derived from the IR excess emission that 60% of the youngest PMS stars in their sample are surrounded by circumstellar disks. In a similar study, Cohen et al. (1989) examined 72 stars in Taurus-Auriga and concluded that about one third of the stars show evidence for disks. Calculations of emission from circumstellar disks (e.g. Adams et al. 1987; Bertout et al. 1988) demonstrate clear infrared signatures accompanying disks similar to the proto-solar nebula; these calculations provide the underpinnings for the studies cited above, but are not the only means of detecting disk matter.

In general, all optical and IR observations suffer from the shortcoming of being more sensitive to energetic activity, such as accretion, rather than to the amount of disk matter. Excess radiation at visual and NIR wavelengths arises in a hot inner disk or a luminous boundary layer between the disk and the star. The strength of the FIR emission, usually optically thick, thus depends on the disk luminosity and temperature distribution, both strong functions of the energy balance in the disk. To discuss the likelihood of planet formation, it is desirable to measure the total mass in a disk and its spatial distribution. Thermal emission from dust grains entrained in these disks is optically thin at long submm wavelengths and therefore provides an excellent way to measure disk masses directly.

Recently, a very successful survey for 1300μm emission towards 87 pre- main sequence (PMS) stars in the Taurus-Auriga dark clouds has been performed at the 30m MRT (Beckwith et al. 1989). During this survey, 37 stars (42%) have been detected with more than 3σ; the non-detections have 3σ upper limits of about 30mJy which means that further detections might be easily possible with more integration time. 28 of 53 (53%) normal T Tauris and 6 of 21 (29%) weak line or "naked" T Tauris could be measured at 1300μm. These percentages are very similar to those found by Strom et al. (1989) from their study of FIR properties.

It was assumed at the outset that the submm continuum emission originates from heated circumstellar dust grains. For the bright sources HL Tau and DG Tau, this interpretation has been clearly established by Beckwith et al. (1986). There are insufficient data to confirm the assumption of thermal emission from dust for every star, but it can be shown for many in the sample and is quite likely to apply to all the members. Another alternative to thermal re-radiation from dust is free-free emission from the ionized regions near the stars, since most of them do show H_α in emission. Comparing the flux density at 1300μm and the H_α equivalent width of all stars in the sample no correlation between the two quantities can be found, making it unlikely that there is significant contribution of free-free emission to the observed 1300μm flux density. This conclusion is also corroborated by VLA observations (Bieging et al. 1984) who surveyed 44 stars in Taurus for 6cm radio emission; the present sample includes 27 of these objects. 24 of them have 6cm flux densities less than about 500μJy whereas the 1300μm flux density is often above 100mJy. Even in an outflowing wind, free-free emission will not produce such steep spectra which implies again that there is no significant contribution besides thermal emission from dust.

Next it is of interest to investigate the spatial distribution of the dust particles around the star. A simple estimate comes from the conversion of the optical depth at 1300μm into a visual extinction: Taking 100mJy as a typical flux density at 1300μm and adopting 100K as an average dust temperature one derives about 12mag of visual extinction averaged over an area of 1500AU (beamsize of 11"). Obviously, any centrally peaked distribution, e.g. freely falling accretion onto the star or a uniform velocity stellar wind would raise this value by orders of magnitude.On the other hand, only a few of the objects in the sample have A_V as high as 3mag, and none have A_V greater than 7mag. Therefore it is likely that the dust grains are distributed very

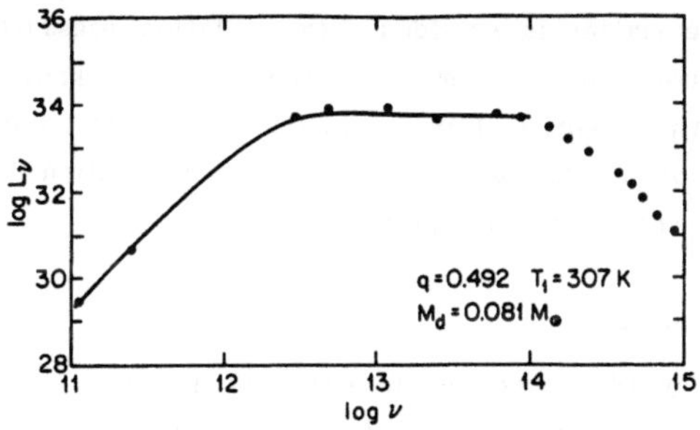

Fig. 2: The energy spectrum of HP Tau is shown to illustrate the two principle regimes of emission. The section between 3 and 100μm shows L_ν is independent of ν. The low frequency emission beyond 200μm is mainly optically thin.

non-uniformly around each star to account for both the low visual extinction and the high 1300μm flux density. The most natural distribution to account for all observations is a highly flattened disk. Because the optical depth within the disk will be rather high even for submm wavelengths one can exclude the rare case where the disk is viewed edge-on; such objects would not have been detected in optical surveys.

As an example Fig. 2 shows the observed and calculated spectrum of HL Tau. The solid line is a fit to the spectrum with $T(r) = T1(r/1AU)^{-q}K$ and the parameters indicated in the figure. The fit is excellent for wavelengths greater than 3μm, where flux from the stellar photosphere and extinction are unimportant. The temperature of the disk at any radius is well-constrained by the FIR emission. The dust grains are warm, ~300K at 1AU from the star. The disk mass is 0.081$M_\odot$ of total interstellar matter. In general, disk temperatures resulting from these fits are closely correlated with the 60μm flux density, demonstrating that the FIR flux is essentially a measure of the disk temperature, the other disk properties being largely irrelevant. Disk temperatures at 1AU exhibit a wide range, from a few tens of degrees to a few hundred degrees with an average value of 120K. The central stars are usually more luminous than 1$L_\odot$ and so the typical disk temperature is smaller than the equilibrium temperature for rapidly spinning small particles exposed to the radiation from the star at 1AU. This is probably even true for the hot disks, although there is considerable uncertainty in the stellar luminosity owing to the unknown contribution from the disk.

Disk masses range from approximately 0.001$M_\odot$ to 1$M_\odot$ or more, the lower limit set by the sensitivity of the observations and the upper value being very uncertain, depending strongly on the choice of unknown disk parameters to fit the spectra. The average mass of about 0.02$M_\odot$ is similar to that assumed for the early solar nebula. These masses refer to the total amount of gas, using a gas to dust ratio of 100. The principle uncertainty in disk mass is given by the choice of κ_ν, the dust opacity at submm wavlengths. A larger than normal value of κ_ν was used for this sample due to the obvious presence of larger than normal dust grains, so that the masses are unlikely to be lower than those derived here. There is certainly enough material to form planets around most of these stars and therefore planetary systems should be relatively common in the Galaxy.

The mass, size and angular momentum of these disks are similar to those associated with the solar nebula during the first stage of planet formation. Because the disks are spatially thin, most of the matter is shielded from radiation pressure and mass-loss from the stars, and it is likely that this matter can survive the known environments around T Tauris during their entire PMS evolution. Since no main sequence stars are observed with disks of $0.01M_{\odot}$ or more, the clear implication is that the circumstellar material must have condensed into planets.

4.2 Large particles around main sequence stars

One of the most spectacular discoveries by IRAS was the detection of an FIR excess around Vega, probably due to large circumstellar dust grains (Aumann et al. 1984). In order to study the evolution of circumstellar disks the submm emission from Vega and other main sequence stars with FIR excess, found during a systematic search of the IRAS Point Source Catalog (Aumann 1985) have been observed at the IRAM 30m telescope with the MPIfR bolometer system at wavelengths of 870 and 1300μm. Such submm observations are extremely difficult and - as a consequence - there are only a few sources which show evidence for circumstellar emission. On the other hand, for an interpretation of the observations it is fortunate that the angular distribution of matter need not be known in these optically thin configurations because i) The temperature of a dust grain and its emission spectrum depend solely on its distance to the star. ii) At least in the case of Vega there is information on the size of the disk from IRAS slit scans. Therefore a sophisticated model will, in particular, yield valuable estimates for the size of the grains.

In order to investigate the grain sizes and their distribution around the star radiative transport calculations as described in Equ.(2) and (3) of Chini et al.(1986b) have been performed. Because of the low optical depth, this practically reduces to calculating the temperatures of the grains in the unattenuated radiation field of the star and their subsequent emission. The grains are assumed to have a standard size distribution $n(a) \propto a^{-3.5}$ of Mathis et al.(1977). Such a distribution, which is equivalent to $dn(m) \propto m-1.8dm$, has been shown to arise from collisional fragmentation (Hellyer, 1970). Absorption and scattering efficiencies Q are calculated from Mie theory. This is not really required for the absorption of stellar light, where $Q_{abs} \approx 1$, but it is necessary to properly calculate Q_{abs} for the emission in the far IR. The optical constants of the dust material are taken from Draine (1985).

In the spectra of Fig. 3 observations for 4 main sequence stars are depicted by asterisks. A model value is explicitly shown by a cross only if it deviates from the observational point by more than 30%. Smaller differences cannot be clearly separated graphically in the figure and are irrelevant in view of the overall inaccuracies. The model always takes into account the observational beam width. The dust is distributed in the models in a spherical shell of constant density. As the configurations are optically thin, distribution of the dust in a disk with the same inner and outer boundary as the shell would result in an identical spectrum as long as the disk absorbs the same fraction of stellar light as does the shell.

The model fit for Vega reproduces the observations very well. The dust cloud lies between 40 and 74AU from the star. Such an extension is compatible with a size at 60μm of 20". The circumstellar dust consists of grains with radii from 18 to 4374μm.

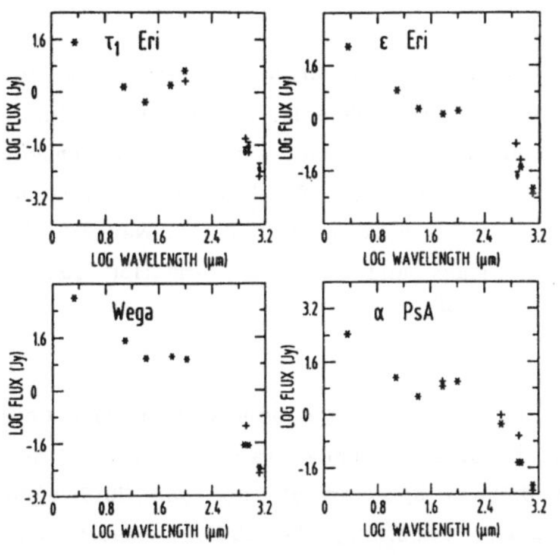

Fig. 3: Observed spectra (asterisks and arrows for upper limits) of main sequence stars with strong far IR excess. Plotted are the fluxes at 2.2μm, in the four IRAS bands and at 870 and 1300μm. The results of the models described in the text are presented by crosses if they deviate by more than 30% from the observations.

The grain temperatures are only mildly size dependent: at a given distance from the star the difference between the largest and smallest particles is less than 10%. The medium grain temperature falls from 140K at the inner edge to 90K at the outer. For a good agreement with the observations, the upper limit a_u of the grain size distribution may be reduced only slightly; reasonable fits can be found down to a_u=486μm. The corresponding mass is then 6.6 $10^{-9}M_\odot$. An open end to a_u, i.e. the existence of particles greater than 1cm, can, on the other hand, not be excluded from the models. They would not change the spectrum because they do not absorb much stellar light. For an $n(a) \propto a^{-3.5}$ distribution the bulk of the mass is locked up in the big particles, whereas the cross section is provided by the small ones. Therefore the dust mass and the upper grain radius are only lower limits. For details see Chini et al. (1989d).

Information on the 60μm size for α PsA and ε Eri comes from Gillett (1985) who gives different values for in-scan and x-cross direction; the numbers are compatible with the sizes of our models. The quality of the 0.87 and 1.3mm fluxes, although made with different beams, does not allow a reliable estimate of the extent of the sources. Our choice for the radius of the outer boundary of the dust cloud around τ_1 Eri is guided by the desire to obtain a reasonable spectral fit. Considering the accuracy of the fitting procedure grains greater than 1cm may also exist around α PsA and ε Eri.

One may summarize the results concerning the circumstellar matter around main sequence stars in the following way: The dust is distributed in a shell where the outer radius is two to three times larger than the inner radius. The total diameter is of order 100AU and seems to be correlated with stellar luminosity in the sense that the dust cloud is larger for brighter stars. Grain temperatures range from 50 to 140K. If the grains, like those of the interstellar medium, have a size distribution $n(a) \propto a^{-3.5}$ or steeper then the diameter of the largest grains $2a_u$ cannot be determined. Our models, where $2a_u$~1cm, give only lower limits to the total dust mass, typically 100 times less than the mass of the Earth. From the analysis of the spectra the presence of km-sized bodies is possible.

4.3 Dust in external galaxies

Despite the great progress submm astronomy has achieved during the past years, data for external galaxies are still very rare in literature. This is, on one hand, due to the extreme faintness of the sources but, on the other hand, also a consequence of their extended emission: weak gradients in the distribution of submm radiation cannot be detected by the differential observing procedure as described above. Nevertheless, some interesting relations between the gas mass and the star formation rate in galaxies have been obtained from existing submm data.

4.3.1 Nonactive spiral galaxies

For a long time M51 was the only spiral galaxy detected at submm wavelengths. Chini et al. (1986a) performed a systematical study of 26 extended galaxies, most of type Sb,c at 350 and 1300μm. All of them had been detected at the four IRAS bands showing increasing spectra from 12 to 100μm. Combining the 1300μm observations, made with a beamsize of 90", and the IRAS data taken from the Point Source Catalogue, the spectra could be interpreted as to originate from two dust components – one, fitting the part between 25 and 60μm with a color temperature of about 50K and another, describing the submm part with a rather cool temperature of 16K. However, the 350μm flux densities, observed with a 30" beam, were by a factor of ten below the value expected from a cool component of 16K, suggesting that the submm emission in these galaxies is extended. Eales et al. (1989) and Stark et al. (1989) also observed some of these galaxies at short submm wavelengths and suggested to fit the spectra between 60 and 1100μm by a single dust component of 30 – 50K. It is obvious that this controversial interpretation of the spectra has major influence on the derived masses in such a way that a single component model of higher dust temperature reduces the total gas mass by a factor of 10 – 100. In fact, there is a tendency that the mass of interstellar matter as determined by Eales et al. (1989) from their submm data is lower than the mass estimated from CO measurements.

The disagreement among the submm data may, at least partially, be due to varying observational procedures; all measurements were performed at comparatively small optical or infrared telescopes with different beam sizes and chopper throws. Undoubtedly, the spatial extent of these spiral galaxies, which is several arcminutes at optical wavelengths, and the low submm surface brightness makes the observations extremely difficult. As the investigation of IRAS maps has shown the size of extended spirals at 60μm is comparable to that in the blue light (Rice et al. 1988). Comparing the integrated IRAS flux densities for the sample of 26 galaxies with those from the Point Source Catalogue, it becomes evident that the energy distributions of the galaxies require significant revision. Likewise, a re-investigation of the submm emission and an enlargement of the sample is necessary for a better understanding of the amount and the temperature of interstellar matter in normal galaxies. Fig. 4 shows the spectral data as currently available for M51 and reflects the large uncertainties as described above.

Nevertheless, following the interpretation by Chini et al. (1986a) there seems to exists a relation between the total luminosity and the total gas content of the form L

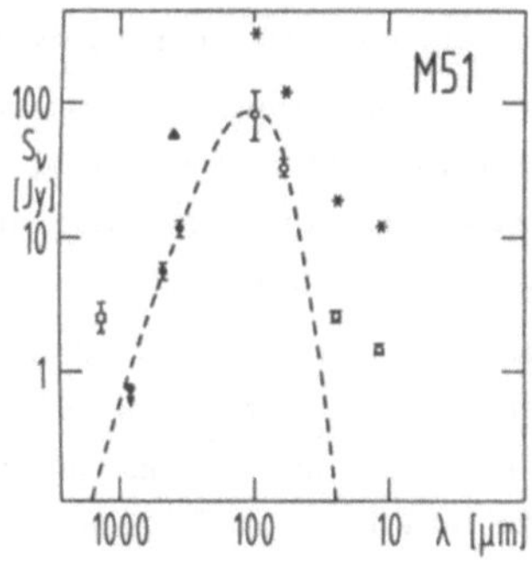

Fig. 4: FIR/submm energy distribution of M 51, demonstrating the difficulties described in the text. Open circles are taken from the IRAS Point Source Catalog, closed circles are observations by Eales et al. (1989); the dashed curve is the fit by Eales et al. of the form $\nu^2 B_\nu(T_d)$ with a dust temperature of 32K. This fit is in contrast to the observations by Chini et al. (1986) (open square) and Smith (1982) (triangle). The total FIR flux densities (asterisks) given by Rice et al. (1989) from integrating IRAS maps also suggest an energy distribution with higher flux densities.

$\propto M^{0.88}$ over the mass range from 3 $10^8 \leq Mg \leq 10^{12}$ $M_\odot$. As the luminosity is considered to be proportional to the star formation rate it seems that the creation of new stars depends almost linearly on the gas content of spiral galaxies; the efficiency of star formation seems to be four times higher in barred and peculiar galaxies than in Sb,c types.

4.3.2 Starburst galaxies

The problem of comparing FIR and submm observations, made with different beamsizes is less severe if one studies distant galaxies where one receives all, or at least a major fraction, of the emission. Krügel et al. (1988a,b) and Chini et al. (1989a) investigated an FIR flux limited sample of starburst galaxies at 870 and 1300μm and showed that their spectra from 60 to 1300μm can be described by emission from a single dust component of 30 ± 5K. Furthermore, these data suggest that the wavelength dependence of dust opacity ß = 2.0 ± 0.2. Converting the amount of dust into a gas mass, values between 10^8 and $10^{11} M_\odot$ were found. In an L vs. M_g diagram starburst galaxies cover the same mass range than normal spiral galaxies, however, the corresponding luminosity is by a factor of about 20 higher (Fig. 5). The reason

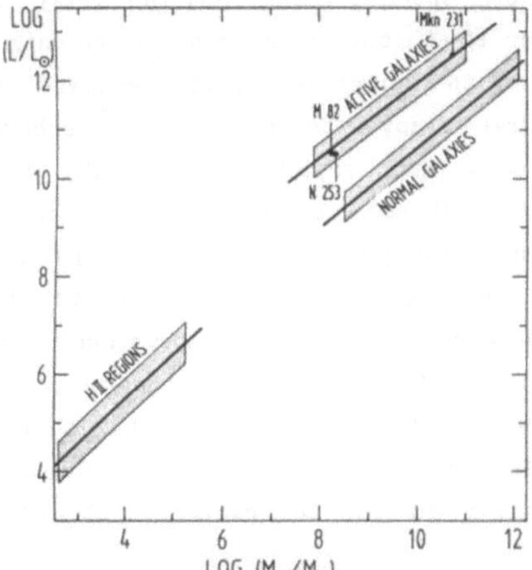

Fig. 5: The relation between total luminosity and gas mass for different classes of objects. The luminosity is derived by integrating the energy distribution from 0.3 - 1300μm, the gas mass is determined from 1300μm data according equation (1). The data for compact HII regions have been taken from Chini et al. (1986c,d).

for this increased efficiency of converting gas into stars is not yet clear but some possibilities are suggestive: i) There might be a higher star formation rate per unit gas mass or ii) the IMF has been shifted towards more massive stars. It should be noted that Krügel et al. (1989) studied the CO emission from these galaxies and derived gas masses which are comparable to those obtained from submm data.

The clear separation of normal spirals and starburst galaxies in terms of their L/Mg ratio is striking. Krügel et al. (1988b) argued that galaxies might exist in two stages, active and non-active. In this dynamical scheme, the gas swings in the gravitational field of the central stars like a piston: The inward motion is caused by gravitational attraction and proceeds through the dissipation of turbulent cloud motion. It is stopped when the gas density is high enough for rapid star formation to occur. The outward motion of the gas is driven by supernova explosions. The period of the burst is determined by the rate at which gas is replenished by mass loss from population II stars (or by the inflow rate of gas from the disk) and by the lifetime of the progenitor stars of supernovae. A typical period is 10^8yr, of which the activity stage comprises some 25% of the time. The transition between starburst and quiescent phase is less than 10% which makes it difficult to observe and might explain the gap between normal and active galaxies in Fig. 5.

4.4 Active galactic nuclei

There is little doubt that quasars do reside in the centers of galaxies. In fact, their spectra increase from optical to FIR wavelengths and thus resemble those of starburst galaxies. Therefore it is not clear where the emission observed by IRAS with comparatively low spatial resolution really comes from. During the controversial discussion about the origin of FIR emission from quasars evidence has grown that such spectra do contain thermal components due to dust radiation.

4.4.1. Radio-quiet quasars

Radio-quiet quasars probably show the most exotic spectra, rising from the X-ray to the FIR regime and showing almost no sign for significant radio emission. Recently, Chini et al. (1989b) presented 1300μm data of all 26 radio-quiet quasars observed by IRAS (Neugebauer et al. 1986). The steep spectral turnover between IRAS and submm data suggests that dust emission on kpc scale may explain most readily the observations in a satisfactory way. In such a thermal model the bulk of dust attains temperatures of about 35K similar to that in starburst galaxies (Chini et al. 1989a). Converting the observed dust emission into a gas mass one obtains 10^8 to $10^{10}M_\odot$ which is also typical for gas rich spirals. Independent HI and CO observations (e.g. Barvainis et al. 1989, Krügel et al. 1989) which confirm the results concerning the gas masses in radio-quiet quasars and starburst galaxies corroborate the thermal interpretation of the FIR/submm spectra. It is therefore likely that the FIR and submm emission from these objects originates from dust in the center of the host galaxies, heated by the active galactic nucleus.

4.4.2. Radio-loud quasars

The systematic study of the nature of radiation from quasars was continued by Chini et al. (1989c) who observed all quasars with steep and flat radio spectra and two or more positive detections in the IRAS bands (Neugebauer et al. 1986). Again, the 870 and 1300μm data played a key role in understanding the origin of FIR/submm emission in these quasars because i) a large gap of two orders of magnitude in wavelength without observations exists for most of the objects. ii) From extrapolating the radio data for steep spectrum quasars to 1300μm one expects extremely weak emission at that wavelength, a fact which on the other hand implies a sharp rise of the spectra towards 100μm.

Fig. 6 shows, as a typical example, the energy distribution of a steep spectrum quasar. The new submm data suggest two clearly distinct components for the steep spectrum sources, consisting of a smooth increasing part between 0.3 to 100μm and another between 1300μm and long radio wavelengths. Several flat spectrum sources also show a steep high frequency radio spectrum (see Fig. 7). The overall spectra from the radio range to the X-ray range of this sample of quasars differ from those of radio-quiet quasars significantly only in the fact that the radio emission is stronger. Because of this similarity, it is tempting to interpret the FIR component of all steep and many flat-spectrum objects again as due to hot dust in a gas rich galaxy, heated by an active nucleus. If one determines the temperature from observations at 60 and 100μm, one finds an average dust temperature of 40 ± 5K in the rest frame of the quasars. The corresponding gas masses are of order 10^8 to 5 $10^9 M_\odot$, the same range as already found for starburst galaxies and radio-quiet quasars by means of the same method. This consistency, which otherwise would be very difficult to explain, can be taken as further support for a thermal origin of the FIR component. In summary, the

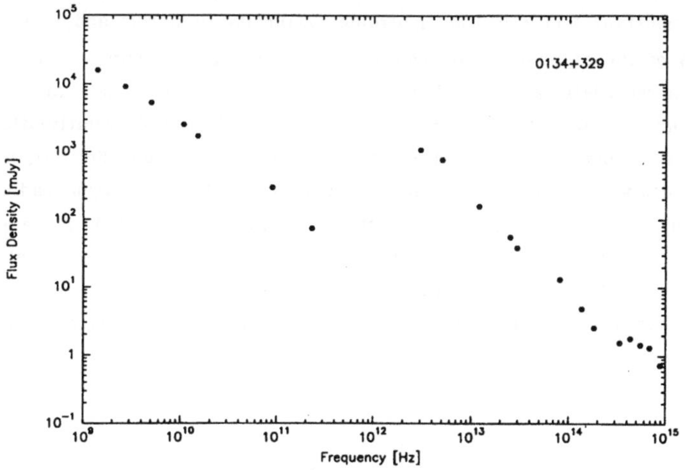

Fig. 6: The energy distribution of a steep spectrum quasar with a new observation at 1300μm. Obviously, the Synchrotron component continues into the submm regime, indicating a separate component at FIR wavelengths. A complete display of all available spectra for this class of objects is given by Chini et al. (1989c) together with the submm data.

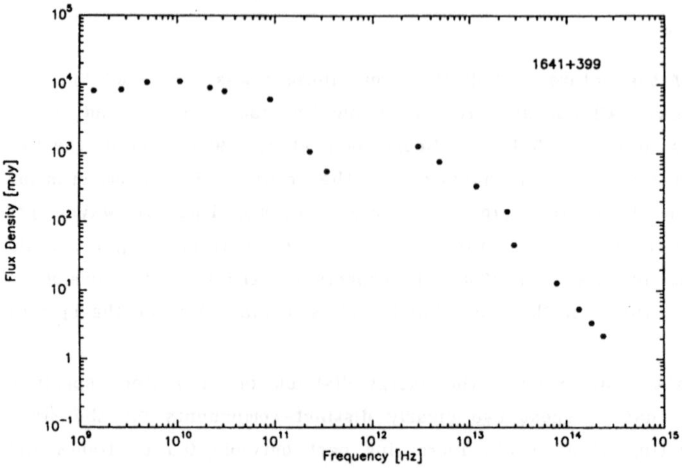

Fig. 7: The energy distribution of a typical flat spectrum quasar with new observations at 870 and 1300μm; more spectra and submm data are given by Chini et al. (1989c). The submm data clearly show a turnover of the flat radio component and suggest an FIR component of separate origin.

submm data corroborate earlier discussions of steep spectrum quasars that suggested already thermal emission from dust in these objects. The indication of a thermal component in some flat-spectrum sources is an unexpected new result obtained from the submm observations.

The submm data of quasars support the well established notion that the orientation of an active nucleus (jet, disk, extended radio lobes) with respect to the observer is an important factor (Barthel 1989) as concerns its appearance as a radio-loud or radio-weak object, as a radio-quasar or a radio-galaxy. Almost certainly, the second important factor is the isotropic luminosity (Fanaroff and Riley 1974). The asymmetries of the extended radio emission of many radio quasars are most readily explained by a moderate amount of relativistic boosting. For small angles between the jet axis and the line of sight to the observer, the (assumed) relativistic jet on one side dominates and can swamp the entire isotropic emission. For larger angles, the radiation becomes weaker and, if not dominated by steep spectrum radio lobes, can thus explain the remarkably large fraction of optically selected quasars with flat or nearly flat radio spectra and also the high fraction of radio-weak quasars with flat or nearly flat radio spectra. The submm data corroborate the argument that all quasars of high isotropic luminosity have a gaseous disk surrounding the active nucleus.

REFERENCES

Adams, F.C., Lada, C.J., Shu, F.H.: 1987, Ap.J. **312**, 788

Aumann, H., Gillett, F., Beichman, C., de Jong, T., Houck, J., Low, F., Neugebauer, G., Walker, R., Wesselius, P.: 1984, Ap.J. **278**, L23

Aumann, H.: 1985, PASP **97**, 885

Baars, J.W.M., Hooghoudt, B.G., Mezger, P.G., de Jonge, M.J.: 1987, A&A **175**, 319
Barthel, P.D.: 1989, Ap.J. **336**, 606
Barvainis, R., Alloin, D., Antonucci, R.: 1989, Ap.J. **337**, L69
Beckwith, S.V.W., Sargent, A.I., Chini, R., Gsten, R.: 1989, A.J. (submitted)
Beckwith, S., Sargent, A.I., Scoville, N.Z., Masson, C.R., Zuckerman, B., Phillips, T.G.: 1986, Ap.J. **309**, 755
Bertout, C., Basri, G., Bouvier, J.: 1988, Ap.J. **330**, 350
Bieging, J.H., Cohen, M., Schwartz, P.R.: 1984, Ap.J. **282**, 699
Booth, R.S., Delgado, G., Hagström, M., Johansson, L.E.B., Murphy, D.C., Olberg, M., Whyborn, N.D., Greve, A., Hansson, B., Lindström, C.O., Rydberg, A.: 1989, A&A **216**, 315
Chini, R., Kreysa, E., Krügel, E., Mezger, P.G.: 1986a, A&A **166**, L8
Chini, R., Krügel, E., Kreysa, E.: 1986b, A&A **167**, 315
Chini, R., Krügel, E., Kreysa, E., Gemünd, H.P. : 1989a, A&A **216**, L7
Chini, R., Kreysa, E., Biermann P.L.: 1989b, A&A **219**, 87
Chini, R., Biermann P.L., Kreysa, E., Gemünd, H.P.: 1989c, A&A (in press)
Chini, R., Krügel, E., Kreysa, E.: 1989d, A&A (submitted)
Cohen, M., Emerson, J.P., Beichman, C.A.: 1989, Ap.J. **339**, 455
Draine, B.: 1985, Ap.J. Suppl. **57**, 587
Eales, S.A., Wynn-Williams, C.G.: 1989, Ap.J. **339**, 859
Elias, J.H., Ennis, D.J., Gezari, D.Y., Hauser, M.G., Houck, J.R., Lo, K.Y., Matthews, K., Nadeau, D., Neugebauer, G., Werner, M.W., Westbrook, W.E.: 1978, Ap.J. **220**, 25
Emerson, J.P., Clegg, P.E., Gee, G., Cunningham, C.T., Griffin, M.J., Brown, L.M.J., Robson, E.I., Longmore, A.J.: 1984, Nature **311**, 237
Fanaroff, B.L., Riley, J.M.: 1974, MNRAS **167**, 31p
Gear, W.K., Brown, L.M.J., Robson, E.I., Griffin, M.J., Smith, M.G., Nolt,I.G., Radostitz, J.V., Lebofsky, L., Veeder, G.: 1986, Ap.J. **304**, 295
Gillett, F.C.: 1985, in F.P. Israel (ed.), Light on Dark Matter, Reidel, Dordrecht, p. 61
Hellyer, B.: 1970, MNRAS 148, 383 Hildebrand, R,H., Whitcomb, S.E., Winston, R., Steining, R.F., Harper, D.A., Moseley, S.H.: 1977, Ap.J. **216**, 698
Jaffe, D.T., Becklin, E.E., Hildebrand, R.H.: 1984, Ap.J. **285**, L31
Kreysa, E.: 1985, Proc. URSSI Intern. Symp. on MM-and Submm-Wave Radioastronomy, Granada/Spain 11.9. - 14.9., p. 153
Krügel, E., Chini, R., Kreysa, E., Sherwood, W.A.: 1988a, A&A **190**, 47
Krügel, E., Chini, R., Kreysa, E., Sherwood, W.A.: 1988b, A&A **193**, L16
Krügel, E., Steppe, H., Chini, R.: 1989, A&A (in press)
Mathis, J., Rumpl, W., Nordsieck, K.: 1977, Ap.J. **217**, 425
Neugebauer, G., Miley, G.K., Soifer, B.T., Clegg, P.E.: 1986, Ap.J. **308**, 815
Rice, W., Lonsdale, C.J., Soifer, B.T., Neugebauer, G., Kopan, E.L., Lloyd, L.A., de Jong, T., Habing, H.J. : 1988 Astrophys.J.Suppl. **68**, 91
Smith, J.: 1982, Ap.J. **261**, 463
Stark, A.A., Davidson, J.A., Platt, S., Harper, D.A., Pernic, R., Loewenstein, R., Engargiola, G., Casey, S.: 1989, Ap.J. **337**, 650
Strom, K.M., Strom, S.E., Edwards, S., Cabrit, S., Skrutskie, M.F.: 1989, preprint (University of Massachusetts)
Telesco, C.M., Harper, D.A.: 1980, Ap.J. **235**, 392
Thronson, H.A., Walker, C.K., Walker, C.E., Maloney, P.: 1987, Ap.J. **318**, 645

Atmospheric Variations in Chemically Peculiar Stars

R. Kroll

Institut für Astronomie und Astrophysik, Universität Würzburg, Am Hubland, D-8700 Würzburg, Fed. Rep. of Germany

The expression 'Chemically Peculiar Stars' was established by *Preston* in 1974 to group together the various types of stars with abnormal chemical composition on the upper main sequence. In this volume *Maitzen* (1989) gives a compilation of the variety of photometric and spectroscopic properties of all subtypes of CP stars. This paper deals exclusively with the classical Ap stars, stars in the spectral range from F to B that show magnetic fields, the CP2 stars in Prestons notation. In these stars we have the possibility to study the interaction of a stellar photosphere with a strong, large scale magnetic field.

1. Abundance Distributions

Most magnetic CP stars show spectral line variations with the rotational period, which is interpreted as a nonuniform distribution of elements at the surface.

Deutsch (1958, 1970) proposed a harmonic analysis method to reconstruct analytically the distribution of elements at the surface from the observed radial velocity and equivalent width variations. Only integral properties of a line are used in this method. *Pyper* (1969) started her abundance mapping by identifying spots at the surface with components in lines. While this approach uses the line profile information, the attachment of the line components introduced some arbitrariness to the method. Similar work was done by *Megessier* (1975). *Khokhlova* (1975) performed a trial and error analysis where an initial guess of an abundance pattern was iterated until the computed line profiles matched the observed ones. Ten years later *Khokhlova* (1985) used a regularizing algorithm method to do the abundance mapping. Finally, *Vogt et al.* (1987) presented a maximum entropy method to find surface detail in stars from line profile variations.

Reviews in Modern Astronomy 2 Editor: G. Klare

Any of these works used a model atmosphere as a base for the abundance mapping that is uniform in its *physical* structure. This seems justified in a first approximation. The intensified metal spectrum gives rise to the backwarming effect, induced by line blanketing. This mechanism steepens the temperature gradient in the atmosphere, but it conserves the effective temperature. If the magnetic field is dipolar, as suggested by the observational work of *Borra & Landstreet* (1977), it is force free and will hence not change the hydrostatic structure and the pressure gradient in the atmosphere.

A slightly closer look shows that this assumptions are problematic. A locally changing metal content introduces a locally changing temperature stratification. This alone would result in different excitation conditions, and different line profiles hence, but even worse, we would have different temperatures at the same geometrical depth. The result would be horizontal energy transfer, so even a change in the local effective temperature is possible.

If the magnetic field deviates from a potential field geometry, as the dipole, the hydrostatic structure is changed by magnetic forces. This scenario has been investigated by *Stepién* (1978) and *Carpenter* (1985). They find that a magnetic force would show its influence as a locally changing effective surface gravity. Even moderate deviations from a dipole could change the effective log g by more than a factor of two, depending on the magnetic field configuration and aspect.

Therefore it seems advisable to check observationally, whether horizontal homogeneity of the thermal and pressure structure in the atmospheres of Ap star might be assumed, or deviations from that must be considered.

2. Temperature Variations

The spectra of CP stars are affected by the steepened temperature gradient and a flux redistribution from the ultraviolet to the visual, due to the crowded lines in the short wavelength region. Most temperature indicators, prooven for normal main sequence stars, like photometric colors, line ratios etc., fail in the case of CP stars. The most reliable method to determine effective temperatures is that of integrated fluxes, established by *Shallis & Blackwell* (1979). However, to apply it, calibrated data from all wavelength regions, especially the ultraviolet are needed. Several attempts have been made to connect photometric indices to effective temperature, *Lanz* (1985) suggest the Geneva $(B2\text{-}G)_0$ index, whereas Mégessier (1988) proposes Strömgren [u-b]. Such methods are valuable for statistical works, but applied to individual stars, the variety of abundance patterns may introduce large uncertainties.

To check local temperature constancy, it is however not necessary to monitor the *effective* temperature, any indicator will do, if it is independent of the chemical composition of the atmosphere. *Kroll* (1987) has shown that the infrared colors of CP2 stars are normal from the J-Filter at 1.25μ up to the longest wavelengths that could be observed with the IRAS satellite. This shows that flux redistribution from the ultraviolet reaches not beyond the visual range. But a little caution is necessary. The flux at a specific wavelength is given by the temperature of a representative layer, say at $\tau = 1$. This point in turn depends on the opacities and hence on the metal abundance. A model calculation, done with Kurucz' program ATLAS8 illustrates the situation. Fig. 1 shows the temperature versus the optical depth at the infrared wavelength of 1.65μ for a model with solar composition and one with ten times the solar metal abundance. At $\tau = 1$ the tem-

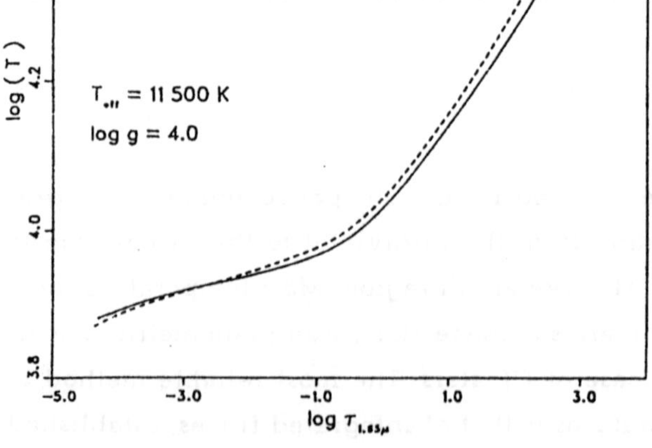

Fig. 1 : Temperature versus optical depth at 1.65μ for model atmospheres with T_{eff} = 10 000 K and log g = 4.0. Solid line : solar metal content, dashed line : ten limes solar metal content

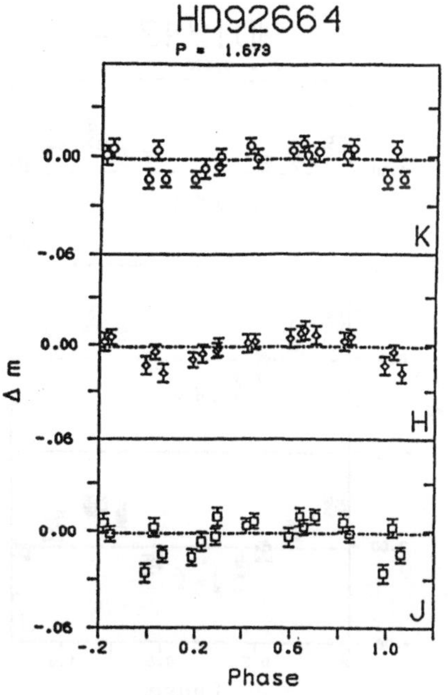

Fig. 2 : IR lightcurve of HR 4185

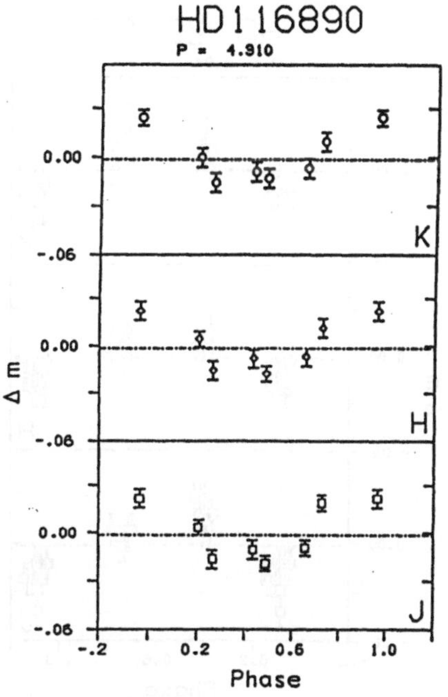

Fig. 3 : IR lightcurve of HR 5066

perature difference amounts to about 300K. We may regard this as an upper limit, because much larger variations in the total metal content would be seen in the spectra, which is not the case.

F. A. Catalano and I have collected IR observations of magnetic Ap stars at the 1m photometric telescope at ESO, La Silla. All of the stars have known photometic periods, so it can be easily checked, whether IR variability occurs or not. We have observed in the J,H and K bands, at wavelengths of 1.25μ, 1.65μ and 2.2μ, respectively. Fig. 2 gives an example of the reached accuracy. HR 4185 (HD 92 664) shows a small, but definite variability with an amplitude in the order of 0.01 mags. Somewhat larger is the amplitude in the case of HR 5066 (HD 116 890). In both stars the line blanketing mechanism, as discussed above, may explain the observed variations.

The star α Dor (HD29305) has a period very close to three days. Fig. 4 shows that the peak to peak variation is about 0.05 mags. Even larger - about 0.06 mags is the peak to peak variation in CU Vir (HD124224, Fig. 5). This star is of special interest, because it is perhaps the fastest

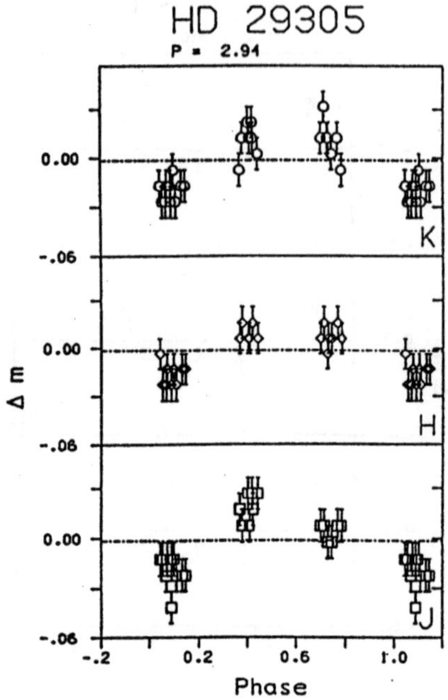

Fig. 4 : IR lightcurve of α Dor

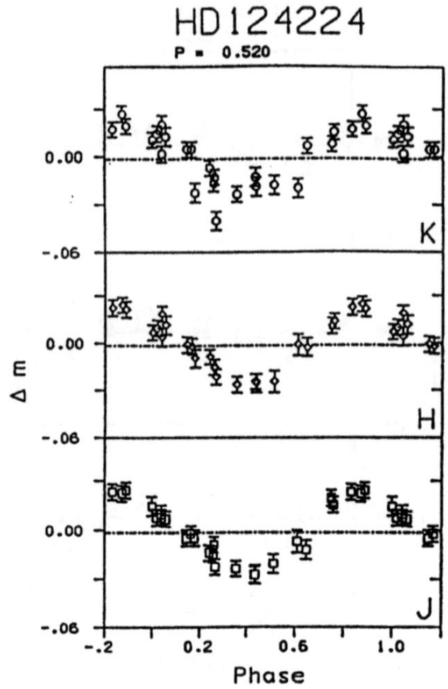

Fig. 5 : IR lightcurve of CU Vir

rotating magnetic Ap star known, with vsini = 100 km/s. In all three filters the variation has in any case the same amplitude and the same phase, which is consistent with a thermal variation in the atmosphere. Pure line blanketing is not sufficient to explain thermal variations of that amount. We must therefore consider effective temperature in the atmosphere of magnetic Ap stars as a *local* quantity. Horizontal energy transport is a possible explanation for this behaviour.

3. Pressure Variations

Looking for pressure variations is more complicated than for thermal variations because there is no simple photometric way to detect them. The H_β index may be used, but the interpretation is ambiguous, because of metallic lines around H_β. But the *profile* of the H_β line contains the information we look for. For stars earlier than about A0, the influence of electron pressure, which may be expressed as effective log g in models, is dominant to that of temperature on the line formation. Further, Balmer-profiles are at a large extent independent from the metal abundance in

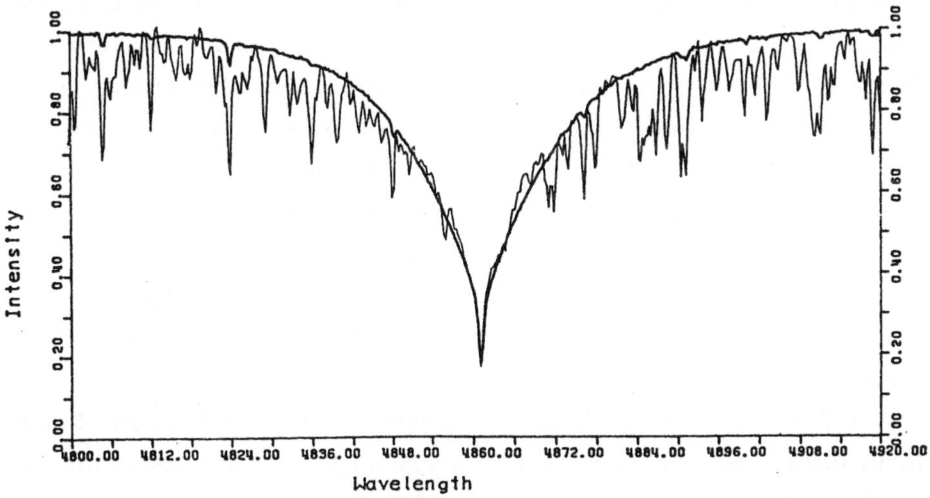

Fig. 6 : Spectrum around H_β of Vega (thick line) and the CP2 star β CrB (thin line)

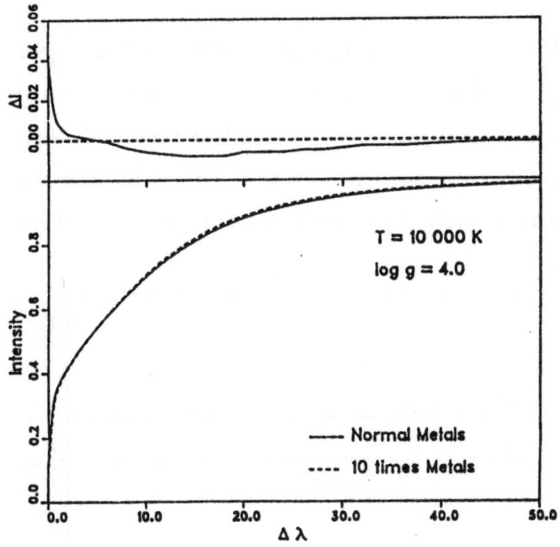

Fig. 7 : Theoretical H_β profiles for a star with solar metal content (solid line) and ten times solar metal content (dashed line). The upper part shows he difference of both profiles

the atmosphere. This can be seen in a superposition of the two spectra in Fig. 6 , one is the H_β line of Vega, the other that of the outstanding CP star β CrB. The metallic spectrum of the two stars is dramatically different, but the H_β line remarkably similar.

The similarity of the H_β line of metal rich and normal atmospheres is also seen in Fig. 7, which compares two profiles calculated with Kurucz' programs ATLAS8 and BALMER, one with an atmosphere with solar metal content, one with ten times more metals. The difference between the two profiles never exceeds 2% of the local intensity except for the very core of the line, were it reaches 4%. NLTE, rotation and other effects would presumably dominate in the profile, so this would be hardly observable.

It has been tried in the past to look for variations in the hydrogen line profiles with photographic observations. Though sometimes large variations have been claimed, none was verified. The drawbacks of the photographic calibration process might have deceived such variability. A photoelectric observing technique seems necessary to monitor such variations.

The 2.2m telescope at Calar Alto together with a CCD in the Coudé focus turned out to be a very useful system to do such observations, I could use this instrumentation in late 1987. For ten stars phase covering observations of the H_β profile with a dispersion of 17 A/mm (0 .2 A/pixel) could be collected.

The integral properties of the line - equivalent width and Doppler width - are not suitable to study variability. Uncertainties in determining the local continuum would cause additional errors and the spectral resolution would be lost. The method I used assumes that if there is line variation, the fundamental frequency will dominate. Hence sine fits were

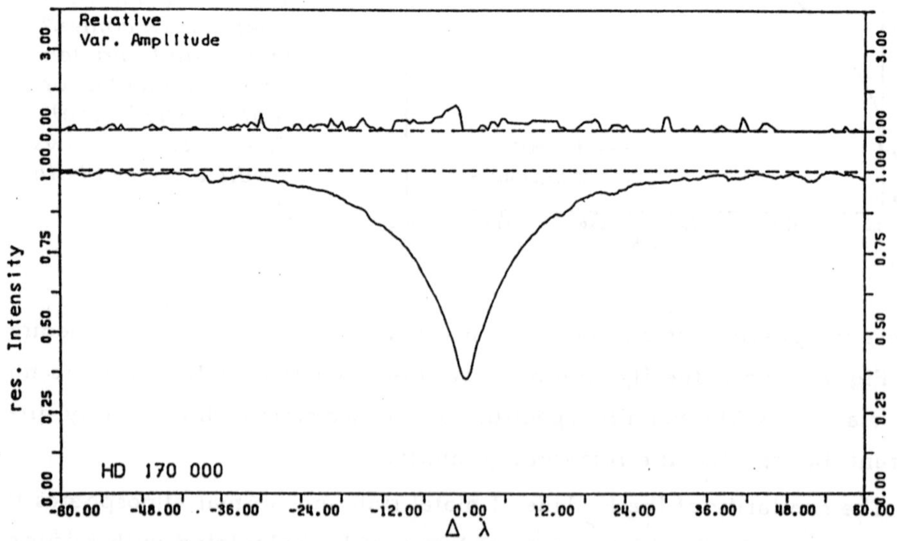

Fig. 8 : Mean H_β profile of φ Dra together with the local variability amplitude. The line is constant

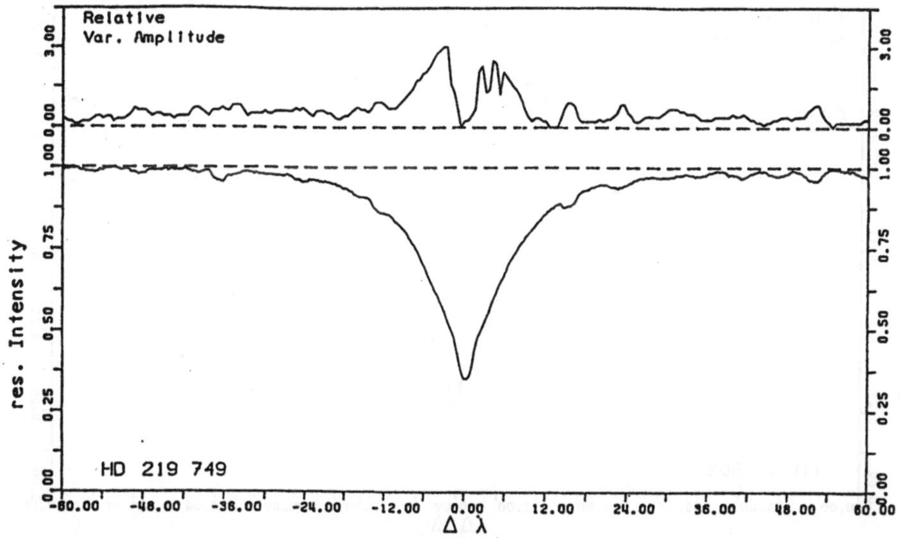

Fig. 9 : Mean H_β profile of HR 8861 and local variability amplitude

tried to the spectra at fixed wavelengths. The relative amplitude is plotted in Figs. 8 to 11 together with the mean H_β profile.

An example of a constant profile shows Fig. 8 . The star HD 170 000 (φ Dra) was suspected be *Musielok* (1984) to have a variable H_γ line. At least for` H_β such variability cannot be confirmed.

The stars HD 219 749 (HR 8861, Fig. 9) and HD 11 503 (γ Ari S, Fig. 10) show profile variability that reach 3% or almost 5%, with a maximum approximately 10 A from the center of the line. The amplitude then drops again to the line core. This pattern is also seen in HD 19 832 (56 Ari, Fig 11), although with smaller amplitudes.

The variability of H_β in magnetic Ap stars seems to have a characteristic f ingerprint: rising amplitude in the wing of the line, drop to the avery center. If the line profile is insensitive to the metal content of the satmosphere, as shown, it depends only on local temperature and electron pressure. It is informative to see, how theoretical models with varying temperature or varying pressure (equivalent to log g in terms of models) behave. Fig. 1 2 shows the H_β variation of a model star with a constant log g = 4.0 and effective temperature varying from 10 000 to 10 400 K, consistent to what the IR observations allow. The line variation is less than 2% all over the line, rises in the wings and stays almost constant up the the core of the line.

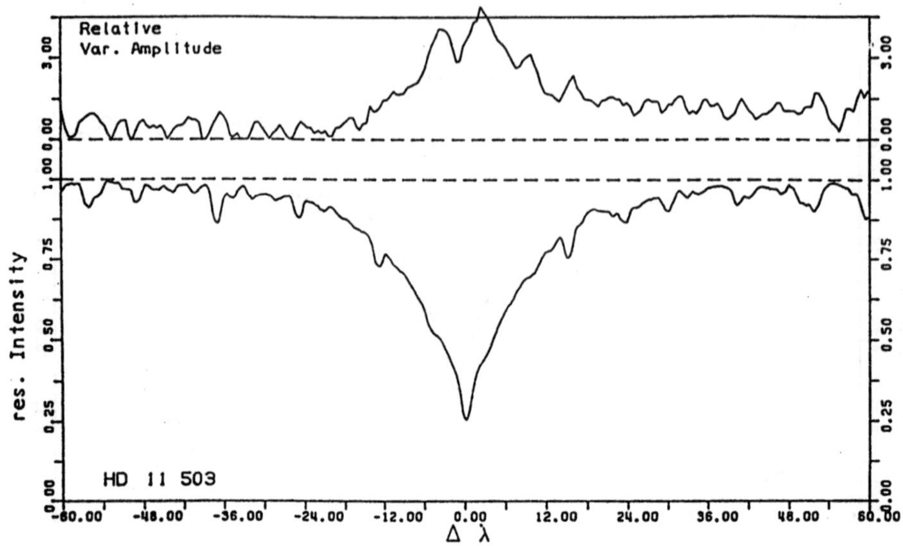

Fig. 10 : Mean H_β profile of γ Ari S and local variability amplitude

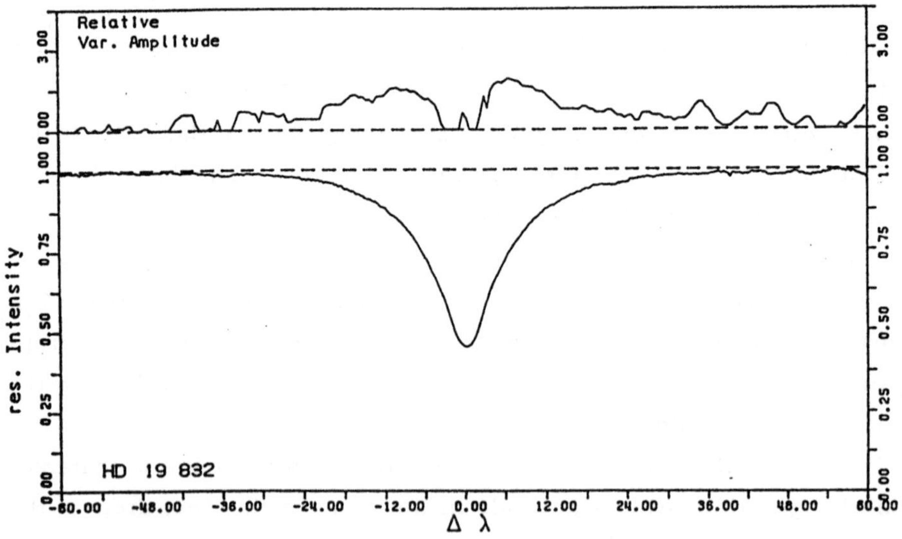

Fig. 11 : Mean H_β profile of 56 Ari and local variability amplitude

Very different from that is the H_β variation of a model star with constant effective temperature (10 000 K), but a changing log g. If we introduce a variation in log g from 3.6 to 4.0, we get, as seen in Fig. 13, a variation in H_β up to about 4%. Moreover we find the maximum of the variability amplitude about 10 A from the center and a significant drop to the core, closely matching the observed variations.

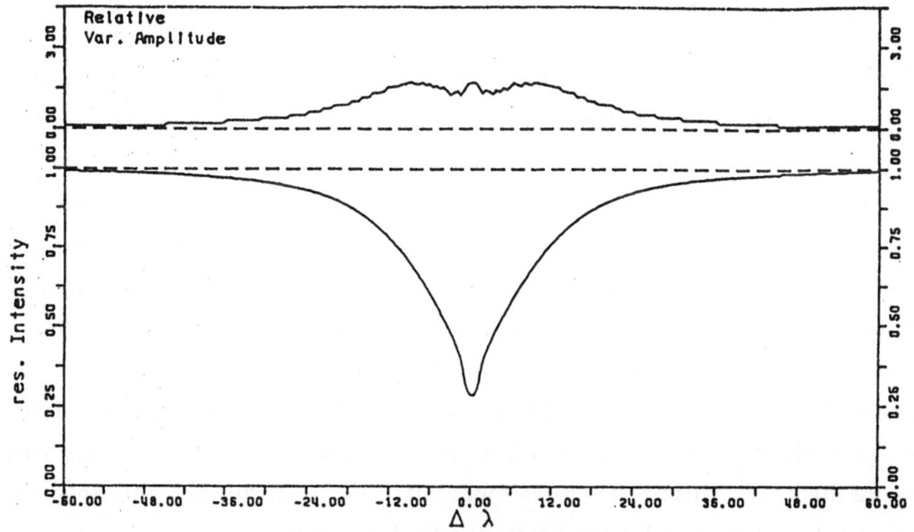

Fig. 12 : H_β variability of a model with log g = 4.0 and changing effective temperature from 10 000 to 10 400 K

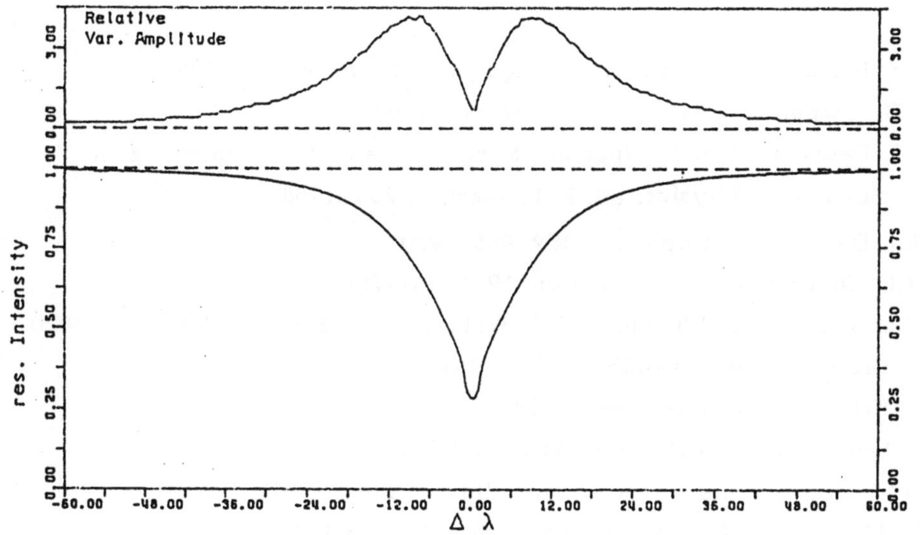

Fig. 13 : H_β variability of a model with T_{eff} = 10 000 K and log g changing from 3.6 to 4.0

The similarity of the observed variations with variations in models with varying log g indicates that we actually have to face local pressure local. This may be understood as evidence for the presence of non force-free magnetic fields.

4. Future Perspectives

Until now, there are no stars, for which we have both, H_β *and* IR measurements. This would be extremely valuable, especially if it could be connected to the magnetic aspect, if the magnetic field variation is known.

The most optimistic forecast of the resolution reached with the completed VLT are .1 milli arc seconds in the visual. The probably closest Ap star is α Cir, at a distance of approximately 20 pc. Assuming a radius of three solar radii, it will show a disk of 1 milli arcsec. Seeing surface structures directly therefore seems not totally impossible within the next ten years. However, the spots in the atmospheres of Ap stars are *opacity* spots, so they will have much less contrast than sunspots, for instance. This will make the observations of course more difficult, it might be that, unfortunately, the VLT will be too small for that purpose.

References

E.F. Borra, J.D. Landstreet : Astrophys. J. Suppl. **42**, 421 (1980)

K.G. Carpenter : Astrophys. J. **289**, 660 (1985)

A.J. Deutsch : IAU Symposium 6, Electromagnetic Phenomena in Cosmical Physics, ed. B. Lehnert, p. 209 (1958)

A.J. Deutsch : Astrophys. J. **159**, 985 (1970)

V.L. Khokhlova : Soviet Astron. **19**, 576 (1975)

V.L. Khokhlova, J.B. Rice, W.H. Wehlau : Astrophys. J. **307**, 768 (1986)

R. Kroll : Astron. Astrophys. **181**, 315 (1987)

R. Kroll : Thesis, Göttingen (1988)

T. Lanz : Astron. Astrophys. **144**, 191 (1985)

H.M. Maitzen : Rev. Mod. Astron, this volume (1989)

C. Mégessier : Astron. Astrophys. Suppl. **72**, 551 (1988)

B. Musielok : Acta Astron. **34**, 387 (1984)

G.W. Preston : Ann. Rev. Astron. Astrophys. **12**, 257 (1974)

D.M. Pyper : Astrophys. J. Suppl. **18**, 347 (1969)

K. Stepién : Astron. Astrophys. **70**, 509 (1978)

S.S. Vogt, G.D. Penrod, A.P. Hatzes : Astrophys. J. **321**, 496 (1987)

Chemically Peculiar Stars of the Upper Main Sequence

H.M. Maitzen

Institut für Astronomie, Universitätssternwarte,
Türkenschanzstr. 17, A-1180 Wien, Austria

1. INTRODUCTION

At the beginning it seems necessary to outline briefly the problem of terminology caused by the historical growth of knowledge on these stars. Students who want to enter bibliographical activities will be first confronted with the question where to look for them e.g. in *Astronomy and Astrophysical Abstracts*. Currently, they have to be told that information is contained under the following entries:

Am Stars, Ap Stars, Bp Stars, Helium-Rich Stars, Helium-Weak Stars, Hg-Mn Stars, Peculiar Stars, Silicon Stars, Spectrum Variables and as of 1987/2 also *CP Stars* (=chemically peculiar stars) which is the proper abbreviation for the whole group of the fore-going list of upper main sequence stars with chemical peculiarities. Quantitatively the largest body of publications relates the Ap Stars in the classical sense, i.e. the group the prototype of which - α^2 Canum Venaticorum - had been detected already at the end of the past century by Maury (1897).

The sequence of peculiarity detections of α^2 CVn was: peculiar line strengths - variations of line strengths - quadrature of line variability with radial velocity changes - photometric variability with same periodicity and coincidence of extrema. This way by 1914 the main characteristics of this classical Ap star were already known, except the existence of a global strong magnetic field which was first discovered for a member of this group, 78 Vir, by Babcock (1947).

Morgan (1933) showed in his spectral classification study that besides the group of Ap stars which share the properties of α^2 CVn (but separate into color dependent subgroups, the most numerous one being the Silicon stars) another group of peculiar stars has to be established, i.e. the Mercury Manganese stars, which differ from the α^2 CVn stars by their non-variability (a few mildly variable objects were reported by Schneider, 1987) aside from their spectral features.

Next in the historical record come the Am Stars (Titus and Morgan, 1940) with their discrepancy between Ca and metallic line spectral types and nonvariability. Like the HgMn stars Am stars are non-magnetic.

It had been recognized already at that time that the appearance of spectral peculiarities was not restricted to spectral type A, but extended to at least late B-type stars when dealing with Silicon and HgMn stars whence the term 'Bp' was introduced producing some confusion because peculiarities among earlier B-stars might not be related to the underlying phenomenon. Some authors prefer therefore to use the more complicated term *Bp-Ap* stars in order to avoid ambiguity.

Investigations on two further peculiarity groups, the Helium strong and Helium weak stars started much later by MacConnell et al. (1970) and Garrison (1967), resp.; this way

Reviews in Modern Astronomy 2 Editor: G. Klare

nearly the whole range of B-type stars was covered with peculiar objects. In order to emphasize the common characteristics, e.g. the fact that spectral viz. chemical peculiarities relate only the surface of the stars, Preston (1974) introduced the new term 'CP' and proposed four subgroups:

CP1 = Am stars, CP2 = magnetic (or classical Ap) stars, CP3 = HgMn stars and CP4 = He-weak stars. This is why many researchers in this field prefer to speak about CP-stars. On the other hand, a weak point in this grouping is the inadequate treatment of the hot Bp stars. Recognizing that Preston's classification scheme has a touch of an odd-even effect with respect to the existence of magnetic fields (CP1 and CP3 nonmagnetic, CP2 magnetic) and taking into account that in both groups of Helium abnormal stars objects with strong magnetic fields had been found (Borra and Landstreet, 1979; Borra et al. 1983), Maitzen (1984) proposed an extension of Prestons CP-scheme (CP4 and CP5 for He-weak and CP6 and CP7 for He-strong stars).

No explicit subject heading is available at present for a group called λ Bootis stars, which have attracted considerable interest only recently (see eg. Baschek and Slettebak, 1988). They have to be looked up under 'peculiar stars' and are characterized by the weakness of their metallic lines, hence a spectral appearance rather opposite to CP1 (=Am) stars.

2. MEETINGS, REVIEW PAPERS, CATALOGUES

Hallmarks of the evolution of knowledge on CP-stars were the big meetings (together with their proceedings) at Vienna 1975 (IAU Coll. No. 32), Liège 1981 (23th Colloque de Liège) and Crimea 1985 (IAU Coll. No. 90). Two other recent meetings deserve attention: on 'Magnetic stars' at Nizhnij Arkhyz in Oct. 1987 and on 'Elemental abundance analyses' at Lausanne in Sep. 1987.

Regular meetings of workshop type are held by the European Working Group on CP-stars short before the ESO-deadline for application for observing time. The bulletin 'A peculiar newsletter' has been issued roughly twice a year (editors H.Hensberge and W.van Rensbergen) since the founding of this group in 1979 and serves also the IAU working group on Ap stars.

Aside from review papers in the proceedings of these meetings two further recent reviews should be mentioned: the book 'The A-type stars: problems and perspectives' with exhaustive treatment of all aspects of peculiarity (including Bp-stars) by S.C.Wolff (1983) and the article 'Recent progress in CP stars detection and classification' by R.Faraggiana (1987).

Most comprehensive is the catalogue work accomplished by P.Renson at the Institut d'Astrophysique de Liège since many years. It contains more than 6000 stars, half of which are CP1 objects. This catalogue together with a list of bibliographical references is still privately circulated, but will soon become available through the 'Centre des Données Stellaires' (Strasbourg).

3. SPECTRAL RANGE AND TEMPERATURE DOMAIN OF CP STARS

As outlined in Section 1 the spectral range was progressively enlarged from peculiar A to B-type stars. The hot limit of the CP-phenomenon seems to be at temperatures ($T_e = 29000K$) where mass loss is not yet very large, whereas the cool end is marked by spectral types where convection is not very important ($T_e = 7000K$). This domain represents the appearance of stable atmospheres as prerequisite for the development of chemical peculiarities.

4. RELATIONSHIP BETWEEN PECULIAR AND NORMAL STARS

It is important to ask whether the peculiar stars form a population well separated in their behaviour from normal stars, since this should be a remarkable constraint for theories explaining the origin of peculiar stars. Evidence has accumulated that there is a continuous transition of peculiarity degrees from the normal level. Maitzen and Vogt (1983) show (see Fig.1) that the distribution of their photoelectric measurements of peculiarity (using the Δa-index introduced by Maitzen, 1976) for a large sample of CP2-stars detected spectroscopically by Bidelman and MacConnell (1973) has a rather sharp cutoff at the detection level for peculiarity, especially for the hotter CP2-stars. This means that stars are lacking in the transition zone to normality just because the spectroscopic technique (based on classification dispersion) did not discover less peculiar stars.

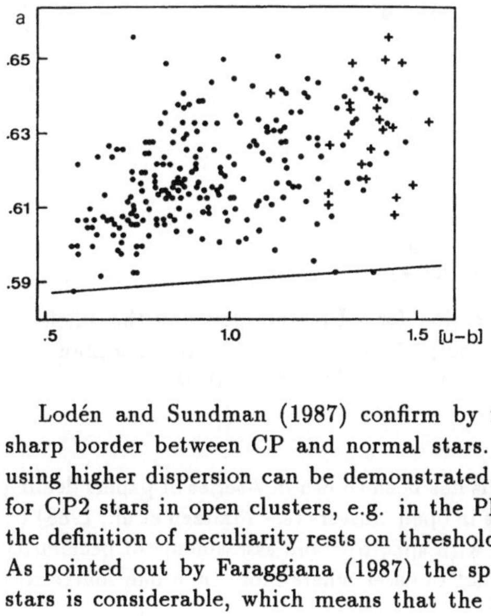

Fig.1: The photometric peculiarity index a vs. $[u-b]$ for hot CP2 stars (from Maitzen and Vogt, 1983). Units are magnitudes. Straight line represents normal stars.

Lodén and Sundman (1987) confirm by fine-classification recently that there is no sharp border between CP and normal stars. That new peculiar stars can be found by using higher dispersion can be demonstrated in a series of papers devoted to the search for CP2 stars in open clusters, e.g. in the Pleiades (Abt and Levato, 1978). Therefore, the definition of peculiarity rests on threshold levels defined by instrumental limitations. As pointed out by Faraggiana (1987) the spread of metallicities among normal A-type stars is considerable, which means that the processes in the atmospheres are not fully understood.

5. DETECTION AND FREQUENCY

If one wants to investigate the frequency of peculiar stars it is necessary to define a level above which peculiarity is assumed. In spectroscopy this is usually done by using a specific dispersion, mainly classification dispersion (about 100 A/mm and IIaO). The spectroscopic detection technique suffers from a number of drawbacks: it is rather time consuming concerning both the observation and the evaluation and it depends on the subjective capability of classifying people whether a star will be regarded as peculiar or

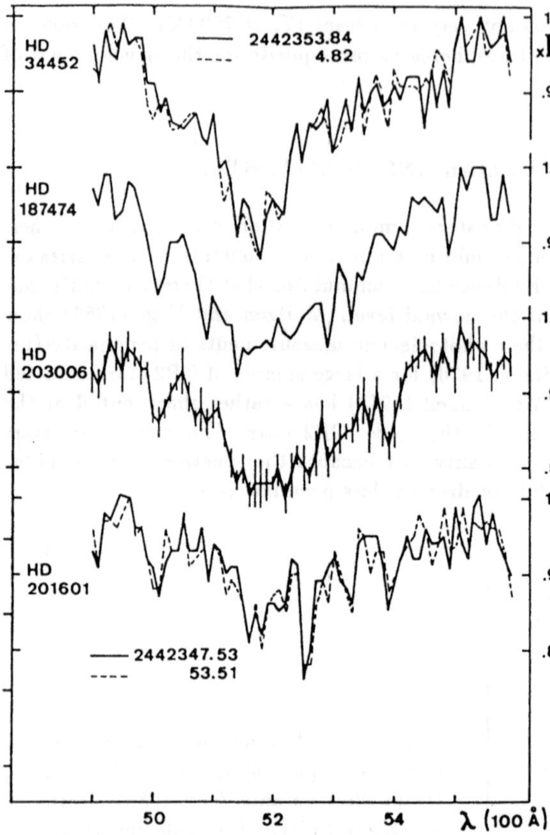

Fig.2: Spectrophotometry of the λ5200 region for CP2 stars covering the whole temperature domain (top: hot, bottom: cool). Ordinates refer to the continuum set to 1 for the bordering intervals. From Maitzen and Muthsam (1980).

not, if its peculiarity is not outstanding. This has been shown in a series of papers dealing with the photometric detection of CP2 stars in open clusters (see Maitzen et al., 1988) by comparing the impersonal Δa-index values with spectroscopic assessments of peculiarity (which by the way were contradictory in a lot of cases where more than one source was available).

Photometric detection of peculiarity resides on broad band features in the spectra of CP-stars, most prominent being those at λ4100 (CP2 and CP1 stars) and λ5200 (only CP2 and part of the CP4 stars). The latter is displayed in Fig. 2. CP3 and the hotter Helium abnormal stars cannot be detected this way. The UBV filters are too broad for efficient detection of CP-stars. Better results can be obtained with the Strömgren *uvby* system: the hotter CP2-stars (=Silicon stars) lie to the left of the hot branch in the c_o vs. $b-y$ diagram. This 'bluening' of $b-y$ is due to the influence of the λ5200 depression on the y-magnitude. However, this diagram applies only for unreddened stars, because the dereddening procedure produces $b-y$ for normal stars which are redder than peculiar ones with the same c_o index.

The influence of the enhanced line blocking around λ 4100 on the m_1 index of the Strömgren system is another tool to pick out CP2, but also CP1 stars. This technique had already been used in the pioneering study by Cameron (1967).

A photometric index specially tailored for the λ5200 depression has been devised by Maitzen (1976). He compared the flux at the center of the depression (filter measurement g_2) with the mean of y and g_1 (located at λ5015). The resulting values increase slowly from normal unreddened B to A-type stars defining the level of normality. The Δa index is defined as deviation from this normality level at the concerning $b - y$ and is significantly positive for CP2 stars, but not for nonmagnetic CP-stars (which may have only a weak statistically positive Δa). Maitzen (1984) has shown that also the cooler CP4-stars, if magnetic, exhibit moderate positive Δa values. Maitzen and Vogt (1983) have amply demonstrated the capability of Δa for singling out CP2 stars. They compared its performance also to the peculiarity index $\Delta(V1-G)$ of the Geneva photometric system which through the wavelength of V1 samples much of the λ5200 depression. A combination of the Geneva system with the Vilnius system has been proposed by North et al. (1982) which could yield a better effect for λ5200 since Geneva-$V1$ is replaced by Vilnius-Z (5160 A).

Δa-photometry has meanwhile been applied to an unprecedentedly large sample of open cluster stars (in 70 clusters, 33 published - see Maitzen et al. 1988) in order to survey the frequency of magnetic chemically peculiar stars in the approximate spectral range B5 - A5 above a given peculiarity threshold, i.e. 0.012 mag in Δa. An example for such a search within an open cluster is given in Fig. 3. A comparison with the CP2-objects in the catalogue of CP-stars in open clusters of Niedzelski and Muciek (1988) compiled from the literature (spectroscopic and photometric sources) results in significantly lower numbers of Δa peculiar stars which can be easily explained by the unprecise definition of spectroscopic peculiarity threshold and classification errors.

Evaluation of the cluster Δa survey is still under way, but a preliminary inspection of the available data does not support the finding of Abt (1979) claiming an increase with age of CP2 stars in open clusters. Critical points in his study are the small number of clusters (n=14) and the use of two different dispersions for spectroscopic detection of peculiarity.

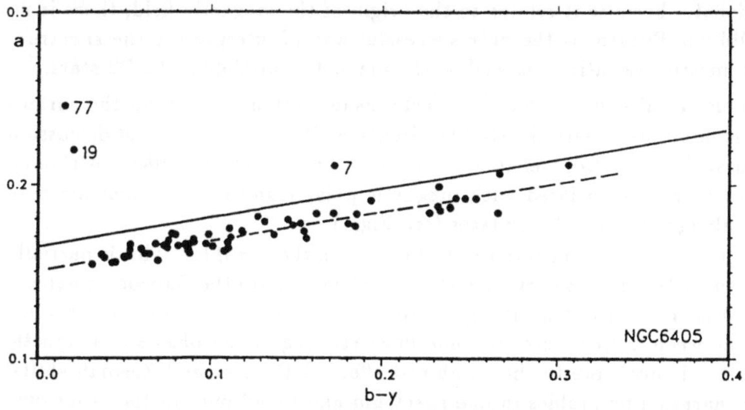

Fig.3: Search for CP2-stars in NGC 6405. Dashed line is the locus of normal stars, full line indicates the peculiarity threshold. From Maitzen and Schneider (1984).

Our data residing on a fixed peculiarity threshold exhibit an average frequency of about 6 percent CP2 stars in the concerning temperature domain, which is slightly less than the average for field CP2 stars in Wolff (1983). There is an indication of intrinsic differences of the CP2 content from cluster to cluster.

Finally, one can hope to study clusters at larger distances in different galactic fields, with the progress of the CCD-technique even out to the Magellanic clouds, in order to compare the appearance of CP2 stars in regions with different metallicities and large scale magnetic fields.

A new field for the application of the Δa index has recently opened by the study of Maitzen and Pavlovski (1989). They found that despite the problem of proper spectroscopic definition of the group of λ Bootis stars (see e.g. Faraggiana, 1987) their Δa values are predominantly negative to such an extent that this index will be used for detection studies and one can hope to consolidate the group together with more spectroscopy.

6. VARIABILITY

6.1 SPECTROSCOPIC

Deutsch (1947) performed an exhaustive study on spectrum variations of CP2 stars which turned out to have periods typically of several days and were most prominent in the lines enhanced in CP2 stars (Sr, Cr, Eu, to a lesser extent Si). Out of phase variations (e.g. Cr, Eu in HD 125248) occur in some stars. The great diversity of CP2 spectral behaviour has become apparent since.

Line variations in quadrature to the radial velocity variations can be successfully explained by the *Rigid Oblique Rotator* model as well as the observed variations of the magnetic field. It was developed by Stibbs (1950) who basically proposed that the (dipolar) magnetic field axis is inclined to the rotation axis. The variable elements are concentrated in patches at the magnetic poles. As the star rotates the observer (which does not look pole-on to the rotational axis) will notice a light house effect: magnetic polar areas together with their element patches alternate in their appearance on the visible hemisphere.

Although the physical background for the origin of the magnetic field inclination is still a matter of debate, because it relates to the origin of the magnetic field, there is no doubt that the Oblique Rotator is the only successful way of interpreting the spectrum (and related photometric) variations as well as the magnetic variability of CP2 stars.

After the Oblique Rotator model had been established, attempts to map the surface distribution of elements started with the study of Deutsch (1958). A very recent discussion of mapping stellar surfaces by Doppler imaging is given by Rice et al. (1989). Although abundance patches may be associated with magnetic polar regions, some elements may form patches also along the magnetic equator (see Fig.4).

Variability is not only seen in line intensities, but also in their *profiles*. This is partially due to variable *Zeeman broadening* in stars with strong fields. With the Babcock magnetic field analyzer (producing two spectra with opposite circular polarization components) line width differences between both spectra become most striking at the phases of magnetic *crossover* (=polarity change) due to the combined effect of Doppler and Zeeman shifts. This yields rather narrow line profiles in one spectrum and broad ones in the other one. The crossover effect is another important confirmation for the validity of the Oblique Rotator model.

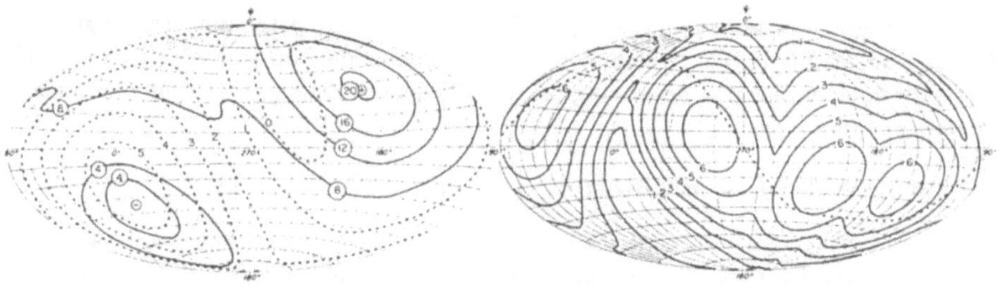

Fig.4: Mapping of peculiar elemental distributions and magnetic field on the surface of α^2 CVn (from Pyper, 1969). Left: solid curves indicate magnetic field, dashed curves local equivalent widths of rare earths. Right: solid curves refer to iron-peak elements, dashed curve to magnetic equator.

Spectrum variations have also been found for the Helium abnormal stars. Together with their magnetic field variations this evidences the presence of Oblique Rotators also among these very hot CP stars (see e.g. Hensler, 1979).

Recently, variations in the UV were reported for the C IV profiles in Helium abnormal stars using IUE-spectra by Shore et al. (1987). They are explained by mass loss along narrow cones above the magnetic poles producing a *magnetically controlled corotating stellar wind*.

6.2 PHOTOMETRIC VARIABILITY

Although the first light curve (in integral light) of a CP star had been obtained already at the beginning of this century by Guthnick and Prager (1914), extensive surveys of photometric variability had to wait for improvements of photoelectric technology. Pioneering observations were carried out by Rakos (1962), followed by Stepien (1968), both in the UBV system. The extrema of the light variations are generally in phase with those of the spectrum and magnetic field, their amplitudes are usually of the order of a few hundredths of magnitude, only exceptionally exceeding 0.1 mag. Amplitudes are often strongly wavelength dependent. As an example Fig. 5 shows the complicated change

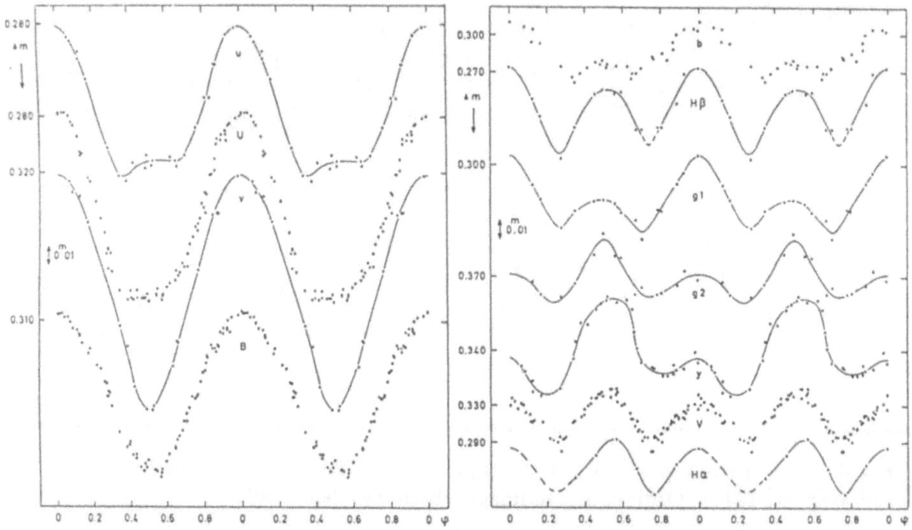

Fig.5

of light curve shapes in the optical region for HD 125248 from the multicolour study of Maitzen and Moffat (1972), which visualizes the appearance of both single and double wave variations in one star depending on the wavelength region chosen for photometric variability observations.

6.3.2 SHORT PERIOD VARIABILITY

Since some CP stars lie within the instability strip δ Scuti type variability with periods of the order of a few hours could be expected. On the other hand, Kreidl (1987) discussed 20 candidates accumulated in 30 years concluding from published and own photometric data that only few possible Ap stars remain that might show consistent δ Scuti variability.

Rapidly growing importance, however, has encountered variability with periods of several minutes during the last ten years. The spectroscopically very exotic HD 101065 (Przybylski's star) was the first cool CP2 stars where such *oscillations* were discovered (Kurtz, 1978). Its principal period is 12.14 minutes and its amplitude 0.02 mag (see Fig.6) which is rather large since other members of the group of rapidly oscillating CP stars exhibit variations of the order of millimagnitudes and less! All stars of this group are at the cool border of the CP-domain.

The fact that the oscillation amplitude varies with the rotational phase led Kurtz (1982) to the *Oblique Pulsator Model* according to which the symmetry axis for the oscillations is the magnetic, not the rotational axis of the star, and strongest oscillations occur at the magnetic poles. In the amplitude spectrum this will give rise to a triplet of frequencies. From their amplitudes the line of sight angle and the obliqueness angle as well as the ratio between minimum and maximum of the effective magnetic field can be derived.

The high overtone oscillations should show up also in very minute amplitude *variations of radial velocity*. So far only one object shows convincing evidence for radial velocity oscillations, so that one can really "see" the pulsations, i.e. HR 1217 with a maximum

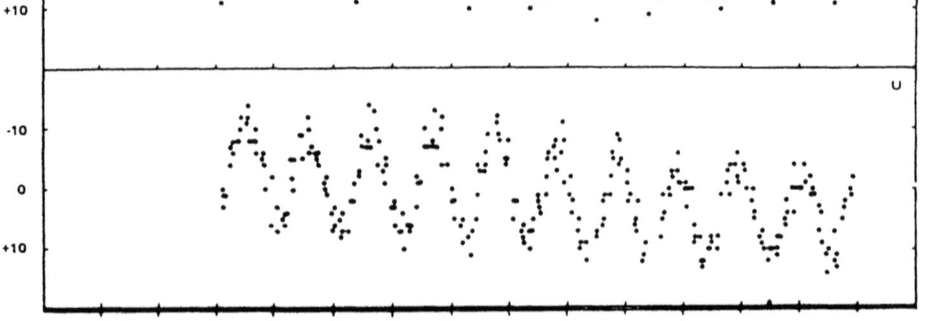

Fig.6: Short period variability in *B* (top) and *U* of HD 101065 (from Kurtz and Wegner, 1979). Ordinates: millimags, abscissae: days/1000.

amplitude of 400 m/s with parallel photometric variations of 6.8 millimags. Both oscillations are stongly modulated down to lower than 130 m/s and 1.8 mmags (Matthews et al., 1988). Controversial results about RV variations for other rapidly oscillating CP2 stars is discussed by Weiss and Schneider (1989). The importance of such very time consuming photometric and spectroscopic variability searches rests on the possibility of identifying pulsation modes from which information on the stellar interior can be derived (for an overview see Däppen et al., 1988). No mechanism has been found so far for exciting these high overtone oscillations.

Changes of this kind have also been detected in the satellite UV, first by Molnar (1973) who recognized the existence of *null wavelength regions* where no photometric variation is present at all.

The physical explanation for photometric variations is *variable line blanketing*, thus line blocking and backwarming of the continuum, rather than local temperature changes over the stellar surface (Wolff and Wolff, 1971).

For the *detection of variability* and observation of lightcurves it is important to choose a filter system which is sensitive to maximum variations. Filters should not be too broad and should be matched to spectral regions where variations can be expected to be specifically large. The *Strömgren* or *uvby*- system is well adapted to this needs, since the amplitude for hotter CP-stars is usually large in u, whereas cool CP2-objects may be conspicuously variable only in v due to strong and variable line blocking.

Work on rotational variability of CP2-stars as well as on Helium abnormal stars is progressing steadily. An overview of this field can be found in Maitzen (1987) and also in the report of the European Working Group on CP-stars (Mathys et al., 1989).

The shape of the light curves can be satisfactorily represented by a sinusoid and its first harmonic (North, 1984). Taking into account the blanketing explanation for photometric variability this indicates that *only two important spots, or a ring* with abundance enhancements appear on the CP surface.

Photometric variability (but also spectrum variations) depends also on a reasonable inclination of the line of sight to the rotational axis of the star observed. But even in case of looking pole-on an Oblique Rotator should exhibit observable variability: the magnetic field will be observed mainly through its transverse components which are linear polarized with the polarization vector describing a 360 degrees rotation during one cycle. Borra and Vaughan (1976) have found this in the case of β Coronae Borealis.

6.3 PERIODS

6.3.1 ROTATIONAL

Although the bulk of CP2 stars are slow rotators with periods of the order of several days, some stars, e.g. CU Vir, rotate nearly as fast as normal ones. On the other hand, there are extremely slow rotators, the slowest one being γ Equulei with a period of about 70 years! The relationship between *vsini* and periods derived from either photometric, spectroscopic or magnetic observations confirms that the latter are rotational periods. An important support for the validity of the Oblique Rotator model even in case of extremely long periods comes from Rice (1988) who observed and discussed the Rare Earths maximum of HR 465 in 1982 establishing a period of 21 years.

The question of *rotational braking* is still controversial. North's (1987) photometric variability study in clusters and associations confirms that hot magnetic stars undergo no braking on the main sequence. Glagolevski (1987) recently argues that braking occurs both before and on the MS. We are still lacking a comprehensive search for variability among the cooler CP2 stars in open clusters which should help to more clearly constrain the influence of any braking mechanism on the MS. Since fainter stars and longer periods are involved the feasability of such a programme suffers from technical reasons.

7. ABUNDANCES, ATMOSPHERIC STRUCTURE

The peculiar spectral appearance of CP2 stars exhibits a bewildering variety of individual patterns so that grouping of these stars becomes the more complex the higher spectral resolutions are employed. For resolutions higher than 10 A/mm *no two Ap stars are exactly alike* (Wolff, 1983). Identifications of lines are difficult in CP2 stars because of the enormous crowding in the optical spectra, but even more in the UV. Leckrone (1986) demonstrates (see Fig. 7) the dramatic gain in spectral resolution from the *International Ultraviolet Explorer (IUE)* to the *High Resolution Spectrograph (HRS)* on the *Hubble Space Telescope (HST)* from which CP-researchers will especially benefit when applying the Statistical Wavelength Coincidence Technique (Hartoog et al., 1973) which was recently refined by Ansari (1987) by including theoretical line strengths. Leckrone (1986) shows also convincingly using theoretical profiles of the Hg II λ 1942 line how it changes with changing isotopic mixture and that such changes are only accessible by the HST HRS.

Cowley (1976) reported the first results for the presence of a number of elements in CP1, CP2 and CP3 stars confirming the dissimilarity of their peculiar spectra. Very recent line identifications (useful for further bibliographical searches) have been performed for the very sharp lined CP2 star HD 110066 by Adelman and Adelman (1988) and Bidelman et al. (1988) offering interesting possibilities of comparisons.

Abundance analysis has been the subject of a unique workshop held in 1987 at the University of Lausanne. Its main aim was to homogenize data reduction and to compare the results of different model atmosphere codes using the same input data. In order to tackle the problem for easier cases, no magnetic CP-star (i.e. CP2) was chosen for this purpose. For the CP1 and CP3 star chosen small abundance differences resulted from different codes. Generally the insufficiency of atomic data for lines appearing in CP stars was deplored. Line blanketed model atmospheres for CP2 stars have been the subject of a series of papers by Muthsam and Stepien; in the latest one dealing with HD 221568 Stepien and Muthsam (1987) state that models with variable chemical composition are not able to reproduce the observed light variations particularly between the Balmer jump and 5000 A. Although 900 000 lines were included, it was felt that the lack of Rare Earths lines might be the reason for this result.

Muthsam (1979) can, however, throw light by his model atmosphere calculations on the influence of blanketing on the atmospheric structure. A CP star atmosphere mimics that of a normal star with higher effectice temperature, and the increase of temperature with optical depth is steeper than in normal stars. The total flux of a CP star is therefore lower than that of a normal star with same UBV colours. Faraggiana (1987) warns that the Paschen continuum is in no way a suitable base for determining the effective temperature.

Using UV data it is possible to derive bolometric magnitudes. T(eff) can be obtained by the method of Shallis and Blackwell (1979) who used the flux in the unperturbed infrared region. Stift (1974) calculated CP2 radii from bolometric magnitudes an T(eff). From the Oblique Rotator model one can statistically infer radii from $v \sin i$ and rotational periods. The radii obtained this way are somewhat larger than corresponding to the

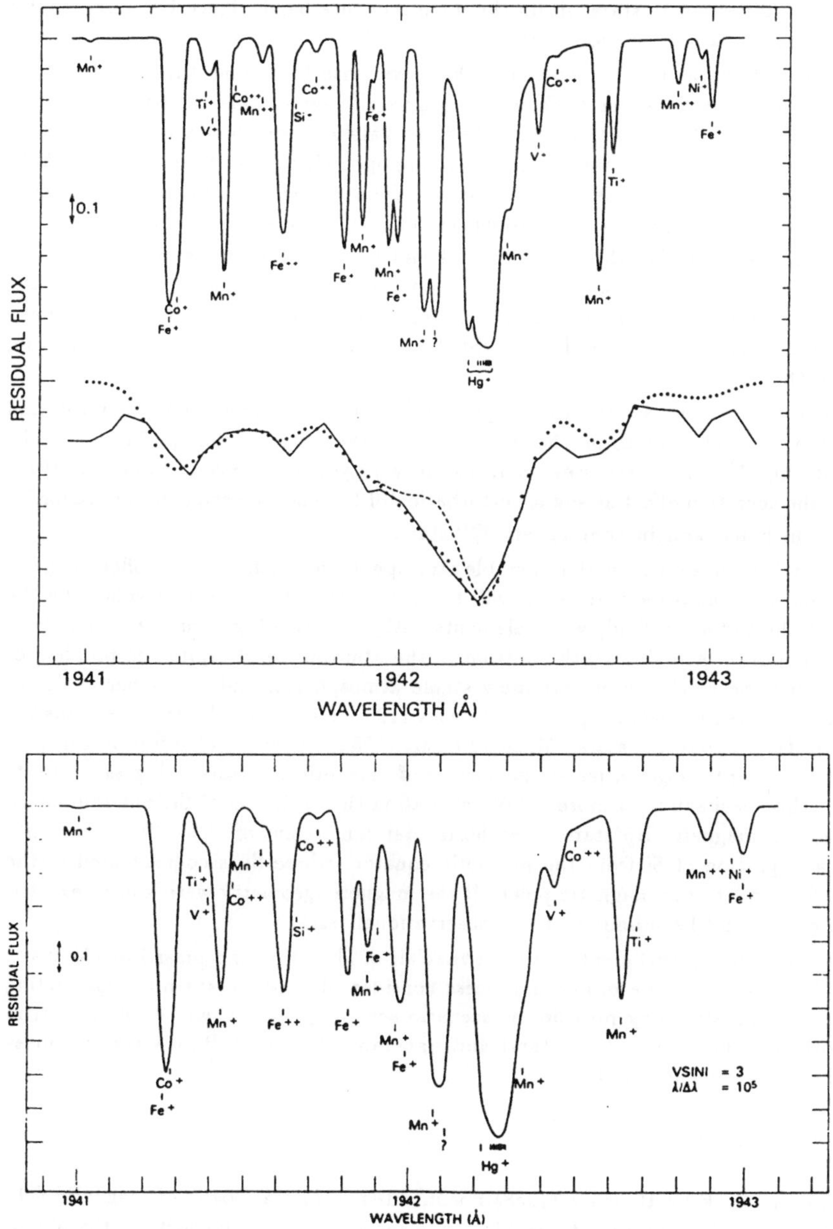

Fig.7: Upper panel:

top: Synthesized spectrum of the CP3 star ι CrB,

bottom: IUE observation (solid curve) and IUE-broadened theoretical computations (other curves).

Lower panel:

theoretical spectrum of ι CrB broadened to the spectral resolution of the High Resolution Spectrograph of HST. From Leckrone (1986).

ZAMS, but still pertain to stars within the main sequence band. Thus, CP-stars exhibit the evolutionary status of unevolved stars.

The origin of the abundance anomalies has given rise to various theories, the most important of which is *Radiative Diffusion*. It is generally agreed that the peculiarities are confined to the outer layers of the stars. The time scale for producing anomalies must be short (million years) since they are found in young open clusters and associations. Wolff (1983) concludes from a discussion of theories involving nuclear processing at the surface of CP stars that the majority of CP properties cannot be explained by them.

The *Magnetic Accretion* theory by Havnes and Conti (1971) relies on a nonnuclear process, i.e. selective capture of heavy element ions which penetrate deeper in the rotating magnetosphere and, after a further ionization, become definitely trapped and diffuse down onto the polar caps. On the other hand, light elements are deflected and carry momentum away from the star.

Although this process sounds very appealing because in principle both slow rotation and peculiar polar abundance patches are explained, the main problem is the time scale, because strong abundance excesses occur also in very young CP stars. Michaud (1976) considers the accretion effect as slight perturbation of the main mechanism, i.e. radiative diffusion, which also acts in nonmagnetic CP-stars.

Necessary for diffusion to work is a stable atmosphere in which elements offering many bound-bound or bound-free transitions will be driven upwards in the atmosphere by the pressure of the radiation field, while elements with few transitions for the existing radiation distribution will sink to the bottom of the atmosphere. Despite the problem of demonstrating the existence of extremely stable atmospheres, and a number of deviations from predicted abundance patterns, Cowley (1981) states that diffusion explains the majority of facts about CP stars. Michaud et al. (1981) showed that diffusion can produce abundance inhomogeneities at the surface of magnetic CP stars. Megessier (1984) developed this mechanism in more detail calculating the evolution of Silicon abundance distribution in magnetic Bp stars. The horizontal component of the diffusion velocity leads to a migration of Si from the magnetic equator (where Si is concentrated in the youngest Si stars) to the magnetic poles. If the magnetic geometry were known exactly, this prediction could be submitted to an observational test.

On the other hand, turbulent motions (Vauclair et al., 1978) and mass loss (Michaud et al., 1983) may alter the diffusion process considerably. Moderate mass loss acting together with diffusion may provide the scenario according to Michaud et al. (1983) in which both overabundances (Am-stars) and underabundances (λ Bootis stars) appear during the main sequence life time.

8. MAGNETIC FIELDS

The pioneering work by Babcock (1958) documented in his catalogue of magnetic field measurements was based on photographic coudé spectra. The nonlinearity of the photographic emulsion emphasizes absorption line cores which leads to an inaccurate determination of Zeeman shifts. Moreover, lack of full compensation of phase shifts introduced by oblique reflections in the coudé light path causes smaller values of the measured effective (=longitudinal) fields since admixtures of the opposite circular polarization component reduce the Zeeman shifts. Wolff (1983) therefore summarizes that the old photographic technique has merits only in case of determining the sign of the magnetic field and its period. Moreover, it is restricted to bright and sharplined stars.

The photoelectric technique by Angel and Landstreet (1970) measures the circular polarization in the wings of a Balmer line (1 percent for 10 kgauss) and can be applied also to fast rotators, which means that fields can also be obtained for CP-stars even in the temperature range of early B type stars (i.e. for the Helium abnormal stars). Results can be found in Borra and Landstreet (1980) and the catalogue of Didelon (1984).

While upper limits for CP1 and CP3 stars are about 200 gauss, which is the value the effective field must exceed to be measurable photographically, the typical H(eff) values of CP2 and Helium abnormal stars are of the order of kgauss, the maximum value being 20 kgauss for Babcock's star HD 215441.

Many of the photographic H(eff) curves are strongly anharmonic while their photoelectric counterparts look essentially sinusoidal. This is probably due (Hensberge et al., 1979) to the inhomogeneous distribution of metallic lines (which yield the field in the photographic case) while hydrogen is equally distributed over the surface.

Important information on the magnetic geometry can in principle be derived from a comparison of the longitudinal field with the (scalar) surface field H_s which depends linearly on the width of the Zeeman pattern (=splitting of the σ components) and not on the direction of the field. For the case of a centered dipole field H_s equals 80 percent of the polar field when the star is seen pole-on, while equator-on it reduces to 64 percent. As a consequence, such a field shows a *double wave* variation during the rotation cycle if H(eff) reverses its sign. Since most of the H_s curves show only one maximum the pertaining magnetic fields must be *hemispherically asymmetric.* They can be represented by a dipole with the center of symmetry displaced from the center of the star, in the most general case, investigated by Stift (1975), magnetic and rotational axes even do not cross each other (=off axis decentered dipole).

Admixture of a modest quadrupole field to the dipole seems to explain the magnetic curves in a number of cases. Much better insights into magnetic field geometries are to be expected from a long term programme carried out by G. Mathys at ESO with the CASPEC spectrograph at the 3.6 m telescope offering the possibility of measuring with high S/N not only line shifts, but also *line profiles* (see Mathys et al., 1989).

A very recent highlight related to the magnetic fields of CP-stars is the detection of *gyrosynchrotron radiation.* Philipps and Lestrade (1988) using VLBI measurements for two magnetic He-strong stars derive an emitting zone of 6 stellar diameters and a brightness temperature of 10^9 K. No gyroresonant radiation at 6 cm could, however, be found for 13 CP2 stars from VLA observations made by Willson et al. (1988), whereas Drake et al. (1987) detected such an emission for 5 CP stars (out of 34) which are all He abnormal stars and have strong fields. It is consistent with gyrosynchrotron emission from continuously injected, mildly relativistic particles trapped in the magnetosphere.

As to the *origin of the magnetic fields* Moss (1989) very recently reviewed the existing theories:

- The *battery* theory was originally proposed by Biermann (1950) and later modified by Dolginov (1977): chemical inhomogeneities cause circulation currents and can maintain toroidal fields in addition to the background poloidal field produced by another mechanism. No observationally testable predictions are made.

- The *dynamo* theory is based on the existence of a contemporaneous dynamo operating in the convective core of the magnetic stars.

- The *fossil* theory has two variants: the magnetic field is either the slowly decaying relics of the frozen-in interstellar magnetic field (a minute part of which has survived the star formation process) or of a dynamo acting in the pre-main sequence phase.

According to Moss (1989) is is not yet possible to decide conclusively whether the contemporaneous dynamo or the fossil field theory do describe the nature of CP magnetic fields properly. Nevertheless, he gives rather clear preference to the latter one:

A major problem for the dynamo theory is the transport of field from the core to the surface. No mechanism has been found so far to provide this in at least five million years. Furthermore, if core fields are limited by equipartition with the kinetic energy of convection which is a plausible assumption, it will be difficult to explain surface fields as strong as those observed.

The fossil theory has the advantage of a wide range of possible initial conditions. While previous modelling of magnetic field structure was based on a fairly homogeneous distribution of the magnetic flux, Moss (1989) argues that after the Hayashi phase any field surviving is likely to be arranged in flux tubes ("ropes"). His calculations demonstrate that surface fields may me distributed over much of the surface (and not concentrated into discrete spots) even if the interior fields are quite inhomogeneous. Urgently required is a larger and more precise body of field geometry observations in order to establish deviations from axisymmetry.

9. REFERENCES

Abt,H.A.:1979, Astrophys.J.230, 485.

Abt,H.A.,Levato,H.:1978, Publ.Astron.Soc.Pacific 90, 210.

Adelman,S.J.,Adelman,C.J.:1988, Mon.Not.R.Astron.Soc.235, 1361.

Angel,J.R.P.,Landstreet,J.D.:1970, Astrophys.J.160, L147.

Ansari,S.G.:1987, Astron.Astrophys.181, 328.

Babcock,H.W.:1947, Astrophys.J.105, 105.

Babcock,H.W.:1958, Astrophys.J.Suppl.3, 141.

Baschek,B.,SLettebak,A.:1988, Astron.Astrophys.207, 112.

Bidelman,W.P.,Cowley,C.R.,Roemmelt,J.C.:1988, Publ.Obs.Univ.Michigan 12,23.

Bidelman,W.P.,MacConnell:1973, Astron.J.78, 687.

Biermann,L.:1950, Z.Naturf. 5a, 65.

Borra,E.F.,Landstreet,J.D.:1979, Astrophys.J.228, 809.

Borra,E.F.,Landstreet,J.D.:1980: Astrophys.J.Suppl.42, 421.

Borra,E.F.,Landstreet,J.D.,Thompson,I.:1983, Astrophys.J.Suppl.53, 151.

Borra,E.F.,Vaughan,A.H.:1976, Astrophys.J.210, L145.

Cameron,R.C.:1967, Georgetown Obs. Monogr. No.21.

Cowley,C.R.:1976, in: Physics of Ap Stars, eds. W.W.Weiss,H.Jenkner, H.J.Wood, Vienna, p.275.

Cowley,C.R.:1981, Astrophys.J.246, 238.

Däppen,W.,Dziembowski,W.A.,Sienkiewicz,R.:1988, in: Advances in helio- and asteroseismology, IAU Symp.No.123, eds. J.Christensen-Dalsgaard, S.Frandsen, Reidel, p. 233.

Deutsch,A.J.:1947, Astrophys.J.105, 283.

Deutsch,A.J.:1958, in: Electromagnetic Phenomena in Cosmical Physics, IAU Symp.No.6, ed. B.Lehnert, Cambridge, p.209.

Didelon,P.:1984, Astron.Astrophys. Suppl.55, 69.

Dolginov,A.Z.:1977,Astron.Astrophys.54, 17.

Drake,S.A.,Abbott,D.C.,Bastian,T.S.,Bieging,J.H.,Churchwell,E.,Dulk,G., Linsky,J.L.:1987, Astrophys.J.322, 902.

Faraggiana,R.:1987, Astrophys.Space Sci.134, 381.
Garrison,R.F.:1967, Astrophys.J.147, 1003.
Glagolevskij,Yu.V.:1987, Soob.Sp.Astr.Obs.54, 73.
Guthnick,P.,Prager,R.:1914, Veröff.K.Sternw.Berlin Babelsberg 1, 1.
Hartoog,M.R.,Cowley,C.R.,Cowley,A.P.:1973, Astrophys.J.182, 847.
Havnes,O.,Conti,P.S.:1971, Astron.Astrophys.14, 1.
Hensberge,H.,van Rensbergen,W.,Goossens,M.,Deridder,G.:1979, Astron. Astrophys.75, 83.
Hensler,G.:1979, Astron.Astrophys.74, 284.
Kreidl,T.J.:1987, in: Stellar Pulsation, eds. A.N.Cox,W.M.Sparks,S.G. Starrfield, Springer, p. 134.
Kurtz,D.W.:1978, Inf.Bull.Var.Stars No.1436.
Kurtz,D.W.:1982, Mon.Not.Roy.Astr.Soc.200, 807.
Kurtz,D.W.,Wegner,G.:1979, Astrophys.J.232, 510.
Leckrone,D.S.:1986, in: Upper Main Sequence Stars with Anomalous Abundances, IAU Coll.No.90, eds. C.R.Cowley,M.M.Dworetsky,C.Mégessier, Reidel, p.109
Lodén,L.O.,Sundman,A.:1987, Astron.Astrophys.8, 351.
MacConnell,D.J.,Frye,R.L.,Bidelman,W.P.:1970, Pub.Astron.Soc.Pacific 82, 730.
Maitzen,H.M.:1976, Astron.Astrophys.51, 223.
Maitzen,H.M.:1984, Astron.Astrophys.138, 493.
Maitzen,H.M.:1987, Hvar Obs. Bull. 11, 1.
Maitzen,H.M.,Moffat,A.F.J.:1972, Astron.Astrophys.16, 385.
Maitzen,H.M.,Muthsam,H.:1980, Astron.Astrophys.83, 334.
Maitzen,H.M.,Pavlovski,K.:1989, Astron.Astrophys.219, 253.
Maitzen,H.M.,Schneider,H.:1984, Astron.Astrophys.138, 189.
Maitzen,H.M.,Schneider,H.,Weiss,W.W.:1988, Astron.Astrophys.Suppl.75, 391.
Maitzen,H.M.,Vogt,N.:1983, Astron.Astrophys.123, 48.
Mathys.,G.,Maitzen,H.M.,North,P.,Hensberge,H.,Weiss,W.W.,Ansari,S.,Catalano, F.A., Didelon,P.,Faraggiana,R.,Fuhrmann,K.,Gerbaldi,M.,Renson,P.,Schneider, H.:1989, The Messenger 55, 41.
Matthews,J.M.,Wehlau,W.H.,Walker,G.A.H.,Yang,S.:1988, Astrophys.J.324, 1099.
Maury,A.C.:1897, Harvard Ann.28, 96.
Mégessier,C.:1984, Astron.Astrophys.138, 267.
Michaud,G.:1976, in: Physics of Ap Stars, eds. W.W.Weiss,H.Jenkner,H.J. Wood, Vienna, p.81.
Michaud,G.,Mégessier,C.,Charland,Y.:1981, Astron.Astrophys.103, 244.
Michaud,G.,Tarasick,D.,Charland,Y.,Pelletier,C.:1983, Astrophys.J.269, 239.
Molnar,M.R.:1973, Astrophys.J.179, 527.
Morgan,W.W.:1933, Astrophys.J.77, 330.
Moss,D.:1989, Mon.Not.R.astr.Soc.236, 629.
Muthsam,H.:1979, Astron.Astrophys.73, 159.
Nidzielski,A.,Muciek,M.:1988, Acta Astron.38, 225.
North,P.:1984, Astron.Astrophys.Suppl.55, 259.
North,P.:1987, Astron.Astrophys.Suppl.69, 371.
North,P.,Hauck,B.,Straižys,V.:1982, Astron.Astrophys.108, 373.
Philipps,R.B.,Lestrade.J.F.:1988, Nature 334, 329.
Preston,G.W.:1974, Ann.Rev.Astron.Astrophys.12, 257.
Pyper,D.M.:1969,Astrophys.J.Suppl.18, 347.
Rakos,K.D.:1962, Lowell Obs.Bull.5, 227.

Rice,J.B.:1988, Astron.Astrophys.199, 299.
Rice,J.B.,Wehlau,W.H.,Khokhlova,V.L.:1989, Astron.Astrophys.208, 179.
Schneider,H.:1987, Hvar Obs.Bull.11, 29.
Shallis,M.J.,Blackwell,D.E.:1979, Astron.Astrophys.79, 48.
Shore,S.N.,Brown,D.N.,Sonneborn,G.:1987, Astron.J.94, 737.
Stepien,K.:1968, Astrophys.J.154, 945.
Stepien,K.,Muthsam,H.:1987, Astron.Astrophys.185, 225.
Stibbs,D.W.N.:1950, Mon.Not.Roy.Astr.Soc. 110, 395.
Stift,M.J.:1974, Astron.Astrophys.34, 153.
Stift,M.J.:1975, Mon.Not.Roy.Astr.Soc.172, 133.
Titus,J.,Morgan,W.W.:1940, Astrophys.J.92,256.
Vauclair,G.,Vauclair,S.,Michaud,G.:1978, Astrophys.J.223, 920.
Weiss,W.W.,Schneider,H.:1989, Astron.Astrophys., in press.
Willson,R.F.,Lang,K.R.,Foster,P.:1988, Astron.Astrophys.199, 255.
Wolff,S.C.:1983, The A-type stars, NASA SP-463, p.33.
Wolff,S.C.,Wolff,R.J.:1971, Astron.J.76, 422.

Dynamics and Structures of Cometary Dust Tails

K. Beisser

Dr. Remeis-Observatory, Sternwartstr. 7,
D-8600 Bamberg, Fed. Rep. of Germany

1) Introduction

According to the Whipple-Model (Whipple 1950, 1951), the nucleus of a comet consists mainly of water ice with embedded dust grains. On the nucleus' approach to the sun around the perihel of its orbit, the ice is vaporizing and so the enclosed dust particles are set free. They are dragged along by the gas flow due to the gas pressure. In a distance of about 20 nucleus radii, the density of the gas has decreased so much that the gas flow does no longer control the motion of the dust particles. Hence, they move through the solar system as free particles.

2) The Motion of Cometary Dust Particles

After decoupling from the gas flow, the dust particles are subject to various forces: solar gravitation, radiation pressure, Poynting-Robertson force, collisions with solar wind particles, neutral particle pressure, Coulomb force (as the particles are charged by photo-ionisation due to the solar UV-radiation), Lorentz force (Schwehm 1980) and cometary gravitation. According to an estimation by Schwehm (1980) the dominating of these forces are the solar gravitation and the radiation pressure, the latter being relevant to the motion of particles in the size range of about 10^{-2} µm to 10 µm (Burns et al., 1979).

These particles form the cometary dust tail. Each particle independently follows its own orbit around the sun. So, to model the dust tail, the orbits of a number of different particles with various emission conditions have to be taken into account and the loci of all these

Reviews in Modern Astronomy 2 Editor: G. Klare

particles at a given observation time t_{obs} have to be constructed.

Constricting the flight periods of the regarded particles to less than 100 days, all forces except the solar gravitation and radiation pressure may be neglected. As the typical velocity of an interplanetary body is of the order of 50 km/s, the radiation pressure $\underline{F}_R$ may be taken in its nonrelativistic form:

$$\underline{F}_R \quad (\langle\sigma_{pr}\rangle / r^2) \; \underline{e}_r \tag{1}$$

$\langle\sigma_{pr}\rangle$: cross section of the particle for radiation pressure averaged over the solar spectrum
r : distance from the sun
$\underline{e}_r$: unit vector radially outward from the sun.

Because this force is proportional to r^{-2}, it can be expressed in terms of the gravitational force $\underline{F}_G$:

$$\underline{F}_R =: -\beta \underline{F}_G; \qquad \beta >= 0; \quad \beta = \beta(r). \tag{2}$$

Thus, the resulting force $\underline{F}_T$

$$\underline{F}_T = \underline{F}_R + \underline{F}_G = (1-\beta)\underline{F}_G =: \mu \underline{F}_G; \qquad \mu <= 1 \tag{3}$$

is equal to the gravitational force with a gravitation constant modified by the factor μ.

μ contains the cross section $\langle\sigma_{pr}\rangle$, which depends on the material and the size of a particle. In general, it cannot be calculated analytically. Here, the particles are assumed to be spheres. Then the cross section and consequently the ß-value can be determined following the Mie-theory (see e. g. Rohde 1978). Results for ß-values of a number of different materials over a wide range of sizes have already been published (Burns et al. 1979, Schwehm 1980, Eaton 1984, Lamy and Perrin 1988).

So, the motion of a particle can be tracked using modified Keplerian mechanics. From it, the orbital elements of each particle can be derived and so its position at t_{obs} is obtained. For a particle with given ß-value, its orbit is determined by the initial conditions: emission position $\underline{r}_{em}$ and emission velocity

$\underline{v}_{em}$ relative to the sun. Whereas $\underline{r}_{em}$ is simply given by the nucleus position at the emission time t_{em}, $\underline{v}_{em}$ consists of three constituents: the velocity $\underline{v}_{em,c}$ of the comet relative to the sun at t_{em}, the velocity $\underline{v}_{em,d}$ of the dust relative to the nucleus at the decoupling from the gas flow, and the velocity $\underline{v}_{em,r}$ due to the nucleus rotation around an assumed axis. Whereas $\underline{r}_{em}$ and $\underline{v}_{em,c}$ can be determined easily from the comet's orbital elements, the decoupling velocity $\underline{v}_{em,d}$ can be calculated by hydrodynamic methods (e. g. Finson and Probstein 1968, Gombosi et al. 1984) but is in general not very well known and, like $\underline{v}_{em,r}$, is found by matching observations in a trial and error procedure.

3) The Generation of Synthetic Dust Tail Images

Several computer codes have been developed to model cometary dust tails in different ways and under various assumptions and confinements, e. g. by Finson and Probstein 1968, Kimura and Liu 1977, Fulle 1987 or Keller and Richter 1987.

The algorithm used in this paper is intended mainly for general studies of connections between emission characteristics and the dust tail appearance and also to check results obtained in other ways rather than to get fits to observations. In it, the nucleus is assumed to be a sphere with a given radius, a rotation axis and a rotation velocity together with a list of emission point sources. A relative efficiency E_i is assigned to each source. The total number of particles around β_0 emitted around the time t_0 is given by the function $N(\beta_0, t_0)$. Their emission velocity is described by the function $v_{em}(\beta_0, t_0)$. To account for the activity fluctuations due to the variable irradiation onto an emission point, N and v_{em} are multiplied by a weight function W(ZD) of the sun's zenith distance ZD at the respective emission point. To calculate the amount of light scattered into the observer's direction by a particle, a scattering function σ (β_0, SA) must be given (SA: scattering angle).

To actually model a dust tail, the above mentioned parameters and functions have to be prescribed. Then, for a large number of particles, their position relative to the nucleus at the observation time t_{obs} is calculated and the corresponding image pixel is determined. The contribution of each particle to its target pixel is

$$(L_0/r^2)*N(t_{em},\beta_{em})*W(ZD)*(\sigma(\beta_{em},SA)/\Delta^2)*E_{em}*dt*d\beta \quad (4)$$

L_0 : Luminosity of the sun
Δ : Distance between particle and earth

Finally, an isophote map of the produced data is created and plotted.

4) Examples for Dust Tail Structures

As an example for the results obtained with the computer code described above, some synthetic images of comet Halley's dust tail on March 8.35, 1986 are presented. For all images, the scattering cross section is assumed to be the geometric cross section which is convenient using $\beta<=0.8$ according to the results of Burns et al. (1979). The isophotes are given in arbitrary units.

Around that day, Halley has shown distinct structures in its dust tail (Beisser and Boehnhardt 1987). In Fig. 1, an isotropic emission of the dust with a vanishing emission velocity $v_{em,d}$ from a non-rotating nucleus is taken which is a quite frequent assumption. In the computed image, no structures similar to the observed ones are present.

Adopting a rotation period of 2.17 days around an axis as given by Sekanina and Larson (1986), one can get a tail structure as shown in Fig. 2. There, it is assumed that one prominent dust source exists at the equator of the nucleus whose activity is modulated by W=cos(0.75*ZD) if positive or W=0 if negative, thus restricting the emission to a solar zenith distance from the source of 120 degrees. $v_{em,d}$ is still set to zero. This leads to a somewhat closer agreement with the observations but still gives not a good reproduction.

If additionally an emission velocity distribution in the form given by Sekanina and Larson 1986 is assumed with a maximum velocity $v_{em,d}$ of (400/r) m/s which is

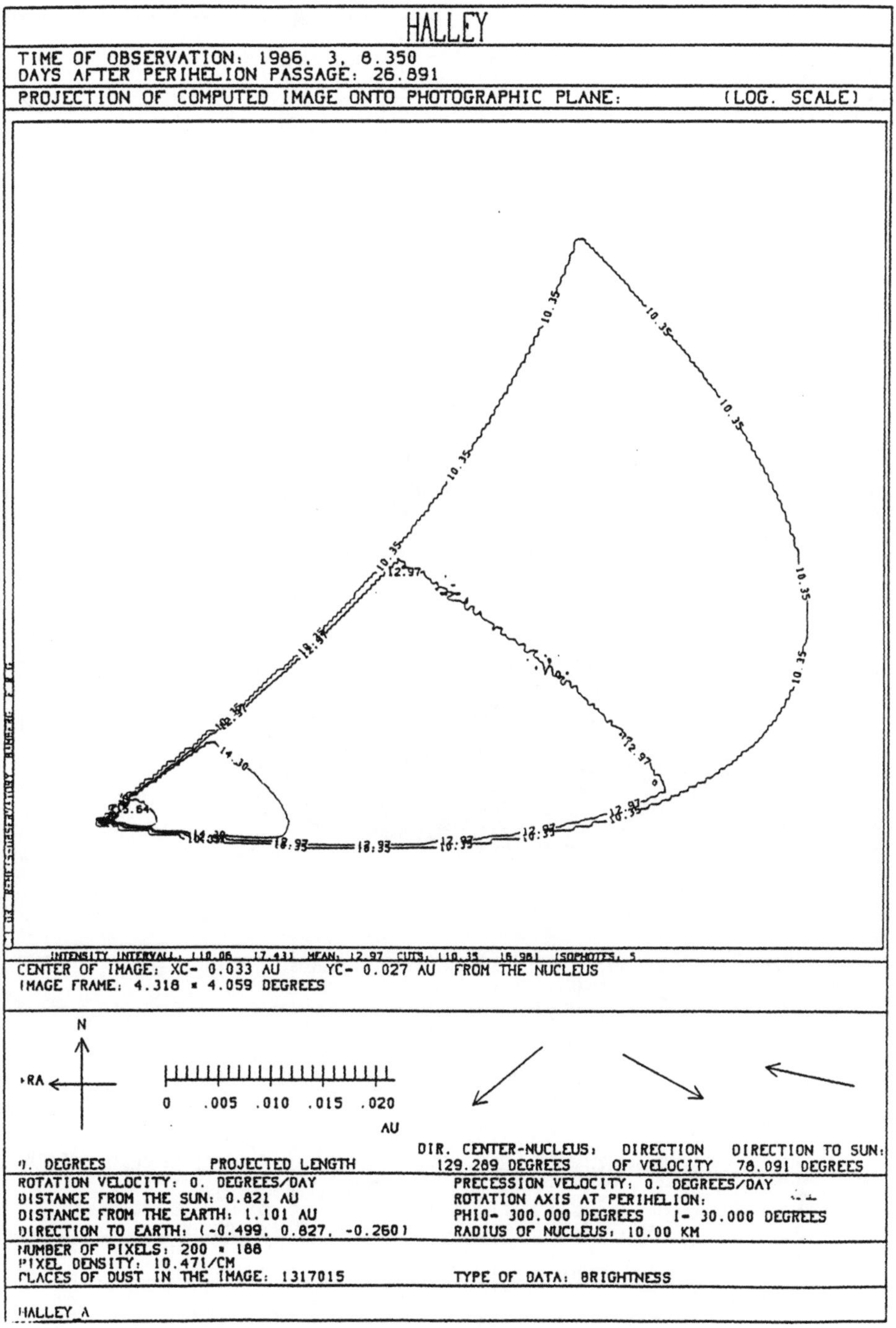

Fig. 1: Synthetic isophotes of Halley's dust tail.
$N = \beta^{+0.5} * r^{-2}$; $v_{em,d} = 0$;
See text for further explanations.

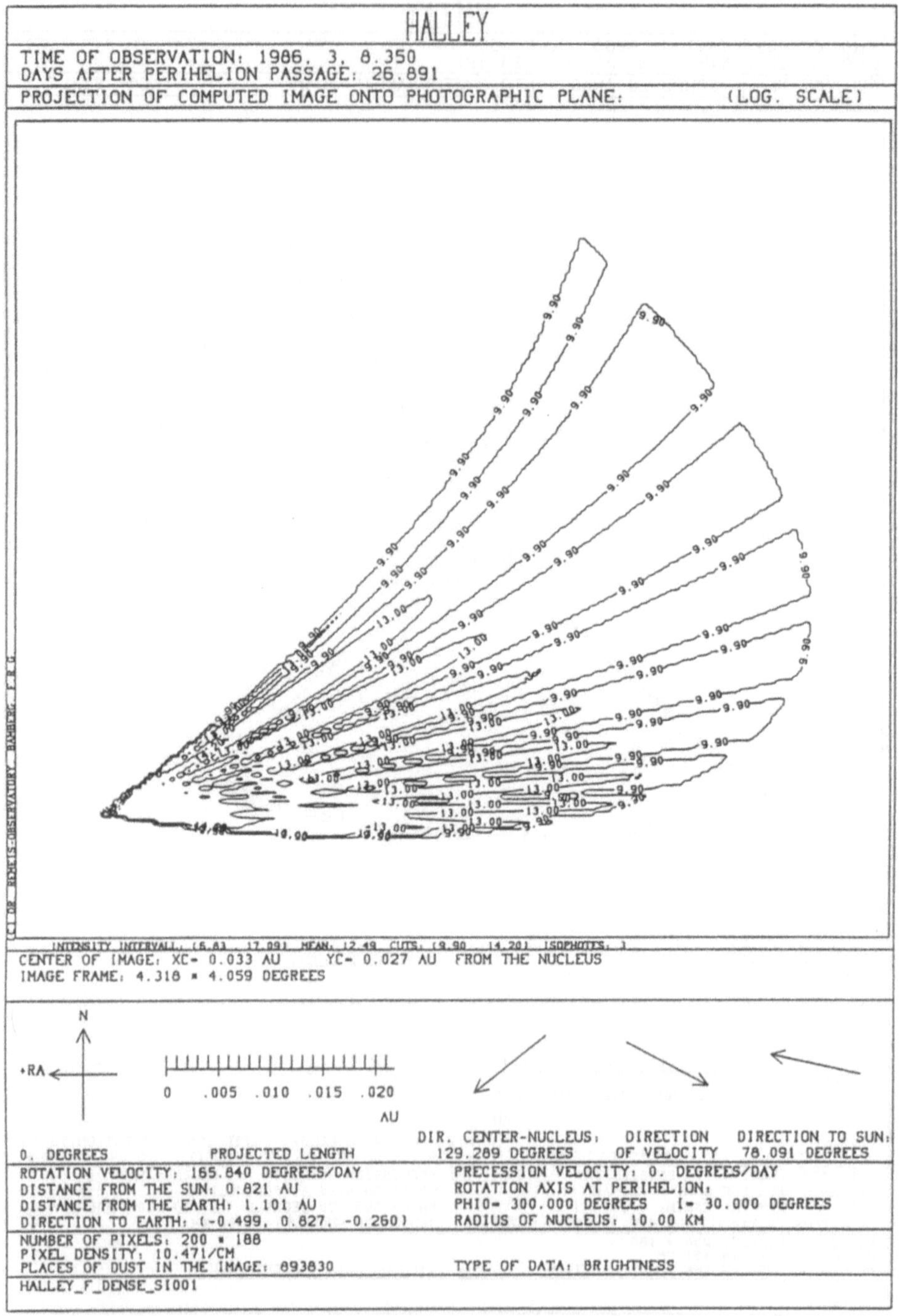

Fig.2: Synthetic isophotes of Halley's dust tail.
$N = (\beta^{+0.5} * r^{-2}) * W$; $v_{em,d} = 0$;
See text for further explanations.

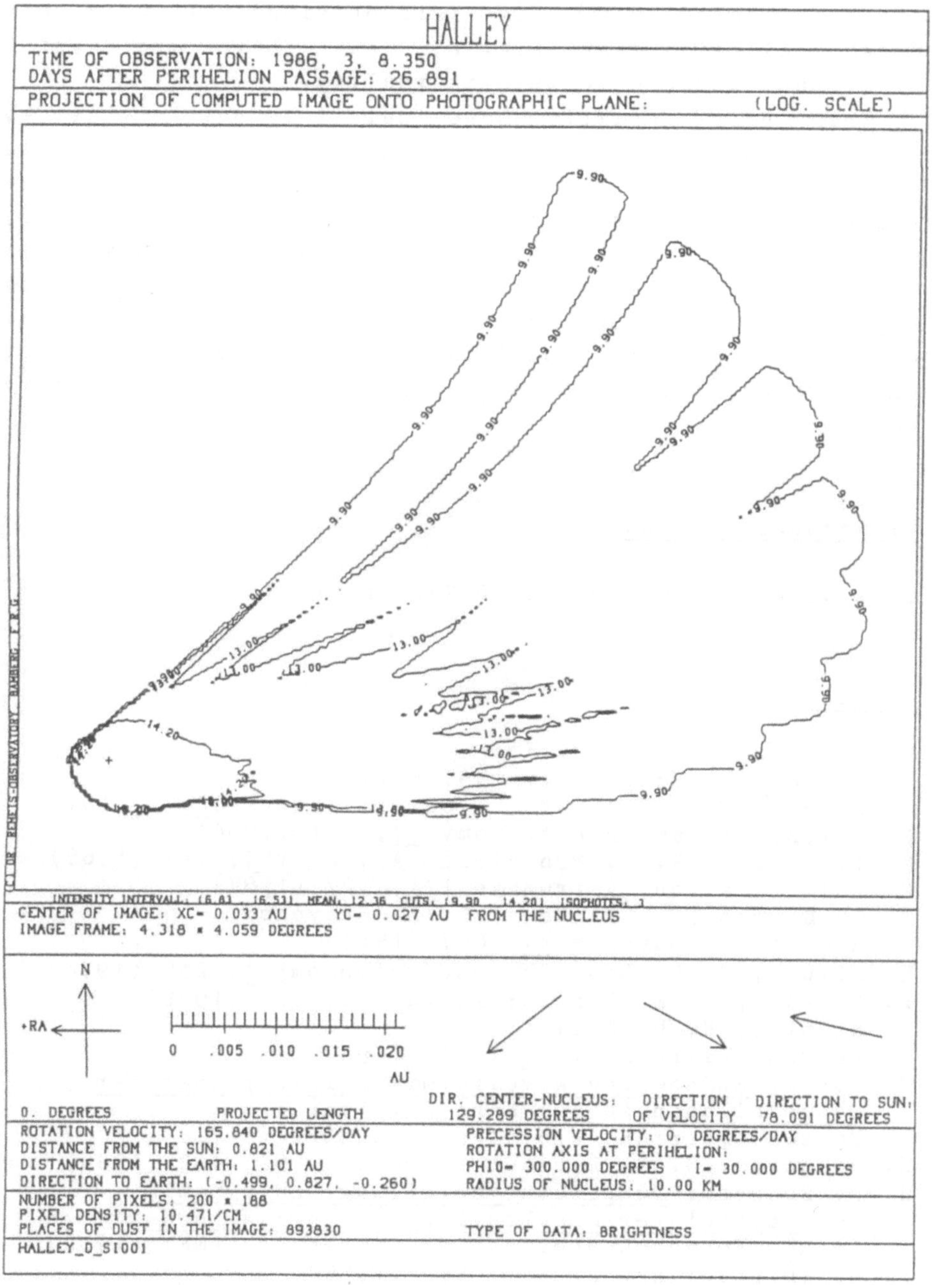

Fig. 3: Synthetic isophotes of Halley's dust tail.
$N = (\beta^{+0.5} * r^{-2}) * W$;
$v_{em,d} = ((1/400)+B*\beta^{-0.5})^{-1}/r$ m/s * W;
$B = 1.24*10^{-4}$ s/m;
See text for further explanations.

also modulated by W, Fig. 3 is obtained. It is a bit more realistic than Fig. 2, but there are still sigificant disagreements with reality.

This brief example shows that it is not trivial to obtain a synthetic image that makes an accurate fit to observations . On the other hand, it can be seen that shape and structure of a cometary dust tail are very sensitive to the emission conditions of the dust near the nucleus, like emission velocity or thermal inertia represented by W. So, the dust tail acts like a "map" of cometary activity, and further study towards a better understanding of this "map" may be useful for cometary research.

5) Acknowledgements

This work is sponsored by DFG-grant Ra 136/13-2.

6) References

K. Beisser, H. Boehnhardt:
Astrophys. Space Sci. 139, 5 (1987)
J. A. Burns, P. L. Lamy, S. Soter: Icarus 40, 1 (1979)
N. Eaton: Vistas in Astronomy 27, 111 (1984)
M. L. Finson, R. F. Probstein: Ap. J. 154, 327 (1968)
M. Fulle: Astron. Astrophys 171, 327 (1987)
T. I. Gombosi, A. F. Nagy, T. E. Cravens:
Rev. of Geophysics 24, 667 (1984)
H. Kimura, C.-P. Liu: Chinese Astronomy 1, 235 (1977)
P. L. Lamy, J.-M. Perrin: Icarus 76, 100 (1988)
K. Richter, H. U. Keller:
Astron. Astrophys. 171, 317 (1987)
M. Rohde: Aufheizung kugelförmiger interplanetarer Staubteilchen in der Nähe der Sonne;
Forschungbericht W78-33 an das BMFT (1978)
Z. Sekanina, S. M. Larson: Astron. J. 92, 462 (1986)
G. H. Schwehm: Temperaturverteilung und Dynamik interplanetarer Staubteilchen in der Nähe der Sonne; Forschungsbericht W79-40 an das BMFT (1980)
F. L. Whipple: Ap. J. 111, 375 (1950)
F. L. Whipple: Ap. J. 113, 464 (1951)

Automated Data Analysis

D. Teuber

Astronomisches Institut, Westfälische Wilhelms-Universität Münster, D-4400 Münster, Fed. Rep. of Germany

1 Introduction

Automating a task is assigning a sequence of operations to an *automaton*, which repeats this sequence upon request without need for further human interaction. The automaton in our case is the digital computer. The operations carried out by the computer are algorithms for analysing data sets. During the years the complexities of algorithms and data structures have grown as is illustrated in **Fig. 1**. The stages of simple *calculations* and *image processing* are fully developed today. *Object description*, the first step towards *automated data analysis*, is being explored. In the future, *model interpretation* will be supported by computers.

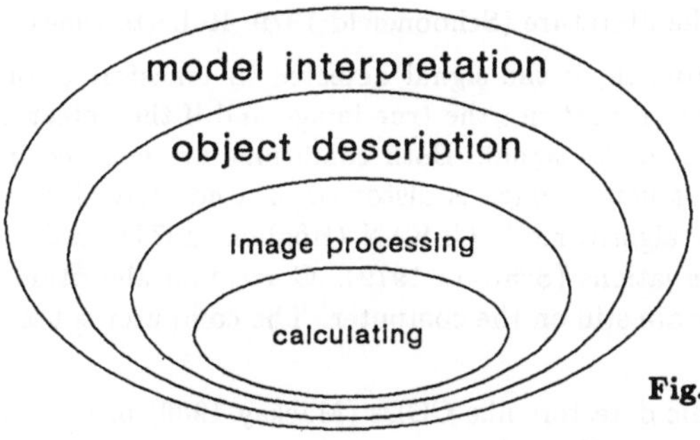

Fig. 1. Digital data processing in astronomical data analysis

Automated analysis on the level of astronomical objects makes survey projects feasible where millions of objects have to be examined, a task no individual nor group of individuals could accomplish without computerised data processing. Survey projects enable us to investigate large structures and to deduce global parameters not apparent in small samples. The experience reported in this article has been gathered with the Muenster Redshift Project (MRSP, Schuecker et al. 1989).

Automated data analysis requires that current knowledge about the physical world is transferred to the computer to serve as starting-point for the analysis. Strategies of analysis must be carefully examined to be able to construct an appropriate hardware and software environment (Teuber 1988).

2 Automating Calculations

In the early days, computers were used to assist in calculating. The astronomer decided which feature of the light distribution in the telescope or density distribution on a photographic plate could be interpreted as an astronomical object. He also decided how to derive numerical parameters for these objects.

The measuring machines ranged from simple, mechanical aids, like ruler and compasses, through high-precision electro-mechanic comparators, like Grant Machines (Rickard et al. 1975). The resulting numbers were fed into the computer to yield astrophysical quantities.

All machines had only one task: to help the astronomer measure in a more objective way.

3 Automating Image Processing

3.1 Digitising Images

Interacting with signals through human senses only, is feasible as long as light distributions are obtained by direct imaging. As soon as measuring methods are introduced where the detector signal represents a transformation of the image, the astronomer can no longer use his bare eyes only. Before he can perceive an image, the signal must be fed into a computer which transforms it into an image. Aperture synthesis is a technique of indirect imaging first applied in radio astronomy, intensity and speckle interferometry are used in the range of optical and IR wavelengths. Principles and problems connected with indirect imaging are discussed in the literature (Schoonefeld 1979; Roberts 1984).

Indirect imaging would cause no problem, if the signal detected in transformation space carried all information necessary to reconstruct the true image and if the detector was an ideal device which adds no noise to the signal. Both conditions are violated in real applications. As a result the reconstructed image is distorted in a non-trivial way. A prominent example of reconstruction algorithms is CLEAN (Högbom 1974), mainly used in the field of radio synthesis observations (Schwarz 1979). Correction algorithms are applied to the raw images while they are still on the computer. The computer is used to both reconstruct and process images.

The introduction of imaging electronic detectors like CCDs (Mackay 1986) promoted the trend towards digital image processing. The electric output signal can easily be converted into a computerised image: positional information is digital by design, and the intensity signal can easily be converted from analog to digital coding.

Images from photographic plates are also digitised. A variety of microdensitometers are available for this purpose. Construction principles and performance characteristics have been reported in conference proceedings (De Jager and Nieuwenhuijzen 1975; Sedmak et al. 1979; Stobie and McInnes 1983; Klinglesmith 1984). Practical hints are given by Andrews and Hunt (1977). Widespread are PDS microdensitometers. They are mainly delivered as low cost steel machines. But a few granite machines capable of scanning plates up to $20'' \times 20''$ were also built. These PDS 2020 microdensitometers have been equipped with laser interferometers (Kinsey 1984) for accurate position measurement, and with high-speed logarithmic amplifiers (Teuber et al. 1987) for reliable photometry in order to meet the high performance requirements of survey work.

3.2 Image Processing Systems

General purpose systems were raised for image processing. The landmarks are listed in **Tab. 1**. Many more systems exist, which were built for particular needs. They are

Tab. 1: Astronomical image processing systems

acronym	name	application [developer]	system
VICAR	Video Image Communication And Retrieval	satellite images [JPL] (Castleman 1979)	IBM, DEC
TV	"Tololo-Vienna" System	SIT-vidicon images [CTIO] (Albrecht 1979)	DEC
IHAP	Image Handling And Processing	general purpose image processing [ESO] (IHAP 1989)	HP
MIDAS	Munich Image and Data Analysis	general purpose image processing [ESO] (MIDAS 1989)	VAX/VMS UNIX
AIPS	Astronomical Image Processing System	radio astronomy and general purpose image processing [NRAO] (Wells 1984)	UNIX
IRAF/ SDAS	Interactive Reduction and Analysis Facility	Hubble Space Telescope data processing [NOAO] (IRAF 1989)	VAX/VMS UNIX

not listed here, though they influenced the development of the more widespread systems. **Fig. 2** shows the typical structure of a software and hardware environment for interactive image processing.

The development of appropriate human/machine interface hardware, such as high resolution graphics and color displays with interactive cursor facilities, permits the astronomer to extract parameters by interacting with the computerised image through computer programs.

When a set of astronomical images can be processed by the same procedure, processing can be automated: data and algorithms are both represented on the computer, and the reduction can be executed as batch job without user interaction. The procedure is developed by the astronomer, often as a result of long and tedious interactive data reduction sessions. The responsibility for chosing input images which are suitable for the process remains with the astronomer.

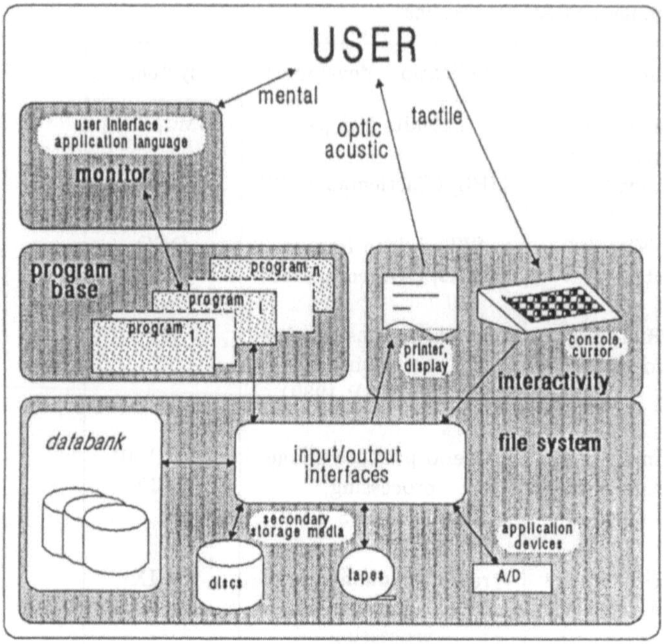

Fig. 2. Structure of a system for interactive and batch image processing

4 Automating Data Analysis

Measuring is no problem for an automaton: it has been designed to perform this operation. Measuring may be a problem for human beings, because they lack the skill or do not have the senses to perceive the signals. Deciding which measurement must be made and which result must be retained is difficult for an automaton, while it comes naturally to the human being who has developed faculties for decision making. If one wants the computer to carry out decisions, one has to understand how human beings proceed and how their procedures can be transferred to the computer. It is obvious that automated data analysis cannot be implemented from scratch, interactive data analysis is needed to recognize problems and to visualise results. Thus, a complete interactive image processing system is a prerequisite.

4.1 Expert Systems

Human beings use knowledge. Human beings can learn, i.e., enhance their knowledge so they can perform better the next time. Human beings use experience in form of heurisms and analogies to other problem domains. Human beings reason (at least they should) about what they do. They can try different approaches to solve a task. Human beings can explain to others why they act in a particular way. Introductions to methods of representing these human faculties on the computer are given by Winston (1984) and Charniak and McDermott (1986).

Human beings can cope with uncertainty. Often a decision is made on the basis of several uncertain features and the resulting classification is chosen to be the one that violates the facts least. A method to describe this process in mathematical form is the Fuzzy Set Theory (Pal and Majumder 1986; Zadeh 1973).

The human faculties mentioned above are often considered prerequisites for intelligence. The branch of science investigating the problem of transferring these faculties to the computer is said to study *Artificial Intelligence* (AI). Automated data analysis in astronomy makes use of methods developed for AI. Examples are the use of Fuzzy Sets in morphological studies of galaxies (DiGesu 1988), and rectification of spectra (Schuecker 1988).

Systems which use AI methods to cover a complete problem domain are often called *expert systems.* An early example of an expert system from the field of medicine is MYCIN (Buchanan and Shortliffe 1985). The MRSP uses expert systems for a variety of problems, among them are spectral classification, quasar detection and galaxy redshift determination from low-dispersion spectra (Gericke 1988; Schuecker 1988; Schuecker et al. 1989; Teuber et al. 1989). Thonnat and Bijaoui (1989) apply an expert system in rule-based morphological classification of galaxies.

4.2 Classification

Classification is a method of grouping objects with respect to a particular problem domain, neglecting minor differences within and ignoring all differences outside the problem domain (Niemann 1983). Classification is a key technique to gain information on all levels of analysis. The result of the classification process is the assignment of an object to a class. Classification is based on object features. Features must be extracted and coded.

Various statistical methods for feature extraction are discussed in detail by Murtagh and Heck (1987) in the context of Principal Components Analysis, Cluster Analysis and Discriminant Analyis. Often, transformations are used which yield a finite set of coefficients, such as Fourier and Walsh transformations. Widely used in astronomy is the calculation of moments (Martin and Lutz 1979; Stobie 1980). Sophisticated algorithms for object detection and star/galaxy separation (Jarvis and Tyson 1981; Heydon-Dumbleton et al. 1988; Maddox et al. 1988; Horstmann 1988) are based on this technique.

Sometimes features are blended. In these cases they may be enhanced by heuristically selected algorithms. The only condition is that these algorithms do not influence the measured value of the feature. Differential filters may be used to restore spectra (Schuecker et al. 1986). Adaptive filtering (Richter 1978) has been applied in photometric mapping of galaxies (Capaccioli et al. 1988).

Features are coded in numerical or syntactical terms (Gonzalez and Thomason 1978). Aside from a few exceptions (Balestreri et al. 1979) features of astronomical objects are described numerically.

After the features have been determined, they are fed into a classifier. A classifier is the algorithm which, on the basis of the feature values, assigns the object to a specific group. A good classifier is expected to yield optimum probabilities for correct classification, wrong classification and rejection.

One type of classifier is the linear discriminating function. Linear discriminating functions are applied in star/galaxy separation (Maddox et al. 1988; Horstmann 1988). They can easily be implemented, but problems arise when classes overlap.

In this case non-linear classifiers yield better results. A classical example of a non-linear classifier is the Bayes classifier. Here, classification relies on 'a priori' knowledge concerning the object (e.g. frequencies of object types) and on 'a posteriori' knowledge (e.g. the 'probability density function' giving the probability of a feature value of a given class). The Bayes classifier assigns the object to the class with highest joint probability of all relevant feature values. Bayes classifiers have been applied in selecting galaxies from the IRAS Point Source Catalogue (Adorf and Meurs 1988) and in the classification of M-type stars (Schuecker et al. 1986).

If the performance of a classifier improves during operation, the classifier is called trainable. For Bayes classifiers, training is the improvement of a priori and a posteriori knowledge. The additional knowledge may be gained under the control of an expert (supervised learning) or by a stand-alone iterative process (unsupervised learning). Cluster analysis is a typical example for the latter. Various classification methods are described by Murtagh (1988). A comprehensive compilation of automation techniques in MK spectral classification is given by Kurtz (1985).

5 Implementing Systems for Automated Data Analysis

5.1 Databanks and Databases

The information gained through the analysis process must be presented in understandable form. Former generations had to live with printed catalogues and atlases and had to find cross references to related objects, phenomena, and authors by intuition and diligence. Modern astronomers find their catalogues and atlases stored on computer media, and data retrieval demands that they write programs to perform 'search and compare' operations.

Large collections of data are referred to as *databases.* The structures of databases reflect the algorithms which generated the data in given programming environments. The data cannot be exploited by algorithms from different application domains, because these algorithms differ and the structures are incompatible.

Data bank systems (Codd 1970) were developed to organise data in a way which permits efficient conversion into the different structures required by the application software and which assures optimal access to the underlying hardware. Ullman (1980) gives a general introduction to data bank techniques. Data bank systems were first implemented in *hierarchical* and *network* architectures. *Relational data banks* (RDB) are most promising for the future.

Data bank systems have three components: database, data bank management system (DBMS), and query language. The database is configurated with the DBMS. The query language is used to specify which data will be accessed by an application. The *Structured Query Language* (SQL) has been announced standard for RDB.

Prominent examples of astronomical data bases are SIMBAD (Egret and Wenger 1988) and STARCAT (Richmond et al. 1988). Large databases are generated with each new space-borne telescope. Data from these observatories must be pre-processed by a computer to remove telemetry signals and can subsequently be easily archived in computer readable form. Examples are the International Ultraviolet Explorer (Benvenuti 1982) and the Hubble Space Telescope (Lubow 1988). Earth-bound observations are archived in a less stringent manner, and not all databases are permanent, some restricting the lifetimes of data to a few years.

While the simplicity of data acquisition and storage offered by modern data processing systems is intriguing, it should not be forgotten that data held in the same database may come from different sources. Information on the generating processes of the data must be included to assure that data sets compiled from the database are meaningful. Heck and Manfroid (1988) discuss aspects of homogeneity in photometric databases. Similar investigations are needed for other databases as well.

Exchanging data between observatories is difficult, not only because the data structures differ, but also because data coding differs. In the past, images and catalogues were stored in almost any format. Only in recent years, standards like FITS have been enforced to ease the exchange of data (Ochsenbein 1985).

5.2 Programming Languages

In automated data analysis, objects cannot be represented by isolated numbers. They must be represented by data structures which reflect their properties and relations within the problem domain. As a consequence, the same object parameter may be represented in different codes, e.g. in a syntactic code for one class of objects and in a numerical code for others. Algorithms for this parameter must be able to abstract from its representation (Shaw 1984). In addition, certain objects can only be manipulated in certain ways. Thus, operations must have different effects on objects of different classes. The programming language must be capable of including such operations.

If the computer is to do something, one must tell 'him'. The usual way is to write a program in a programming language for which the computer is equipped with a compiler. Applications in astronomy today are dominated by number crunching operations and the language FORTRAN is undoubtably the appropriate favorite. In recent times, the language C gains increasing importance, because computer systems are now frequently equipped with UNIX operating systems; the close link between UNIX and C is well known (Kernighan and Ritchie 1977). C may become widely used, when major image processing systems (MIDAS, IRAF) are distributed in C.

C offers many degrees of freedom, which can either lead to a well-structured, readable or to a nasty 'do-all-in-one-statement' code. Consequently, documentation standards and rules for programming style must be enforced even at small astronomical sites.

Programming languages other than FORTRAN and C are not as widespread, but are used in restricted application domains and in instrumentation development.

FORTRAN and C belong to the class of *procedural languages*. In procedural languages one must describe explicitly how each data element is to be processed. In FORTRAN, only primitive data structures (multidimensional arrays) are included. In C, pointers ease the definition of complex structures such as trees.

From the regime of data bank languages came the impact of *non-procedural languages*. Here, one must not specify explicitly how each data element is to be processed, the computer operates on whole data sets.

Another approach to data processing was introduced with the *functional languages*. Operators are defined independently of data types. An example of such a language is LISP (LIst Processing Language, Winston and Horn 1981). LISP, like FORTRAN, was developed in the late 1950s. It had then little success, because of its high demand on computing power. Today, LISP is widely used in programming for non-numerical applications.

The programming languages listed so far regard data as passive structures which are manipulated by active algorithms. In *object-oriented* languages this separation is suspended. An *object* comprises both, the data and the algorithms which can be applied

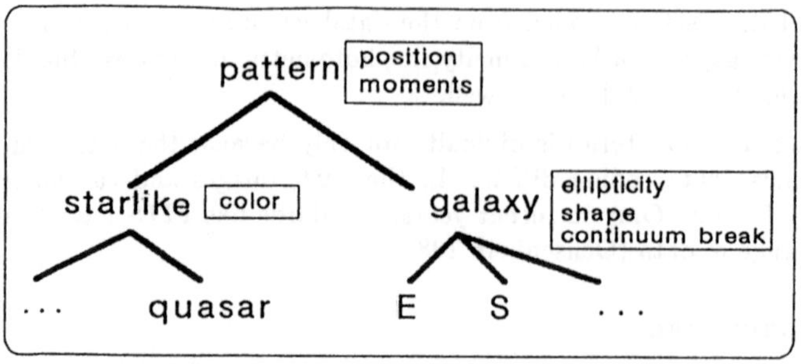

Fig. 3. Representing astronomical objects in a hierarchy

to the data. The description of an algorithm is called *method.* A method is activated when an object receives an appropriate *message.* A message specifies the addressee (identifier of the object) and the name of the requested method, and supplies arguments for the method.

Objects with the same structure are said to belong to the same *class.* A class is also treated as an object. A new object is generated by sending a message to a class. The new object, called *instance* of the class, inherits the properties of its class. Thus, hierarchical relationships among objects can easily be represented. **Fig. 3** shows an example.

Identical messages can be sent to objects which belong to different classes. The objects will select the appropriate methods for application to their data. The message 'classify spectrum' may be interpreted by objects of class STARLIKE as well as by objects of class GALAXY.

From the hierarchical structure of object-oriented languages it is apparent that object-oriented programming is a way to ease maintenance and to promote re-usability of program code. Once a method has been implemented for a class, it can be used by every instance of that class. A comprehensive discussion of selecting the right programming language is given by Bolkart (1987). SMALLTALK was one of the first object-oriented languages (Ingalls 1981). LISP has been enhanced to exhibit object-orientedness. Stroustrup (1987) introduced C++ for object-oriented programming in C-environments. C++ is chosen to become the new implementation language for the MRSP support software GAME (Teuber 1990).

5.3 Computing Hardware

Automated data analysis calls for high instruction rates. Wells (1986) discusses parallel computers for this purpose. Recently RISC machines appeared on the market which make explicit parallelism superfluous for some applications (Teuber 1990). Neural networks represent a computer architecture which deviates from the deterministic behaviour of the above processor types and promise very high processing speed. Currently, true neural network processors are not available. They are, however, emulated on computers with classical architectures.

Neural networks are based on processing elements which imitate the neurons of the human brain. A comprehensive report on how neural networks could be used in astronomy is given by Adorf (1989). **Fig. 4a** illustrates the principle. A neuron has a specific number of optionally weighted inputs x_i. If the sum of the inputs exceeds some threshold a_i, the binary output switches from 'off' to 'on' state. A Hopfield-net (**Fig. 4b**) represents the

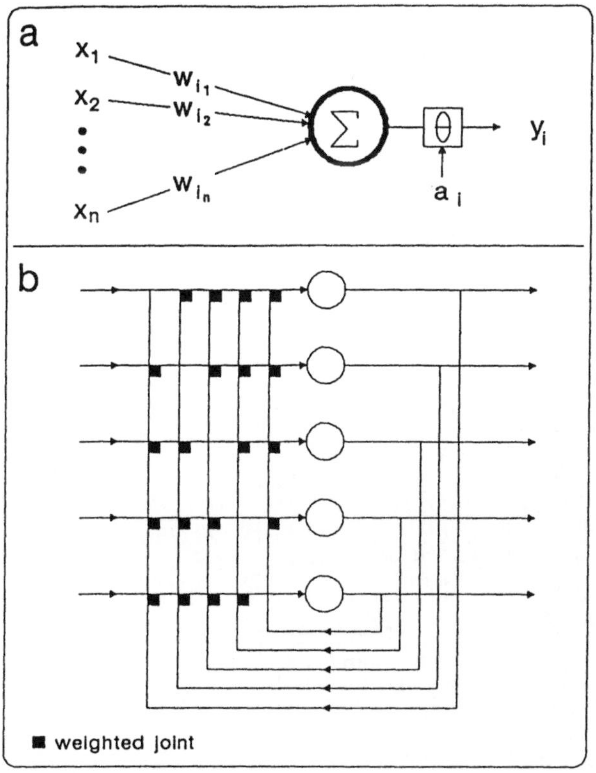

Fig. 4. Adapted from Adorf (1989). **(a)** Artificial neuron with inputs $x_1 \ldots x_n$ coming e.g. from other neurons are multiplied with weights $w_{i_1} \ldots w_{i_n}$, summed up and finally thresholded using bias a_i. **(b)** Hopfield network with five neurons.

case where the input of each neuron is connected to the output of every other neuron and to the input state. All other network topologies are special cases of this one. Neural networks have been applied to scheduling observing runs on the Hubble Space Telescope (Johnston 1989) and to sky mapping using the IRAS point source catalogue (Adorf and Meurs 1988).

6 Automated Data Analysis with MRSP

MRSP aims at deriving three-dimensional distributions of matter in the universe. It is discussed by Schuecker et al. (1989).

Fig. 5 illustrates the processing sequence for the field of an ESO-SRC sky atlas plate. Methods for automated data analysis are applied during all stages of the analysis process: image segmentation, star/galaxy separation, spectral classification and redshift determination. The MRSP is mainly concerned with extragalactic objects, but has to evaluate all objects in the region under study for this purpose. The resulting data sets will be available to the interested user after they have been transferred into an INGRES data bank system (INGRES 1988) from where they can be remotely accessed in standardised formats. Data are presently stored in a local format suitable for MRSP purposes.

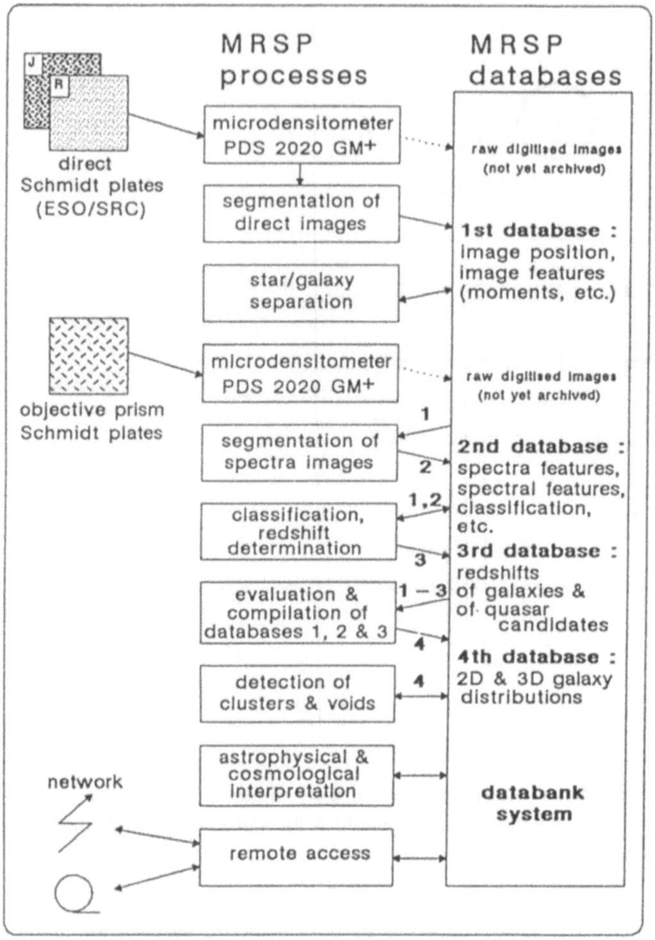

Fig. 5. MRSP analysis system

7 Concluding Remarks

Automated data analysis assists the astronomer in the decision making processes applied for extracting astronomical information from data. Automated data analysis is the step between image processing and model interpretation. It is built on human expertise in the astronomical problem domain as well as in the implementation domain. Tools developed in AI are applied (classification, expert system). Programming languages and computers are chosen to fulfil the increasing requirements. Expert systems have begun to appear in astronomy. Data banks permit the astromical community to share the large body of resulting information.

8 Acknowledgement

The author and his coworkers at Muenster thank the Deutsche Forschungsgemeinschaft for generous support of the MRSP.

References

1 H.-M. Adorf, E.J.A. Meurs: In *Large-Scale Structures in the Universe*, W.C. Seitter, H.W. Duerbeck, M. Tacke, eds., Lecture Notes in Physics, Vol. **310**, Springer, Berlin (1988), p. 315

2 H.-M- Adorf:: In *Knowledge-Based Systems in Astronomy*, A. Heck, F. Murtagh, eds., Lecture Notes in Physics, Vol. **329**, Springer, Berlin (1989), p. 215

3 R. Albrecht: In *Image Processing in Astronomy*, Proc., G. Sedmak, M. Capaccioli, R.J. Allen, eds., Osservatorio Astronomico di Trieste, Trieste (1979), p. 155

4 H.C. Andrews, B.R. Hunt: *Digital Image Restauration*, Signal Processing Series, Prentice-Hall (1977), p. 106

5 M. Balestreri, A. Della Ventura, G. Fresta: In *Image Processing in Astronomy*, Proc., G. Sedmak, M. Capaccioli, R.J. Allen, eds., Osservatorio Astronomico di Trieste, Trieste (1979), p. 268

6 P. Benvenuti: In *Ultraviolet Stellar Classification*, ESA SP-182 (1982), p. 81

7 W. Bolkart: *Programmiersprachen der vierten und fünften Generation*, McGraw-Hill, Hamburg (1987)

8 B.G. Buchanan, E.H. Shortliffe: *Rule-Based Expert Systems*, Addison-Wesley, Workingham (1985)

9 M. Capaccioli, H. Lorenz, G.M. Richter: In *Astrophotography*, Proc. IAU Workshop, S. Marx, ed., Springer, Berlin (1988), p. 157

10 D.J. Carnochan: In *Automated Data Retrieval in Astronomy*, IAU Coll. 64, C. Jaschek, W. Heintz, eds., D. Reidel Publishing Company, Dordrecht (1982), p. 63

11 K.R. Castleman: *Digital Image Processing*, Prentice-Hall, Englewood Cliffs (1979)

12 E. Charniak, D. McDermott: *Introduction to Artificial Intelligence*, Addison-Wesley, Reading (1986)

13 E.F. Codd: *A Relational Model of Data for Large Shared Data Banks*, Comm. ACM 13, 6 (1970), p. 370

14 D. Egret, M. Wenger: In *Astronomy from Large Databases*, Proc., F. Murtagh, A. Heck, eds., ESO Conference and Workshop Proceedings No. 28 (1988), p. 323

15 V. Gericke: In *Large-Scale Structures in the Universe*, W.C. Seitter, H.W. Duerbeck, M. Tacke, eds., Lecture Notes in Physics, Vol. **310**, Springer, Berlin (1988), p. 235

16 R.C Gonzalez, M.G. Thomason: *Syntactic Pattern Recognition*, Addison-Wesley, Reading (1978)

17 A. Heck, J. Manfroid: In *Astronomy from Large Databases*, Proc., F. Murtagh, A. Heck, eds., ESO Conference and Workshop Proceedings No. 28 (1988), p. 329

18 J.A. Högbom: Astron. Astrophys. Suppl. **15**, 417 (1977)

19 IHAP: IHAP Users Guide, ESO (1989)

20 D.H.H. Ingalls: BYTE, Vol. **6**, No 8 (1981)

21 INGRES: *INGRES, Release 5.0 for the UNIX Operating System*, Relational Technology, Alameda, California (1988)

22 IRAF: IRAF System Handbook, NOAO (1989)

23 C. De Jager, H. Nieuwenhuijzen, eds.: *Image Processing Techniques in Astronomy*, Proc., Astrophys. Space Sci. Lib., Vol. **54**, Reidel, Dordrecht (1975)

24 J.F. Jarvis, J.A. Tyson: Astron. J. **86**, p. 476 (1981)

25 M.D. Johnston: In *Knowledge-Based Systems in Astronomy*, A. Heck, F. Murtagh, eds., Lecture Notes in Physics, Vol. **329**, Springer, Berlin (1989), p. 33

26 B.W. Kernighan, D.M. Ritchie: *The C Programming Language*, Prentice-Hall, Englewood Cliffs (1977)

27 J.H. Kinsey: In *Astronomical Microdensitometry Conference*, D.A. Klinglesmith, ed., NASA Conf. Publ. 2317 (1984), p. 135

28 M.J. Kurtz: In *The MK Process and Stellar Classification*, Proc., R.F. Garrison, ed., David Dunlop Observatory, Toronto (1985), p. 136

29 S.H. Lubow: In *Astronomy from Large Databases*, Proc., F. Murtagh, A. Heck, eds., ESO Conference and Workshop Proceedings No. 28 (1988), p. 373

30 C.D. Mackay: Annual Review of Astronomy and Astrophys., Vol. **24** (1986)

31 R. Martin, R.K. Lutz: In *Image Processing in Astronomy*, Proc., G. Sedmak, M. Capaccioli, R.J. Allen, eds., Osservatorio Astronomico di Trieste, Trieste (1979), p. 211

32 MIDAS: MIDAS Users Guide, ESO (1989)

33 F. Murtagh, A. Heck: *Multivariate Data Analysis*, Astrophys. Space Sci. Lib., Reidel, Dordrecht (1987)

34 F. Murtagh: In *Large-Scale Structures in the Universe*, W.C. Seitter, H.W. Duerbeck, M. Tacke, eds., Lecture Notes in Physics, Vol. **310**, Springer, Berlin (1988), p. 308

35 H. Niemann: *Klassifikation von Mustern*, Springer, Berlin (1983), p. 3

36 F. Ochsenbein: Mem. S. A. It., Vol. **56**, No. 2-3 (1985), p. 363

37 S.K. Pal, D.K. Dutta Majumder: *FUZZY Mathematical Approach to Pattern Recognition*, Wiley Eastern Ltd., New Delhi (1986)

38 A. Richmond, T. McGlynn, F. Ochsenbein, F. Romelfanger, G. Russo: In *Astronomy from Large Databases*, Proc., F. Murtagh, A. Heck, eds., ESO Conference and Workshop Proceedings No. 28 (1988), p. 465

39 G.M. Richter: Astron. Nachr., **299**, 283 (1978)

40 J. Rickard, W. Nees, F. Middleburg: ESO Bulletin No. 12 (1975), p.5

41 J.A. Roberts, ed.: *Indirect Imaging*, Proc., Cambridge University Press, Cambridge (1984)

42 C. van Schoonefeld, ed.: *Image Formation from Coherence Functions in Astronomy*, Proc. IAU Coll. 49, Astrophys. Space Science Library, Reidel, Dordrecht (1979)

43 P. Schuecker, H. Horstmann, C.C. Volkmer: In *Data Analysis in Astronomy II*, V. DiGesu, L. Scarsi, P. Crane, J.H. Friedman, S. Levialdi, eds., Plenum Press, New York (1986), p. 109

44 P. Schuecker: In *Large-Scale Structures in the Universe*, W.C. Seitter, H.W. Duerbeck, M. Tacke, eds., Lecture Notes in Physics, Vol. **310**, Springer, Berlin (1988), p. 142

45 P. Schuecker, H. Horstmann, W.C. Seitter, H.-A. Ott, R. Duemmler, H.-J. Tucholke, D. Teuber, J. Meijer, B. Cunow. These proceedings (1989)

46 U.J. Schwarz: In *Image Formation from Coherence Functions in Astronomy*, Proc. IAU Coll. 49, Astrophys. Space Science Library, Vol. **76**, Reidel (1979), p. 261

47 G. Sedmak, M. Capaccioli, R.J. Allen, eds.: *Image Processing in Astronomy*, Proc., Osservatorio Astronomico di Trieste, Trieste (1979)

48 M. Shaw: In *On Conceptual Modelling*, M.L. Brodie, J. Mylopoulos, J.W. Schmidt, eds., Springer, New York (1984), p. 49

49 R.S. Stobie: In *Applications of Digital Image Processing in Astronomy*, Proc., D.A. Elliot ed., SPIE Vol. **264**, (1980), p. 208

50 R.S. Stobie, B. McInnes, eds.: *Workshop on Astronomical Measuring Machines*, Occasional Reports of the Royal Observatory, Edinburgh (1983)

51 B. Stroustrup: *The C++ Programming Language*, Addison-Wesley, Reading (1987)

52 D. Teuber, R. Budell, C.C. Volkmer, M. Hiesgen: Sterne, **63** (1987), p. 9

53 D. Teuber: In *Large-Scale Structures in the Universe*, W.C. Seitter, H.W. Duerbeck, M. Tacke, eds., Lecture Notes in Physics, Vol. **310**, Springer, Berlin (1988), p. 323

54 D. Teuber, P. Schuecker, H. Horstmann: In *Knowledge-Based Systems in Astronomy*, A. Heck, F. Murtagh, eds., Lecture Notes in Physics, Vol. **329**, Springer, Berlin (1989), p. 53

55 D. Teuber: (New computer hardware for the MRSP), to be published (1990)

56 M. Thonnat, A. Bijaoui:: In *Knowledge-Based Systems in Astronomy*, A. Heck, F. Murtagh, eds., Lecture Notes in Physics, Vol. **329**, Springer, Berlin (1989), p. 107

57 J.D. Ullman: *Principles of Database*, Computer Science Press, Rockville, MD (1980)
58 P.H. Winston, B.K.P. Horn: *LISP*, Addison-Wesley, Reading (1981)
59 P.H. Winston: *Artificial Intelligence*, Addison-Wesley, Reading (1984)
60 D.C. Wells: In *Data Analysis in Astronomy I*, V. DiGesu, L. Scarsi, P. Crane, J.H. Friedman, S. Levialdi, eds., Plenum Press, New York (1985), p. 195
61 D.C. Wells: In *Data Analysis in Astronomy II*, V. DiGesu, L. Scarsi, P. Crane, J.H. Friedman, S. Levialdi, eds., Plenum Press, New York (1986), p. 315
62 L.A. Zadeh: IEEE Trans. Syst., Man and Cybern., Vol. **SMC-3**, p. 28 (1973)

MIDAS

P. Grosbøl

European Southern Observatory, Karl-Schwarzschild-Str. 2,
D-8046 Garching, Fed. Rep. of Germany

The Munich Image Data Analysis System (MIDAS) made by the European Southern Observatory is reviewed with emphasis on the new portable version which supports both VMS and UNIX systems. The general facilities available are described including the current support and distribution policy. The hardware and software needs for installing MIDAS are summarized with indications of the requirements for a minimum system on which it is possible to perform data reductions.

1 Introduction

Image processing has a long tradition in ESO where digital data reductions were introduced together with the first digital detector systems used at the La Silla Observatory. The developments of ESO's first Image Handling And Processing system IHAP was started around 1975 by F. Middelburg. It was implemented on HP 1000 computers under the HP-RTE operating system which was used for data acquisition at telescopes. By 1978, IHAP was in regular use mainly for reduction and analysis of spectral data from *e.g.* the IDS detector or direct images from photographic plates.

With the introduction of new 32-bit mini-computers in 1980/81 (*e.g.* VAX 11/780 from Digital), it was decided to start the development of a second generation image processing system which especially should make it easy to add application software. The design of the Munich Image Data Analysis System MIDAS was started in 1981 by Klaus Banse (see Banse *et al.* 1983). The implementation was done mainly in VAX-FORTRAN with usage of VAX/VMS specific features. This version of MIDAS was in general usage by 1985 at which time also regular releases to other institutes were made.

Starting around 1983, workstations consisting of local processor, storage and display became available. These systems offer very interesting features for interactive image processing. They mostly run under the UNIX operating system and could therefore not be used directly with MIDAS. Due to this fact and the general acceptance of UNIX as a vendor independent operating system, a new portable version of MIDAS was started in 1986. The first official release of this portable MIDAS was made in November 1988 with support of both VMS and general UNIX systems.

This paper describes the general design features of the portable MIDAS system. Available applications and different options for adding new code to the system are discussed. The current distribution and support policy is reviewed. Finally, general hardware requirements for running the MIDAS software are defined.

Reviews in Modern Astronomy 2 Editor: G. Klare

2 The MIDAS System

The main objective of MIDAS is to provide a basic image processing environment for reduction and analysis of astronomical data in which it is easy to add new applications also for non-professional programmers. It consists of a monitor which controls the execution of individual tasks and a large set of image processing applications. The aim for the portable version was to make it available on a wide variety of computers so that user would have a significant choice of hardware configuration also accommodating their other needs. Further, consideration were made to make only the minimum amount of changes at the user and programming interface levels in order to limit the difficulties of migration from the old VMS version to the portable.

2.1 General features

The main features of the portable version of MIDAS are summarized below:

- Support of computers with VMS and UNIX operating system.
- Device independent interfaces to peripherals by use of special libraries *e.g.* AGL for plotting, IDI for image display and TW for terminals.
- Support display and hard-copy standards like X11 and PostScript.
- Support of standard compliers like FORTRAN-77 and C.
- Full support of data exchange in FITS format.
- On-line help, history and logging facilities.
- Flexible command/control language with full flow control and debugging facilities.
- Easy programming environment at different levels.
- Extensive subroutine libraries in FORTRAN and C with access to the MIDAS data base *e.g.* ST- and TB-interfaces.

In orded to port the software between different systems, MIDAS is using a layered structure as show in Figure 1. The only system dependent layer of MIDAS is the OS-interfaces which bind the MIDAS code to the operating system. Standard versions of OS-routines exist for different UNIX implementations and for VAX/VMS. Over this layer, standard MIDAS interfaces for data access *i.e.* SC- and TC-routines are places. These routines and the monitor are the only parts

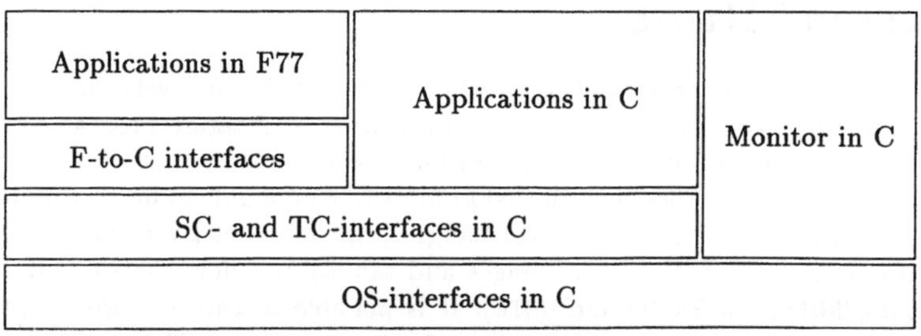

Figure 1: Layered structure of portable MIDAS

allowed to use the OS-levels, except for very special data exchange applications. All normal application programs are using only the MIDAS libraries. Programs written in C can directly use the standard libraries in C while FORTRAN application access them through a FORTRAN-to-C interface which gives the ST- and TB-subroutine libraries. This FORTRAN-to-C interface is compiler dependent and must also be verified during a new installation of MIDAS.

2.2 Data Structures

Five types of data structures are defined in MIDAS, namely: keywords, descriptors, images, tables and catalogs. Keywords and descriptors can store arrays of a given type *i.e.* strings, integers and real in either single or double precision. Keywords are associated to the monitor and provide local variables while descriptors are linked to data files. Images contain data matrices with less than 16 dimensions and sampled at equal step. Table on the other hand can store heterogeneous data (see Grosbøl and Ponz, 1985). Finally, catalogs are lists of data files such as images or tables.

3 MIDAS Applications

The individual commands in MIDAS can be divided in the three main categories depending on their implementation and usage. The first group is the intrinsic monitor commands which are build into the monitor and therefore can be invoked very fast. They consist of simple utilities for manipulation of keywords, debugging etc. Commands can otherwise be classified as basic, being available in all contexts, or application specific that is associated to a given application package.

The basic commands include all general utilities such as simple data manipulation with descriptors, images or tables, image display and graphics utilities, and data exchange facilities. The general data manipulation routines consist of a large variety of commands from simple arithmetic over trivial analysis of statistics to elaborate operations like digital filters, transforms and rebinning. The display commands include a full set of applications for cursor interaction and visualization of image data. Full support of data in FITS format is given by the data exchange tasks which both accept data from standard tape devices and directly from disk. Compatibility with the old IHAP format is also provided.

Application packages contain set of commands which is used in a more specific context such as special reduction procedures for instruments. MIDAS offers extensive packages in the areas of spectral reductions including data in long slit and echelle formats, CCD observations, crowed field photometry (`ROMAFOT`), object search and classification (`INVENTORY`), fitting and modeling of data, astrometry and statistical analysis.

4 Programming in MIDAS

It is possible to program reduction sequences and application at two different levels, namely: a) in procedures using the MIDAS Control Language or b) by writing application programs in either FORTRAN or C using the MIDAS interface libraries for access to data structures.

Procedures are written as normal text files with the list of MIDAS commands to be executed. They have access to all facilities in the MIDAS Control Language such as local variables (*e.g.* KEYWORDS), other data types (*e.g.* descriptors, images and tables) and flow control statements (*e.g.* `IF-THEN/ELSE/ENDIF`, `DO/ENDDO` and `GOTO`). It is possible to call one procedure from another down to 7 levels which provides a good way of structure them. They need not be compiled making them fast and easy to develop especially for packaged reduction sequences which only needs commands already available in MIDAS.

New algorithms can be programmed in the MIDAS environment using standard compiler such as FORTRAN and C. Due to portability reasons, it is recommended to write in standard ANSI FORTRAN-77 or C. It is however possible to use five extension to FORTRAN, namely: a) `IMPLICIT NONE`, b) `INCLUDE`, c) `DO/ENDDO`, d) exclamation mark comments and e) local variable with 16 characters including underscore. These extension are removed by a pre-compiler distributed with MIDAS. Application have access to all MIDAS data structures (*i.e.* keywords, descriptors, images and tables) through an extensive set of interface routines provided for both FORTRAN and C. Further, a number of general facilities are available such as error handling, log-files and file utilities. The size of these libraries have been minimized in order to make them easy to used for astronomers. Such programs can then be executed from the MIDAS monitor and used in procedures. Since they have to be complied and linked, the development is normally slower, however, for computer intensive tasks they give a higher efficiency.

5 Distribution and Support

The system is released twice a year, currently in May and November. Releases are provided free of charge to non-profit astronomical research institutes and distributed in source code including installation procedures and documentation. Institutes wishing to use MIDAS must send a formal request to ESO. Normally, releases are sent to individual user sites, however, agreements can be made so that a single organization in a country receives a master version and than redistributes it to other local sites. Together with the releases, the MIDAS manual and other relevant documentation are updated. Due to the large overhead in producing printed manuals only a limited number are sent to sites while individual users have to request documentation and updates explicitly.

The last version of the original VMS version of MIDAS was used by almost 100 sites around the world but mostly in Europe. The first releases of the portable version (*i.e.* 88NOV and 89MAY) have been requested by approximately 40 sites many of those being new installations on UNIX computers *e.g.* workstations.

The user support for MIDAS consists of documentation, hot-line service and user meetings. The general functionality and the individual image processing commands are explained in the MIDAS User Guide which has two volumes, the first dedicated to the basic system while the second deals with specific application packages. Beside the User Guide , separate documents describe other areas such as installation procedures and programming interfaces. The Hot-line service provides user with fast way of getting answers to questions concerning problems with MIDAS. Such questions can be sent through electronic mail to the MIDAS account using either SPAN or BITNET/EARN. An annual user meeting, normally in April or May, is held in connection with the ESO/ST-ECF Data Analysis Workshop. During this meeting, the current status and developments of MIDAS are presented by the Image Processing Group after which a general discussion takes place. Informations on MIDAS are also published in the MIDAS Memo which appears in each issue of the ESO Messenger.

6 Requirements for running MIDAS

The portable version can be implemented on either VAX/VMS or UNIX systems making a large variety of computer systems able to do data reductions with MIDAS. The basic requirements for installing MIDAS are as follows:

- a computer running either the VAX/VMS (version 4.7 or later) or a UNIX (System V or BSD) operating system typically with 4-8 Mbyte of memory and 200-300 Mbyte of disk storage. The memory size required mainly depends on the amount memory used by the display/window managing system and on the size of data files normally processed. For a X11 window system, 8 Mbyte is sufficient to avoid swap overhead for 512 × 512 images but larger memory is needed for 1024^2 frames. The MIDAS system is distributed in source code which including documentation takes approximately 25 Mbyte. A full MIDAS installation may take up to 100 Mbyte of disk space depending of the type of CPU but can be significantly reduced if not all applications are implemented. Since the operating system and utilities take another 50-100 Mbyte, a system for reduction of images should at least have 300 Mbyte disk storage.
- image and graphics display. MIDAS uses the Image Display Interfaces for all applications which access image display devices and is therefore device independent. It is possible for an institute to write a special set of driver routines for a given device. The window management/display standard for workstations is the X11 system which is the only one officially supported by MIDAS. This system is available from most major vendors *e.g.* DEC, SUN, HP and Apollo. For graphics, all devices supported by the Astronet Graphic Library are available *e.g.* Tektronix 4010 or X11 for terminal display while hardcopies can be done devices using *e.g.* PostScript or HPGL.
- FORTRAN-77 and C compliers. All code are written in either FORTRAN or C and have to be compiled during the installation. Most applications use ANSI FORTRAN-77 with the five extension mentioned above which are removed by a preprocessor. The C code for the monitor and low level interfaces conform to the Kernighan and Ritchie definitions. For VMS systems which does not have a C compiler possibilities of using public domain compilers are being investigated. A few routines are using the NAG library, however, the system can be implemented without.
- data input/output devices. It is necessary to include means to import and export data. This can either be done through a standard 9-track 1/2" tape drive, helical scan tapes or networks. The interfaces to physical devices vary very much from system to system. MIDAS uses a simple I/O interface for accessing such units. Although some version are provided in the standard release, each site normally would have to create their own version. The input and output commands in MIDAS also access data directly from disk which make transfer of data through networks possible.

A wide range of systems fulfill these requirements. The choice of a given configuration may depend on the kind of data to be reduced or other application which should be executed. An extensive set of benchmarks on workstations with the basic portable MIDAS system excluding image display and graphics was done by the Image Processing Group (see Grosbøl *et al.* 1988).

7 Future Developments

Besides addition of new applications, a number of new features for MIDAS is under consideration. These include better support of a distributed data base and introduction of a client-server model for making use of all resources in a Local Area Network (*e.g.* a central compute server). The adoption of a client-server model would further make it possible to introduce different user interfaces tailored to specific application with *e.g.* menus, icons and mouse interaction. This would also include a upgrade of the present syntax of the MIDAS Control Language in order to make it context-free. Finally, support for data acquisition is anticipated in the sense that a MIDAS off-line reduction session should communicate with data acquisition tasks providing a fast view, preliminary reduction facility during observations at La Silla. The exact time table for the implementation of the above mentions features depends mainly on the manpower resources available in the Image Processing Group.

8 Acknowledgements

The IHAP system was designed and developed by the late Frank Middelburg who laid the foundation of the image processing tradition in ESO with his innovative ideas. The MIDAS system was designed by Klaus Banse and implemented by the Image Processing Group in ESO with major contributions by Daniel Ponz, Rein Warmels, Carlos Guirao and Charlie Ounnas. The plotting package in MIDAS is using the Astronet Graphics Library which is kindly made available by the Italian Astronet. Many application packages in MIDAS have been provided by other individual or institutes to whom the MIDAS group is grateful.

References

Banse,K., Crane,Ph., Ounnas,Oh., Ponz,D. : 1983, 'MIDAS' in *Proc. of DECUS*, Zürich, p.87.

Grosbøl,P., Ponz,D. : 1985, 'The MIDAS Table File System' in *Mem.S.A.It.*, **56**, 429.

Grosbøl,P., Banse,K., Guirao,C., Ponz,D., Warmels,R. : 1988, 'MIDAS benchmarks of Work-stations' in *ESO Messenger*, **54**, 59.

The Sun's Differential Rotation*

M. Stix

Kiepenheuer-Institut für Sonnenphysik,
Schöneckstr. 6, D-7800 Freiburg, Fed. Rep. of Germany

Summary. Equatorial zones of the Sun rotate more rapidly than high-latitude regions. Modern observational results of this well-known phenomenon are reviewed. In addition, I review results concerning some probably related phenomena: meridional circulation, the correlation between latitudinal and longitudinal motions on the solar surface, a possible dependence on latitude of the surface temperature, and direct rotational effects upon solar convection. Temporal variations of some of these phenomena, notably the "torsional oscillator", are also mentioned; most of the variations should be either statistical or consequences of the solar magnetic cycle.

The theory of the Sun's differential rotation offers *mean* and *explicit* models. In both kinds of models correlations between certain fluctuating variables, above all the *Reynolds stresses*, maintain the non-uniformity of the angular velocity against diffusive decay.

In the mean, or *mean-field*, models the correlations are related to the large-scale flow by means of simple equations, containing parameters such as the *anisotropic viscosity*, or as the *latitude-dependent coefficient of heat transport*. With the explicit models, on the other hand, an attempt is made to calculate the Reynolds stresses and other correlations ab initio by a numerical integration of the detailed equations of motion.

The aim of all models is not only to explain the rotation observed at the solar surface, but also to predict the form of the angular velocity *within* the convection zone. Only some of the mean models have been successful in this respect (if we take for true the rather conical isorotation surfaces recently produced by helioseismology). The explicit models continue to yield cylindrical isorotation; possible reasons for this are briefly discussed.

1. A Historical Remark

In his "Rosa Ursina", which appeared in 1630, Christoph Scheiner repeatedly mentions that sunspots in close proximity to the solar equator move more rapidly across the solar disc than spots further away from the equator. Figure 1 is an example. In fact, with his large set of observations Scheiner should have been able to derive the law of rotation, i.e. the form of the dependence on latitude of the angular velocity, Ω.

In Scheiner's time the idea of a differentially rotating star was not really appreciated — it was even disputed by Galileo Galilei. Thus, the rotation law was derived for the first time only much later, namely by Carrington (1863) who found $\Omega \sim \sin^{7/4} \psi$, where ψ is the heliographic latitude.

In Chap. 2 of the present paper I shall review some of the more recent observations. The emphasis will be on results rather than on methods. The

* Mitteilung aus dem Kiepenheuer-Institut Nr. 315

Reviews in Modern Astronomy 2 Editor: G. Klare

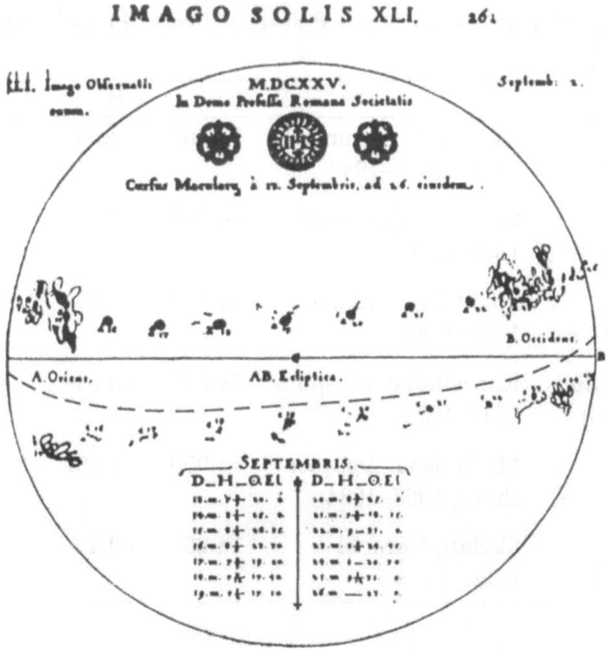

Fig. 1. Sunspot positions, from 13 to 26 September 1625, drawn by C. Scheiner. The northern group clearly moved more slowly than the southern, which is closer to the equator (*dashed line*, added for this reproduction).

latter, together with their possible error sources, have been discussed by Schröter (1985). But I shall also collect results concerning related phenomena such as large-scale velocity fields other than the rotation itself, or as the dependence on latitude of the solar surface temperature. Then, in Chaps. 3 and 4, I shall present current theoretical models, and shall discuss whether (and perhaps why) some of them fail.

A recent review of large-scale solar velocity fields has been given by Bogart (1987). For the present paper, I draw some material from the new monograph by Rüdiger (1989), and from the volume edited by Durney and Sofia (1987). The subject is also treated in Chap. 7 of Stix (1989).

2. Observations

2.1 Rotation Laws

The angular velocity at the solar surface is now commonly written in the form

$$\Omega = A + B\sin^2\psi + C\sin^4\psi \quad . \tag{1}$$

For a number of measurements the constants A, B, and C are listed in Tab.

Table 1. Coefficients for the solar differential rotation (sidereal, in degrees/day).

		A	B	C
Newton and Nunn (1951)	Greenwich, recurrent spots, 1878–1944	14.368	−2.69	
Balthasar et al. (1986)	Greenwich, all spots 1874–1976	14.551	−2.87	
Howard et al. (1984)	Mt. Wilson, all spots 1921–1982	14.552	−2.84	
Hanslmeier and Lustig (1986)	Kanzelhöhe, all spots 1947–1985	14.397	−2.64	
Snodgrass (1984)	Mt. Wilson, Doppler shifts, 1967–1984	14.050	−1.492	−2.606
Timothy et al. (1975)	Skylab, Coronal holes, 1973	14.23	−0.4	

1. Most entries to this table give only A and B because only low-latitude data have been used. The close coincidence of the Greenwich and Mt. Wilson results (those based on *all* spots) is remarkable. Recurrent spots rotate slightly slower, as exemplified by the result of Newton and Nunn (1951). Still slower is the rotation of the solar surface when measured by the Doppler shift of spectral lines, although scattered light in the spectrograph might account for part of the effect (Schröter, 1985). These different rotation rates can be interpreted as an inward increase of the angular velocity: the motion of sunspots is coupled to the motion of the solar gas somewhere beneath the surface, and the coupling of the young spots (contained only in the full samples) occurs in slightly deeper layers, with higher angular velocity, than that of the older recurrent spots. In comparison to the total depth of the convection zone the coupling depth is however small for all kinds of spots, only a few thousand kilometers, as estimated by Schüssler (1987).

Hanslmeier and Lustig (1986) also used *all* spots contained in the material collected at the Kanzelhöhe observatory. Nevertheless, their A and B are closer to the values found by Newton and Nunn (1951) than to those of Balthasar et al. (1986) and Howard et al. (1984). The reason for this is not clear; one possibility might be that, due to changing weather conditions, young spots often could not be followed on the disk during their short life, and that, therefore, the sample is weighted towards the long-lived recurrent spots.

The inferred inward increase of the angular velocity in the outermost part of the Sun's convection zone is not seen in the helioseismological results of Brown et al. (1989), because these authors use only harmonics up to $l = 99$. But it does show up in the work of Hill et al. (1988) who included harmonics up to $l = 995$, and so could investigate the layer immediately below the surface. Conservation of angular momentum in the supergranular flow could be responsible for the effect (Gilman and Foukal, 1979).

The final entry to Tab. 1 demonstrates the much smaller latitudinal gradient of Ω, i.e. the much smaller $|B|$, which is typical for structures of very large scale. For the large-scale magnetic field (which of course governs the shape of the coronal holes referred to in the table) the rotation curve can be directly derived by inspection of consecutive solar magnetic charts (Bumba and Hejna, 1987); it can also most conveniently be obtained by an autocorrelation analysis (e.g. Stenflo, 1989). Hoeksema and Scherrer (1987) determined the rotation rate of the coronal magnetic field calculated from photospheric measurements.

Stenflo (1989) has attributed his rotation rate for the large-scale field to a layer near the base of the solar convection zone, at a depth of $\simeq 2 \times 10^8$ m, where probably all solar magnetism has its origin (e.g. Schüssler, 1983, 1984; Stix, 1987b). By linear interpolation to a surface rotation law he has then constructed the contours of isorotation within the convection zone. The result strikingly resembles the the isorotation curves found by helioseismological methods, and also the isorotation curves *predicted* by some of the mean-field models (cf. Fig. 5 below).

2.2 Meridional Circulation

The term "meridional circulation" generally refers to a large-scale axisymmetric flow which lies in meridional planes, viz.

$$\mathbf{v}_m = (v_r(r,\theta),\ v_\theta(r,\theta),\ 0) \tag{2}$$

in spherical polar co-ordinates (r, θ, ϕ). It is clear that such a flow is capable of transporting angular momentum. In fact, Kippenhahn (1964) has shown that the result of such transport could well be an accelerated equatorial zone as observed, if only the sense of the circulation cell is equatorwards in its upper, and polewards in its deeper part.

Unfortunately there are no safe observational results concerning meridional circulation on the Sun. Generally the uncertainties are of the same order as the velocities actually measured, namely a few meters per second. Doppler shifts of spectral lines indicate a small poleward flow, with a "consensus" of 10 – 20 m/s (LaBonte and Howard, 1982). Topka et al. (1982) found a flow of the same magnitude and direction in latitude zones above 20°, where they used filaments as tracers. Another kind of tracer is provided by the chromospheric Ca II emission network: the proper motion of the individual network mottles occasionally indicated a circulation of, again, a few tens of meters per second, with varying sense (and with very large error bars), but no significant stable pattern has yet been established (Schröter et al., 1978).

Table 2. Meridional circulation, derived from proper motions of sunspots. The symmetric and antisymmetric parts (with respect to the equator) are given separately; ψ is heliographic latitude.

		symmetric	antisymmetric
Tuominen (1942)	Greenwich, 1874–1935	$\simeq 2$ m/s $\lvert\psi\rvert < 16° \rightarrow$ eq. $\lvert\psi\rvert > 16° \rightarrow$ poles	$\simeq 1$ m/s $\lvert\psi\rvert < 16°$ N$\rightarrow$S
Hanslmeier and Lustig (1986)	Kanzelhöhe, 1947–1985		$\simeq 4$ m/s $\lvert\psi\rvert < 35°$ N$\rightarrow$S
Balthasar et al. (1986)	Greenwich, 1874–1976		"tendency" N$\rightarrow$S
Howard and Gilman (1986)	Mt. Wilson 1917–1983	$\simeq 1.5$ m/s $\lvert\psi\rvert < 25° \rightarrow$ eq. $\lvert\psi\rvert > 25° \rightarrow$ poles	1–3 m/s see Fig. 2

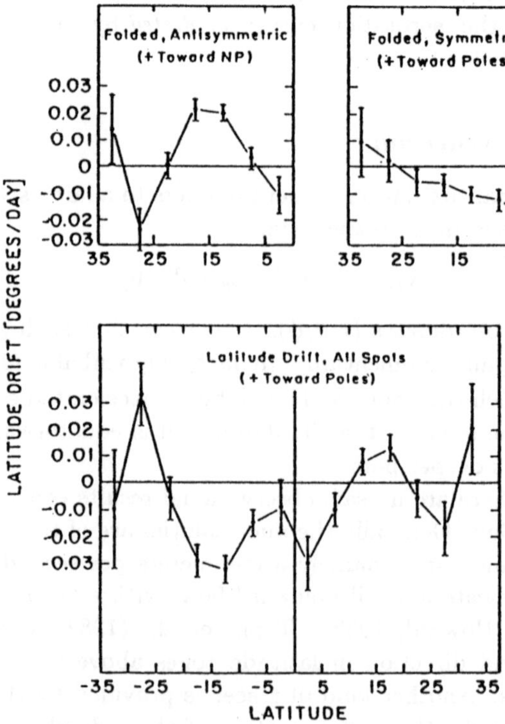

Fig. 2. Average latitudinal proper motion of sunspots observed at Mt. Wilson, 1917–1983. *Top:* symmetric and antisymmetric parts; *below:* full latitudinal dependence. From Howard and Gilman (1986).

The most often used tracers of solar meridional circulation are the sunspots. Table 2 lists the classical work of Tuominen (1942) together with a few modern results, and Fig. 2 illustrates the rather detailed analysis of Howard and Gilman (1986). The latter essentially confirms the finding of Tuominen (1942) of an equatorward flow at low latitude, and a poleward flow at higher latitude. It is remarkable that the flow is not symmetric with respect to the equator. However, large error bars appear in this as well as in all other investigations of the subject.

Perhaps at this state of affairs the conclusion should merely be that theoretical models must not predict steady circulation velocities at the solar surface beyond, say, 10 m/s (cf. also Rüdiger, 1989, p.80). Nevertheless, the establishment of a significant circulation on the Sun remains an interesting observational challenge; it would help to discriminate between the diverse models.

2.3 Giant Cells and Reynolds Stresses

Convection patterns which have scales significantly larger than the supergranulation, but still smaller than the whole Sun, are often called "giant cells". Other names are "giant granulation" which alludes to the possible common nature of granulation and supergranulation, or "global convection" which indicates that such a pattern may pervade the entire depth of the convection zone. The typical scale of giant cells is 10^8 m.

Concerning the Sun's differential rotation, giant cells are of interest for two reasons. The first is that the explicit models described below predict a more or less *regular* pattern of non-axisymmetric global convection; the second is that correlations of the form

$$Q_{\theta\phi} = < v_\theta v_\phi > \ , \tag{3}$$

also called *Reynolds stresses*, play a significant role in the transport of angular momentum.

Doppler as well as tracer methods have been employed in the search of giant cells on the solar surface. Depending on scale (or, equivalently, on longitudinal wave number) measured values or upper limits around 10 m/s have been found, see e.g. the review of Bogart (1987). However, no long-lived regular pattern has yet been detected, although occasional indications do exist. One example is a regular arrangement of solar filaments documented by Wagner and Gilliam (1976), another the velocity pattern observed by Cram et al. (1983) during a 20-day interval in the polar regions of the Sun.

If — in an otherwise random velocity field — non-zero correlations of form (3) exist, then they should have their origin in a *global* force. The Coriolis force is an obvious candidate. In comparison to the other inertial forces the influence of the Coriolis force is measured by the inverse of the Rossby number

$$\mathrm{Ro} = u/2\Omega l \ , \tag{4}$$

where u and l are the velocity and the scale of the flow in the rotating frame

Table 3. Convection cells and Rossby numbers.

	u [m/s]	l [m]	Ro
granulation	1000	10^6	200
supergranulation	500	10^7	10
giant cells	10	10^8	0.02

of reference. For giant cells the Rossby number is smallest, cf. Tab. 3, and we must therefore expect the the largest rotational effect.

Observed Reynolds stresses are listed in Tab. 4. These values are averages over latitude, with the data from the southern hemisphere, where $Q_{\theta\phi}$ is negative, multiplied by -1. The reversal of sign at the equator is a good indication that the solar rotation is the cause for the non-vanishing correlation. The detailed latitude dependence, as derived, e.g., by Ward (1965) and by Gilman and Howard (1984b), shows an approximately linear increase of $|Q_{\theta\phi}|$ on both sides of the equator; it is shown in Fig. 3.

Table 4. Correlation $|Q_{\theta\phi}|$, in units of $10^3(\mathrm{m/s})^2$, between longitudinal and latitudinal velocity conponents.

Ward (1965)	sunspot groups	1.5
Belvedere et al. (1976)	Ca II faculae	4
Schröter and Wöhl (1976)	Ca II network mottles	4
Gilman and Howard (1984b)	sunspot groups	2
	individual spots	0.6
Balthasar et al. (1986)	sunspot groups	2

We may ask whether it is possible to see directly the influence of solar rotation on individual convection cells. Due to their very slow motion, giant cells are difficult to identify. Therefore we may resort to the largest well-documented structure, the supergranulation (although we must be aware that the effect upon giant cells should be much larger, and of more importance for the understanding of the Sun's differential rotation). Kubičela (1973) has found a vortical structure in the supergranular flow which, according to sign and magnitude, could indeed be caused by the Coriolis force. Rimmele and Schröter (1989) found a variation with latitude of the size and velocity of supergranulation cells, while Brune and Wöhl (1982) and Münzer et al. (1989) found a similar variation in the mesh size of the chromospheric Ca II network.

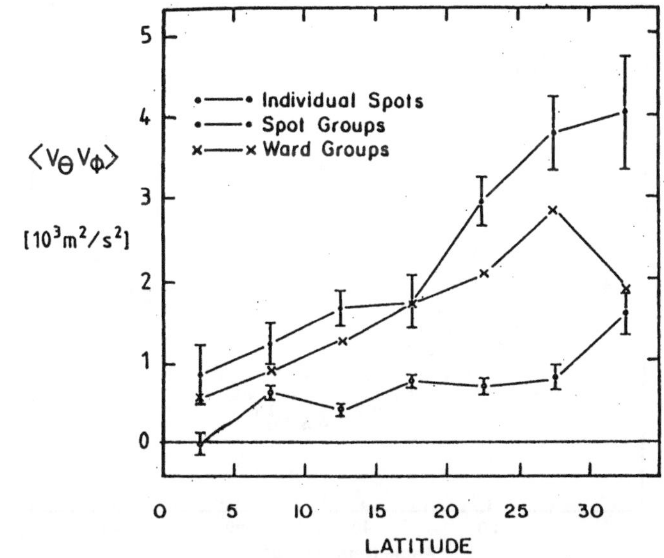

Fig. 3. Correlation $Q_{\theta\phi}$ between longitudinal and latitudinal velocity components, as a function of latitude. After Gilman and Howard (1984b).

2.4 Pole-Equator Temperature Differences

Early measurements, starting in 1933 with Abetti and Castelli, of differences in temperature between the solar equator and the polar regions have been discussed by Beckers (1960). Such a difference,

$$\Delta T = T_{pole} - T_{equator} \quad ,$$

can be obtained by a measurement of polar and equatorial intensities of Fraunhofer lines or of narrow continuum bands in the spectrum. Beckers concluded that there was evidence for a solar-cycle dependent ΔT, of relative magnitude $\simeq -3\%$ during activity minimum and $\simeq +1.5\%$ during activity maximum. Results of similar magnitude were obtained by Plaskett (1970), although this author's conclusion was a *steady* difference, with the poles hotter that the equator by several hundred degrees.

Plaskett found his result attractive because of his interpretation of the Sun's differential rotation as a "thermal wind": the azimuthal flow resulting from the balance between the Coriolis force and the latitudinal pressure gradient maintained by the temperature difference. The amount of ΔT required depends on the depth in the Sun into which the non-uniform angular velocity reaches. Plaskett needed his large ΔT because he considered a layer of only $\simeq 150$ km thickness; Gilman (1968), who included the whole convection zone into his thermal-wind model, found $\Delta T \simeq 4$ K sufficient.

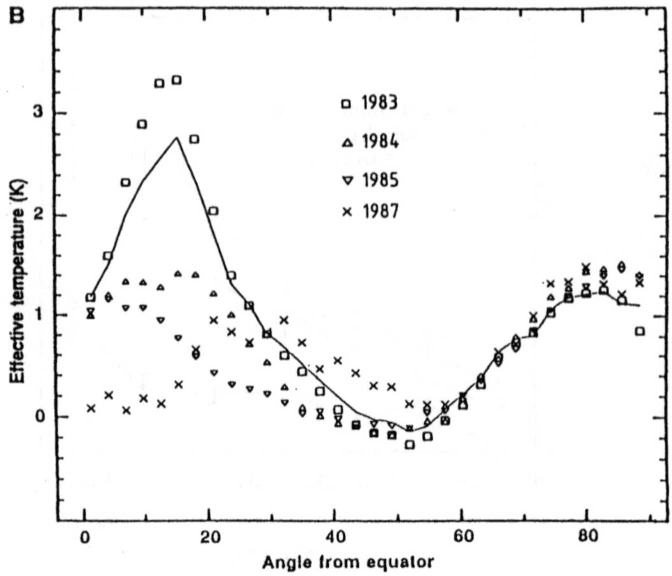

Fig. 4. Effective temperature as a function of heliographic latitude, for the period 1983–1987. After Kuhn et al. (1988).

Most modern determinations of the pole-equator temperature difference yield still smaller ΔT and thus seem to rule out the thermal wind model (which, in any case, is probably too simple for the solar convection zone where the deviation from the *adiabatic* gradient is essential, rather than the temperature gradient itself). Appenzeller and Schröter (1967), Caccin et al. (1970), and Falciani et al. (1974) all give upper limits rather than significant non-zero values. The most stringent limit was set by Altrock and Canfield (1972), who found $\Delta T = 1.5 \pm 0.6$ K, with a possible systematic error of 3–4 K.

In recent years Kuhn et al. (1988), using an instrument originally designed to measure the solar oblateness, determined the effective temperature as a function of latitude. As shown in Fig. 4, they found a cycle-dependent peak (maximum height $\simeq$ 3 K) at low latitude, and a stationary peak (of height $\simeq$ 1.5 K) at higher latitude. While the former probably is a secondary effect of the solar dynamo, the latter, which is of the same magnitude as the ΔT obtained by Altrock and Canfield, might well be related to the mechanism which maintains the differential rotation. If not as *cause* of differential rotation (as in the thermal wind model) a non-zero ΔT may arise as *by-product* since any mechanism which transports angular momentum will also transport heat. Our models must satisfy the constraint set by the smallness of the observed ΔT.

2.5 Variations

The physical quantities described in Sects. 2.1–2.4 above show temporal variations which may be divided into two categories. The first comprises the statistical variations. These might account for some of the diverging results obtained by diverse observational studies, in particular when the observing periods were short. An example could be the meridional circulations derived from proper motions of Ca II mottles in 1974 (Schröter and Wöhl, 1975) and 1976 (Schröter et al. (1978).

To the second category belong the long-term variations, e.g. that of the effective temperature shown in Fig. 4, which are probably — and in some cases certainly — related to the 11-year solar activity cycle. I shall take here the conservative point of view that the solar cycle is a consequence of the differential rotation, and *not* vice-versa. This view mainly rests on an energy argument (e.g. Babcock, 1961). Cycle-related variations should therefore serve as *diagnostic tools* rather than as essential ingredients to the theory of differential rotation. Besides the already mentioned variation of temperature, the most important cyclic variations are the following:

(a) Averaged over latitude, the rate of solar rotation seems to be slightly larger during activity minimum than during activity maximum. This result has been found in sunspot proper motions; it is seen in the data from Mt. Wilson (Gilman and Howard (1984a) and from Greenwich (Balthasar et al., 1986).

(b) Superimposed on the mean latitude profile (1) is the so-called torsional oscillator. It consists of latitude bands of alternatingly faster and slower than average rotation. The bands travel towards the equator. They have been detected in Doppler signals by Howard and LaBonte (1980) and confirmed in a study of sunspot proper motions by Tuominen et al. (1983). In fact Tuominen (1952) already predicted the existence of torsional oscillations connected to the cyclic latitudinal motions which he had found.

The torsional oscillation has been interpreted as a consequence of the cyclic variation of the mean azimuthal Lorentz force which arises in the solar dynamo (Schüssler, 1981; Yoshimura, 1981).

(c) Cyclic variations of the meridional circulation have been reported first by Tuominen (1952), and then by Tuominen et al. (1983) on the basis of the Greenwich sunspot data. The meridional circulation cells seem to propagate towards the solar equator. Tuominen and Virtanen (1984) and Rüdiger et al. (1986) explain this propagation by the meridional components of the mean Lorentz force, in analogy to the above explanation of the torsional oscillator.

Nevertheless the whole subject of a variable meridional circulation remains controversial: Neither Balthasar et al. (1986), nor Howard and Gilman (1986), nor Hanslmeier and Lustig (1986) and Lustig and Hanslmeier (1987) find significant results. Also controversial are the *polewards* drifting narrow circulation cells ("azimuthal rolls") reported by Ribes et al. (1985) and Ribes and Laclare (1988); Wöhl (1988) has criticized these reports.

(d) Radiometers on the Solar Maximum Mission satellite indicate a cyclic variation of the total solar luminosity which is in phase with the activity cycle and has a relative amplitude of $\simeq 0.04\%$ (Willson and Hudson, 1988). Possibly this variation is related to the temperature variation measured at low latitude (Kuhn et al. 1988; cf. Fig. 4).

3. Mean-Field Models

The starting point of any theory for the Sun's differential rotation must be the principle of conservation of angular momentum. This is particularly so in the *mean-field* models where time-dependent fluctuations occur only in the form of averages such as the Reynolds stresses discussed above. In addition to $Q_{\theta\phi}$ let us define

$$Q_{r\phi} = < v_r v_\phi > \quad ; \tag{5}$$

the conservation of angular momentum is then expressed by

$$\mathrm{div}(\rho r^2 \sin^2\theta \, \Omega \, \mathbf{v}_m + \rho r \sin\theta \, (Q_{r\phi}, \; Q_{\theta\phi}, \; 0)) = 0 \quad . \tag{6}$$

where ρ is the density and $\mathbf{v}_m$ the meridional circulation velocity which was already defined. The two terms on the left of (6) describe the transport of angular momentum by meridional circulation on the one hand and by Reynolds stresses on the other hand. Generally both types of transport will occur, but I shall first consider models in which the Reynolds stresses play the more fundamental role.

3.1 Reynolds Stresses and Anisotropic Viscosity

In principle it is now necessary to calculate $Q_{r\phi}$ and $Q_{\theta\phi}$. To this end we should know the fluctuating velocity field. That is, we may either solve the hydrodynamic equations which govern this velocity field (e.g. Rüdiger, 1980; Kichatinov, 1986, 1987), or we may make plausible assumptions concernig, e.g., the size and shape of the individual eddies (e.g. Durney and Spruit, 1979; Durney, 1989a). In any case the result will remain more or less hypothetical: a rigorous solution of the problem of turbulent convection under anisotropic conditions (gravity, rotation) does not exist.

Let me therefore follow Rüdiger (1977, 1980) and write

$$Q_{r\phi} = -\nu_t r \sin\theta \frac{\partial\Omega}{\partial r} + \Lambda_r \sin\theta \, \Omega \quad , \tag{7}$$

$$Q_{\theta\phi} = -\nu_t s \sin\theta \frac{\partial\Omega}{\partial\theta} + \Lambda_h \cos\theta \, \Omega \quad . \tag{8}$$

In this way the Reynolds stresses are divided into diffusive and non-diffusive terms.

The diffusive terms are proportional to the gradient of Ω and describe the smoothing effect of the turbulence upon the large-scale shear. The turbulent diffusivity, ν_t, is of order $ul \simeq 10^9\ \mathrm{m^2/s}$; therefore, the solar convection zone would attain a state of rigid rotation within a few years if non-diffusive Reynolds stresses and meridional circulation were absent. The dimensionless parameter s in (8) indicates that the diffusive effect can be of different strength in the vertical and horizontal directions.

The non-diffusive terms in (7) and (8), called the Λ-effect by Rüdiger, are possible drivers of differential rotation. In the simplest case there is only one parameter, namely Λ_r. If

$$\Lambda_r = 2\nu_t(s-1) \ , \tag{9}$$

then this case corresponds exactly to an anisotropic viscosity tensor ν_{ij} which, in spherical polar co-ordinates, has the form

$$\nu_{ij} = \nu_t \begin{pmatrix} 1 & 0 & 0 \\ 0 & s & 0 \\ 0 & 0 & s \end{pmatrix} \ . \tag{10}$$

Kippenhahn (1963), Cocke (1967), and Köhler (1970) have used this viscosity tensor in their models of the Sun's differential rotation. Their work was based on earlier ideas of Lebedinsky (1941), Wasiutyński (1946), and Biermann (1951).

The anisotropic viscosity models based on (9) and (10) in general also include meridional circulation. One reason is that that the viscosity (10) directly introduces driving terms into the equation of meridional balance. The other is more subtle: suppose there would be no circulation, $\mathbf{v}_m = 0$. In this case one can show that the solution to the problem is an angular velocity which depends only on depth, or $\Omega = \Omega(r)$. The centrifugal force resulting from such Ω generally is not conservative, and thus cannot be balanced by a pressure gradient. A circulation must therefore exist. It transports angular momentum and so establishes the full dependence $\Omega(r,\theta)$.

The result of a recent version (Pidatella et al., 1986) of an anisotropic viscosity model is shown in Fig. 5a. The contours of constant Ω are rather similar to those revealed by helioseismological measurements (Libbrecht, 1988; note that in this case the model preceded the measurement, so it was a real prediction!). In particular, these contours are *not* parallel to the axis of rotation, as they are in the explicit models discussed in Chap. 4, or in some of the earlier models with anisotropic viscosity (e.g. Köhler, 1970). I shall return to this point in Sect. 4.2 below. The surface values of the circulation velocity, as well as of the pole-equator temperature difference, are compatible with the observational results enumerated in the preceding chapter.

On the other hand, the simple one-parameter models cannot contain the whole truth. As Rüdiger has pointed out, the sign of $Q_{\theta\phi}$ contradicts the observed sign if $\partial\Omega/\partial\theta$ is such as observed and $\Lambda_h = 0$; this is evident by mere inspection of (8). Therefore Λ_h must not be zero. Rüdiger and Tuominen (1987) have produced first models which include the additional term.

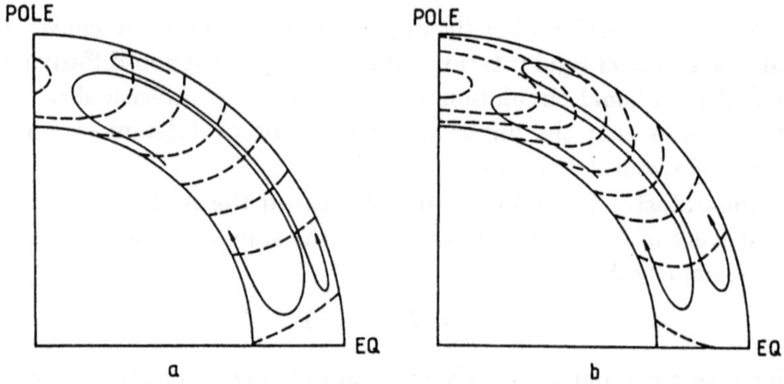

Fig. 5. Surfaces of constant angular velocity (*dashed*), and streamlines of meridional circulation for a model with anisotropic viscosity (a), and a model with latitude-dependent heat transport (b). From Stix (1987a).

3.2 Latitude-Dependent Heat Transport

Another class of mean-field models of the Sun's differential rotation makes use of the energy balance within the convection zone. I have discussed these models together with the anisotropic viscosity models in an earlier review (Stix, 1987a); the presentation here will therefore be short.

As rotation must influence the convective eddies in some way or another (cf. the observational indications mentioned above), it is pausible to assume that the coefficient of convective energy transport, and therewith the convective flux, $\mathbf{F}_C$, depends on latitude (Weiss, 1965; Durney and Roxburgh, 1971). A convenient way to model this dependence is the introduction of a latitude-dependent ratio, $\alpha(\theta)$, of mixing length to scale height (Pidatella et al., 1986). That is

$$\mathbf{F}_C = -\rho T \frac{lv}{2} \nabla S \quad , \tag{11}$$

where $l = \alpha(\theta) H$ is the mixing length; H is the scale height, v the mean convection velocity, and S the mean specific entropy. The balance of the meridional energy flux,

$$\operatorname{div}(\mathbf{F}_C + \mathbf{F}_R) + \rho T \, \mathbf{v}_m \cdot \nabla S = 0 \quad , \tag{12}$$

then leads to meridional circulation, $\mathbf{v}_m$, and to differential rotation, $\Omega(r, \theta)$, as in the models described earlier. The method, results, and also critical remarks can be found in Belvedere and Paternò (1977), Belvedere et al. (1980), Moss and Vilhu (1983), and Pidatella et al. (1986). A recent result is shown in Fig. 5b. The angular velocity distribution is similar to that in Fig. 5a, i.e. quite compatible with the helioseismological result. However, it appears that these models have a tendency of producing too large (as compared to observations) pole-equator temperature differences and circulation velocities. At present I therefore prefer the models based on anisotropic viscosity.

4. Explicit Models

4.1 Concepts and Results

As Tab. 3 illustrates, the most interesting convection cells (as far as solar rotation is concerned) are also the *largest* convection cells. This suggests that the interesting aspects of convection on the rotating Sun may be captured by a numerical integration of the equations of motion. In particular P. A. Gilman and G. A. Glatzmaier have persued this concept in a number of ever more sophisticated models, the "global convection models". The latest of these are described by Glatzmaier (1985) and by Gilman and Miller (1986); Glatzmaier (1987) gives a general review.

Even supercomputers have their limitations. Therefore the numerical grids used so far are still rather coarse. In depth $\Delta r = 10^7$ m, in latitude $\Delta\theta = 2°$ is typical. In longitude, a Fourier representation is normally chosen, with Fourier components up to, typically, $m = 36$ (corresponding to $\Delta\phi = 5°$). These numbers imply that the smaller convection structures visible at the solar surface are not resolved by the numerical code, and that the steep radial gradients of pressure, density, and temperature in the layer just below the surface cannot be modelled correctly. Therefore, solar convection is simulated only out to 90%, resp. 93%, of the solar radius in the models by Gilman and Glatzmaier. Of course small-scale convection may also occur inside these upper model boundaries. But this is not considered explicitly; it is parametrized in the form of a scalar turbulent viscosity, ν_t, and a scalar thermal diffusivity, κ_t. These two parameters are of order lv, cf. (11), or $\simeq 10^9$ m^2/s.

A further simplification, called the anelastic approximation, is the neglect of the time derivative of the density in the equation of continuity. This means that acoustic waves cannot be desribed. Although these waves certainly provide the most important means to determine the internal solar angular velocity observationally, they should be of no relevance to the *origin* of the differential rotation.

The global convection models start out from a reference state in which there is no explicit motion, and all energy is transported by turbulent convection, described as a diffusive process. Global convection sets in when the spherical shell considered is heated at its inner boundary to such a degree that thermal diffusion is no longer sufficient. The whole layer then becomes unstable to convection itself. Perhaps it is worth mentioning that such an instability was also observed in some of the axisymmetric, or mean-field models (Gierasch, 1974; Schmidt, 1982; Chan et al., 1987). Those models then work even without drivers such as anisotropic viscosity.

It should be noted here that the concept of a globally unstable convection zone essentially differs from the way convection is treated in a standard solar model. In the standard model, the efficiency of convection, as expressed by the convection velocity v in (11), adjusts itself according to the requirements of energy transport. More heating at the base of the convection zone would simply mean that v increases, and *not* the onset of global convection.

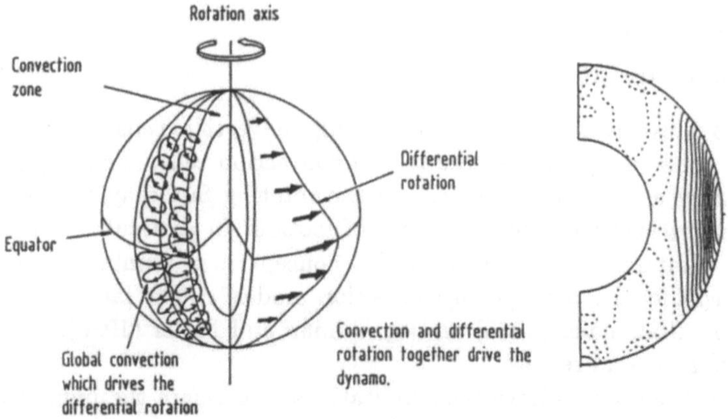

Fig. 6. Global convection model. *Left:* Convection rolls and differential rotation, from Gilman (1983); *right:* Contours of constant angular velocity in a meridional plane (*solid* faster, *dashed* slower than average), from Glatzmaier (1985).

Figure 6 illustrates the main features of the global convection models. In contrast to the meridional circulation of the mean-field models the global convection is *non-axisymmetric.* Under the inluence of the Coriolis force the convection cells take the form of rolls, or "bananas", aligned in the north-south direction. Also due to the Coriolis force, the flow in the rolls is distorted in such a way that angular momentum transport takes place. This leads to the desired differential rotation.

The models yield the correct rotation law at the solar surface, and the correlation $Q_{\theta\phi}$ is such as observed. In addition, the models succeed in generating (axisymmetric) meridional circulation and differences in surface temperature within the limits set by observation; the actual numbers depend on the parameters ν_t and κ_t.

An open question is the existence on the Sun of a regular pattern of non-axisymmetric global convection, i.e. the aforementioned bananas. As said above, the domain to which the numerical code is applicable does not reach the solar surface. The models therefore are screened by a layer of turbulent convection of small scale. The screening of the global pattern is rather efficient for energy flux differences, but inefficient for the differential rotation (Spruit, 1977). For bananas the efficiency of the obscuration depends on their diameter: the smaller the structure, the less is seen at the surface (Stix, 1981; van Ballegooijen, 1986). For the roll diameters actually calculated the predicted velocities are of the same order as the present observational limit of, say, 10 m/s.

The most important result of the explicit models is shown on the right of Fig. 6: the angular velocity is constant on surfaces of, approximately, cylindrical shape. This result is obviously at variance with the heloseismological evidence of nearly conical isorotation surfaces (Libbrecht, 1988). I shall discuss possible reasons for this discrepancy in the following section.

4.2 The Taylor-Proudman Theorem

In rotating systems the occurence of structures (e.g. convection rolls or isorotation surfaces) aligned with the axis of rotation is common. It was predicted by Proudman (1916), and experimentally confirmed by Taylor (1921). In a recent experiment on Spacelab 3 the banana-shaped rolls known from the global-convection models were also reproduced (Hart et al., 1986a,b; Toomre et al. 1987).

Let us see under which conditions the rotational alignment can be expected. For this purpose we consider the motion, $\mathbf{u}$, in a frame of reference rotating with constant angular velocity, $\mathbf{\Omega}_0$. This motion is governed by

$$(\mathbf{u}\cdot\nabla)\,\mathbf{u} = -\frac{1}{\rho}\nabla P - \nabla\Phi + \mathbf{R} - 2\,\mathbf{\Omega}_0\times\mathbf{u} \quad , \tag{13}$$

where P is the pressure, $\mathbf{R}$ the frictional force, and Φ a potential which includes the centrifugal term.

We apply the curl operator to (13). The Coriolis term yields $-\Omega_0\partial\mathbf{u}/\partial z$, where z is the co-ordinate along the axis of rotation, plus a term parallel to $\mathbf{\Omega}_0$. The components perpendicular to $\mathbf{\Omega}_0$ thus become independent of z if *all* the remaining forces in (13) are either curl-free or small in comparison to the Coriolis force.

In all the above models *friction* is treated as a diffusion process, with diffusivity ν_t. Let l be a typical scale of variation of $\mathbf{u}$, say $l = 10^8$ m; the ratio of the friction force to the Coriolis force is then of order 10^{-2} (using $\nu_t = 10^9$ m^2/s and $\Omega_0 = 3\times10^{-6}$ s^{-1}).

The magnitude of the *inertia term*, $(\mathbf{u}\cdot\nabla)\,\mathbf{u}$, divided by the magnitude of the Coriolis force, yields the Rossby number already introduced in (4). For the large-scale flow considered here this number is small, cf. Tab. 3.

Finally, consider the *pressure term*. If the two before-mentioned terms are negligible the curl operation upon (13) gives (for the ϕ-component)

$$2\Omega_0\frac{\partial u_\phi}{\partial z} = \frac{1}{\rho^2}(\nabla P\times\nabla\rho)_\phi \quad . \tag{14}$$

The right hand expression is zero in the barotropic case where $P = P(\rho)$, and of course in the case ρ = const. it also vanishes. Probably the mean-field model of Köhler (1970) produced cylindrical isorotation because of its ρ = const. assumption.

In a rotating system the situation is generally baroclinic: the vectors ∇P and $\nabla\rho$ make a finite angle. Let us eliminate $\nabla\rho$ with the help of

$$\frac{\nabla\rho}{\rho} = \frac{\nabla P}{P} - \frac{\nabla T}{T}\ , \qquad \nabla S = C_P\Big(\frac{\nabla T}{T} - \nabla_a\frac{\nabla P}{P}\Big)\ , \tag{15}$$

and (11); here C_P is the specific heat at constant pressure and ∇_a the double-logarithmic adiabatic temperature gradient. We thus obtain

$$\Omega_0\frac{\partial u_\phi}{\partial z} = \frac{1}{\rho^2 l v T C_P}(\nabla P\times\mathbf{F}_C)_\phi \quad . \tag{16}$$

Because of the hydrostatic equilibrium ∇P is a radial vector in good approximation. Therefore the interpretation of (16) depends on how far the vector of convective energy flux deviates from the radial direction. Durney (1976) argued that large deviations should be unacceptable and that, therefore, u_ϕ and the angular velocity $\Omega \equiv u_\phi / r \sin\theta$ should be constant on cylinders (at least in the lower part of the convection zone); otherwise the emergent flux at the solar surface would not show such high spherical symmetry. However, Spruit (1977) has demonstrated that small-scale convection very efficiently screens any latitude dependence of the energy flux. That is, we may well allow for a strongly inclined $\mathbf{F}_C$ and associated flux differences in the deep convection zone. Then there is no need to assume a cylindrical rotation law.

For the derivation of (16) I have used relation (11) between $\mathbf{F}_C$ and ∇S. Durney (1989b) has pointed out that even if $\partial S/\partial r$ depended strongly on θ this needs not be the case with F_{Cr} because the tranport coefficient, here written as lv, in general depends on latitude also. Thus there is even less reason for cylindrical isorotation. We may strengthen the argument by recalling that on the Sun the transport coefficient might be a tensor rather than a scalar. Then $\mathbf{F}_C$ needs not even be parallel to ∇S!

In all, we may say that either an underestimate of the inertia or friction forces, or a too much simplified treatment of the convective energy transport, or a combination of these reasons, may lead to cylindrical isorotation. The explicit models of Sect. 4.1. should be tested accordingly; I would expect that these models can be modified so to produce the observed results.

Gilman et. al (1989) see an additional possibility to save the explicit models: they suggest to include a transition layer to the solar core. In this view a cycle of angular momentum transport would proceed as follows: at low latitude from the convection zone downwards into the transition layer, within this layer toward the poles, at high latitude back into the convection zone, and then within the convection zone again towards the equator. This latter part would maintain a high angular velocity at lower latitude at *all* depths of the convection zone, that is *conical* isorotation surfaces.

Details of this new model are not yet specified. A general difficulty might be that time scales become long, which means transport processes become inefficient, as soon as the core gets involved. For this reason I would prefer to change the explicit models by improvements *within* the convection zone. But certainly all possibilities must be explored.

References

Altrock, R. C., Canfield, R. C. (1972): *Solar Phys.* **23**, 257

Appenzeller, I., Schröter, E. H. (1967): *Astrophys. J.* **147**, 1100

Babcock, H. W. (1961): *Astrophys. J.* **133**, 572

Ballegooijen, A. A. van (1986): *Astrophys. J.* **304**, 828

Balthasar, H., Vázquez, M., Wöhl, H. (1986): *Astron. Astrophys.* **155**, 87

Beckers, J. M. (1960): *Bull. Astron. Inst. Netherlands* **15**, 85

Belvedere, G., Godoli, G., Motta, S., Paternò, L., Zappalà, R. A. (1976): *Solar Phys.* **46**, 23
Belvedere, G., Paternò, L. (1977): *Solar Phys.* **54**, 289
Belvedere, G., Paternò, L., Stix, M. (1980): *Geophys. Astrophys. Fluid Dynamics* **14**, 209
Biermann, L. (1951): *Z. Astrophys.* **28**, 304
Bogart, R. S. (1987): *Solar Phys.* **110**, 23
Brown, T. M., Christensen-Dalsgaard, J., Dziembowski, W. A., Goode, P., Gough, D. O., Morrow, C. A. (1989): *Astrophys. J.* **343**, 526
Brune, R., Wöhl, H. (1982): *Solar Phys.* **75**, 75
Bumba, V., Hejna, L. (1987): *Bull. Astron. Inst. Czechosl.* **38**, 29
Caccin, B., Falciani, R., Moschi, G., Rigutti, M. (1970): *Solar Phys.* **13**, 33
Carrington, R. C. (1863): *Observations of the Spots on the Sun from November 9, 1853, to March 24, 1861, made at Redhill* (Williams and Norgate, London, Edinburgh)
Chan, K. L., Sofia, S., Mayr, H. G. (1987): in Durney and Sofia (1987), p. 347
Cocke, W. J. (1967): *Astrophys. J.* **150**, 1041
Cram, L. E., Durney, B. R., Guenther, D. B. (1983): *Astrophys. J.* **267**, 442
Durney, B. R. (1976): *Astrophys. J.* **204**, 589
Durney, B. R. (1989a): *Astrophys. J.* **338**, 509
Durney, B. R. (1989b): *Solar Phys.* in press
Durney, B. R., Roxburgh, I. W. (1971): *Solar Phys.* **16**, 3
Durney, B. R., Sofia, S. (eds., 1987): *The Internal Solar Angular Velocity* (Reidel, Dordrecht)
Durney, B. R., Spruit, H. C. (1979): *Astrophys. J.* **234**, 1067
Falciani, R., Rigutti, M., Roberti, G. (1974): *Solar Phys.* **35**, 277
Gierasch, P. J. (1974): *Astrophys. J.* **190**, 199
Gilman, P. A. (1968): *Science* **160**, 760
Gilman, P. A. (1983): IAU Symp. **102**, 247
Gilman, P. A., Foukal, P. V. (1979): *Astrophys. J.* **229**, 1179
Gilman, P. A., Howard, R. (1984a): *Astrophys. J.* **283**, 385
Gilman, P. A., Howard, R. (1984b): *Solar Phys.* **93**, 171
Gilman, P. A., Miller, J. (1986): *Astrophys. J. Suppl.* **61**, 585
Gilman, P. A., Morrow, C. A., DeLuca, E. E. (1989): *Astrophys. J.* **338**, 528
Glatzmaier, G. A. (1985): *Astrophys. J.* **291**, 300
Glatzmaier, G. A. (1987): in Durney and Sofia (1987), p. 263
Hanslmeier, A., Lustig, G. (1986): *Astron. Astrophys.* **154**, 227
Hart, J. E., Glatzmaier, G. A., Toomre, J. (1986a): *J. Fluid Mech.* **173**, 519
Hart, J. E., Toomre, J., Deane, A. E., Hurlburt, N. E., Glatzmaier, G. A., Fichtl, G. H., Leslie, F., Fowlis, W. W., Gilman, P. A. (1986b): *Science* **234**, 61
Hill, F., Gough, D., Toomre, J., Haber, D. A. (1988): IAU Symp. **123**, 45
Hoeksema, J. T., Scherrer, P. H. (1987): *Astrophys. J.* **318**, 428
Howard, R., Gilman, P. A. (1986): *Astrophys. J.* **307**, 389
Howard, R., Gilman, P. A., Gilman, P. I. (1984): *Astrophys. J.* **283**, 373
Howard, R., LaBonte, B. J. (1980): *Astrophys. J.* **239**, L33
Kichatinov, L. L. (1986): *Geophys. Astrophys. Fluid Dynamics* **35**, 93
Kichatinov, L. L. (1987): *Geophys. Astrophys. Fluid Dynamics* **38**, 273
Kippenhahn, R. (1963): *Astrophys. J.* **137**, 664
Kippenhahn, R. (1964): IV. Cent. Nasc. Galileo Galilei, Atti Conv., Tomo 4, ed. by G. Barbèra, Firence, p. 291
Köhler, H. (1970): *Solar Phys.* **13**, 3
Kubičela, A. (1973): in *Solar Activity and Related Interplanetary and Terrestrial Phenomena*, ed. by J. Xanthakis (Springer, Berlin, Heidelberg), p. 123
Kuhn, J. R., Libbrecht, K. G., Dicke, R. H. (1988): *Science* **242**, 908
LaBonte, B. J., Howard, R. (1982): *Solar Phys.* **80**, 361
Lebedinsky, A. I. (1941): *Astron. Zh.* **18**, 10
Libbrecht, K. G. (1988): in *Seismology of the Sun and Sun-Like Stars*, ed. by E. J. Rolfe, esa SP-286, p. 131

Lustig, G., Hanslmeier, A. (1987): *Astron. Astrophys.* **172**, 332
Moss, D., Vilhu, O. (1983): *Astron. Astrophys.* **119**, 47
Münzer, H., Hanslmeier, A., Schröter, E. H., Wöhl, H. (1989): *Astron. Astrophys.* **213**, 431
Newton, H. W., Nunn, M. L. (1951): *Mon. Not. R. astr. Soc.* **111**, 413
Pidatella, R. M., Stix, M., Belvedere, G., Led Paternò, L. (1986): *Astron. Astrophys.* **156**, 22
Plaskett, H. H. (1970): *Mon. Not. R. astr. Soc.* **148**, 149
Proudman, J. (1916): *Proc. Roy. Soc.* A **92**, 408
Ribes, E., Laclare, F. (1988): *Geophys. Astrophys. Fluid Dynamics* **41**, 171
Ribes, E., Mein, P., Mangeney, A. (1985): *Nature* **318**, 170
Rimmele, T., Schröter, E. H. (1989): *Astron. Astrophys.* **221**, 137
Rüdiger, G. (1977): *Astron. Nachr.* **298**, 245
Rüdiger, G. (1980): *Geophys. Astrophys. Fluid Dynamics* **16**, 239
Rüdiger, G. (1982): *Geophys. Astrophys. Fluid Dynamics* **21**, 1
Rüdiger, G. (1989): *Differential Rotation and Stellar Convection* (Akademie-Verlag, Ber-lin)
Rüdiger, G., Tuominen, I. (1987): in Durney and Sofia (1987), p. 361
Rüdiger, G., Tuominen, I., Krause, F., Virtanen, H. (1986): *Astron. Astrophys.* **166**, 306
Schmidt, W. (1982): *Geophys. Astrophys. Fluid Dynamics* **21**, 27
Schröter, E. H. (1985): *Solar Phys.* **100**, 141
Schröter, E. H., Wöhl, H. (1975): *Solar Phys.* **42**, 3
Schröter, E. H., Wöhl, H. (1976): *Solar Phys.* **49**, 19
Schröter, E. H., Wöhl, H., Soltau, D., Vázquez, M. (1978): *Solar Phys.* **60**, 181
Schüssler, M. (1981): *Astron. Astrophys.* **94**, L17
Schüssler, M. (1983): IAU Symp. **102**, 213
Schüssler, M. (1984): in *The Hydromagnetics of the Sun*, ed. by P. Hoyng and M. Kuperus, esa SP-220, p. 67
Schüssler, M. (1987): in Durney and Sofia (1987), p. 303
Snodgrass, H. B. (1984): *Solar Phys.* **94**, 13
Spruit, H. C. (1977): *Astron. Astrophys.* **55**, 151
Stenflo, J. O. (1989): *Astron. Astrophys.* **210**, 403
Stix, M. (1981): *Astron. Astrophys.* **93**, 339
Stix, M. (1987a): in Durney and Sofia (1987), p. 329
Stix, M. (1987b): in *Solar and Stellar Physics*, ed. by E.-H. Schröter and M. Schüssler, Lect. Notes Phys. (Springer, Berlin, Heidelberg), Vol. 292, p. 15
Stix, M. (1989): *The Sun* (Springer, Berlin, Heidelberg)
Taylor, G. I. (1921): *Proc. Roy. Soc.* A **100**, 114
Timothy, A. F., Krieger, A. S., Vaiana, G. S. (1975): *Solar Phys.* **42**, 135
Toomre, J., Hart, J. E., Glatzmaier, G. A. (1987): in Durney and Sofia (1987), p. 27
Topka, K., Moore, R., LaBonte, B. J., Howard, R. (1982): *Solar Phys.* **79**, 231
Tuominen, I., Virtanen, H. (1984): *Astron. Nachr.* **305**, 225
Tuominen, J. (1942): *Z. Astrophys.* **21**, 96
Tuominen, J. (1952): *Z. Astrophys.* **30**, 261
Tuominen, J., Tuominen, I., Kyröläinen, J. (1983): *Mon. Not. R. astr. Soc.* **205**, 691
Wagner, W. J., Gilliam, L. B. (1976): *Solar Phys.* **50**, 265
Ward, F. (1965): *Astrophys. J.* **141**, 534
Wasiutyński, J. (1946): *Astrophys. Norv.* **4**
Weiss, N. O. (1965): *Observatory* **85**, 37
Willson, R. C., Hudson, H. S. (1988): *Nature* **332**, 810
Wöhl, H. (1988): *Solar Phys.* **116**, 199
Yoshimura, H. (1981): *Astrophys. J.* **247**, 1102

Lighting up Pancakes – Towards a Theory of Galaxy-formation

T. Buchert

Max-Planck-Institut für Physik und Astrophysik, Institut für Astrophysik, Karl-Schwarzschild-Str. 2, D-8046 Garching b. München, Fed. Rep. of Germany

Abstract: An extended 'revival'-version of the so-called 'Pancake-theory' is presented and defined on the basis of a short list of assumptions. Aspects of structure formation on large scales as well as concepts to extract small-scale structures like galactic halos together with their large-scale distribution in space are put into perspective.

1. Introduction

Science does not develop at a neat uniform rate. There are long periods when it seems to be stagnating, and short ones when it appears to be striding forward in seven league boots. Modern cosmology today seems to be experiencing one of those short periods which can be summarized by pointing to the need to replace traditional homogeneous model-universes by *inhomogeneous cosmologies* or, in other words, to the huge list of activities towards a *theory of galaxy-formation.* In such a period it is important to sort out those ideas, which really prepare the new stage of cosmological thinking and, at the same time, help to keep progressing results at a simple and essential level. This attitude is not in conformity with the general trend of making things complex resulting in discriminating decisions among models at a stage, where the consequences of the basic assumptions are far from being well understood. I try in this talk to consider building-stones of a galaxy-formation model by tracing all the steps from theoretical grounds (here confined to the framework of Newtonian Gravity) to the prediction of a point-distribution in space, which is ready to be compared with galaxy-distributions on the sky. By 'theory' of galaxy-formation I understand a formation model which predicts, on the basis of a short list of assumptions, the large-scale distribution of luminous matter, initiated by small-amplitude inhomogeneities at the time of decoupling of matter and radiation and evolved according to gravitational instability of a homogeneous and isotropic model-universe. So, at first, only the locations of galaxies and their velocities are predicted, the theory is not concerned with the details of the formation process of individual galaxies and their internal structure. These could of course put in doubt and alter certain

Reviews in Modern Astronomy 2 Editor: G. Klare

assumptions made. We are on the way 'towards' such a theory, but at present no theory exists which uniquely and free of contradictions explains the observed galaxy-distribution on the basis of cosmological assumptions respecting known observational constraints.

The talk is divided into two main sections:

First, basic theoretical assumptions are introduced, which define an extended and clear-cut version of the so-called 'Pancake-theory'. This is done by discussing the spatial scales of inhomogeneity relevant to the galaxy-formation problem. I shall introduce "the scale of mean-isotropy", which concerns properties of inhomogeneous cosmologies as well as galaxy-samples in the large, and "the scale of coherence of structure"; both scales are inherent to assumptions related to *initial conditions*. The former scale will define the upper bound for our considerations, the latter will be the smallest scale of initial perturbations of a homogeneous model-universe related to a typical property of HDM ("Hot Dark Matter")-spectra. Then, I shall introduce the concept of 'generic motions' which is related to the third assumption underlying the *evolution-model*, namely that the model is constructed on the basis of solutions for self-gravitating "dust" (i.e. pressureless matter) describing the mass-dominating matter-content in the universe, a description which is valid on the considered spatial scales far into the non-linear regime of structure formation. Finally, I present details of the constructed 'Pancake-models', which primarily concern metamorphoses of a continuous density-field and its singularities (in particular I shall explain, how "Pancakes" are defined and how they originate e.g. in a two-dimensional medium).

Second, our interest will concern the extraction of a point-process in space from those models. For this purpose, I propose three concepts to 'light up' the Pancakes in the model: The concepts of a "statistically fair sample", a "high-spatial resolution of Pancakes" and a "dynamical thresholding-algorithm" will lead to the prediction of a large-scale distribution of "luminous points" that will be assigned to galaxies.

2. Spatial Scales of Inhomogeneity and Basic Assumptions

The attempt to go all the way from theoretical grounds to observational data can only be successful, if we confine our emphasis to basic assumptions. I hope that from these you will also get a vivid notion of the galaxy-formation problem and its present status. First, I formulate three conceptual assumptions and relate them to the galaxy-formation model to be constructed:

2.1. The "scale of mean-isotropy"

We have several spatial scales in our problem and it is useful to figure out that range of scales which is relevant to the galaxy-formation problem as defined in the introduction. First of all, we introduce the term *large scales*, which in cosmology refers to the small range of scales from several Mpcs (clusters of galaxies) to several tens of Mpcs (superclusters of galaxies). On such scales, the observation of highly isotropic and enormously smooth radiation backgrounds justifies the use of an *absolute reference frame*, relative to which all quantities can be measured; (e.g. inhomogeneities are measured in terms of the density-contrast $\delta = (\rho - \rho_H)/\rho_H$, where ρ_H is the mean-density according to an isotropic expansion). This frame is modelled by Friedmann-Lemaître (FL)-cosmologies, describing trajectories of *fundamental observers*, initially sitting at $\vec{X}$ in such a model-universe:

$$\vec{f}_H = a(t)\vec{X} \quad ; \quad \text{e.g. } a(t) \propto t^{2/3} \text{ in the standard model;}$$
$$X_i : \text{Lagrangian coordinates.}$$

Nowadays, cosmology experiences a fast change of view-points on its way from *homogeneous cosmology*, which pretends to describe the universe in the large, and *inhomogeneous cosmology*, which accounts for the existence of inhomogeneities such as galaxies or clusters of galaxies. In order to keep the FL-reference frame, it is natural, but not necessarily general, to subject *inhomogeneous model-cosmologies* or simply *models* $\vec{f}$ to the following average-principle: Models $\vec{f}$ which include vector-displacements $\vec{F}$ from homogeneity are assumed to obey the following formal principles:

$$\vec{f} = \vec{f}_H + \vec{F} \quad \text{(superposition – principle);}$$
$$\langle \vec{f} \rangle_{\lambda_{\max}} = \vec{f}_H \quad \text{(average – principle).}$$

The spatial average is taken by integrating the vector-field in 'Eulerian space' (comoving with the FL-frame) and dividing by the volume on the scale $\lambda_{\max}$. I emphasize that $\vec{F}$ is not a small perturbation. The density-contrast on the scale of superclusters exceeds the value 1 demanding that $\vec{f}$ should be a non-linear model. To guarantee that $\vec{f}$ is *Newtonian*, the scale $\lambda_{\max}$ should be considerably smaller than the horizon-scale (which today is about 12000 Mpc/h, $h \in [0.5, 1]$); the peculiar-velocities $\dot{\vec{F}}$ relative to the Hubble-velocity should be considerably smaller than the speed of light (today they are of the order of 1000 km/s). At the same time, $\lambda_{\max}$ should be considerably larger than the typical clustering-size of galaxies, i.e. larger than scales we called 'large' previously. This point will face us with the concept of a "statistically fair sample" outlined in section 3.

I define λ_{max} to be the scale 300 Mpc/h. This will be the largest scale in the model. On it we require the mean matter-distribution to obey an expansion law of FL-type. (Note that all scales have to be understood as time-dependent lengths comoving with the background).

In summary, we can only give heuristic reasons, why we fix that scale: First, we want to keep FL-cosmologies as models of the mean-distribution in accordance with observations; I emphasize, however, that we cannot in general expect that averaging any inhomogeneous distribution will yield a homogeneous solution of the gravitational field-equations; (recently this has been addressed in the framework of General Relativity, where it has been shown how mean-quantities are affected by local inhomogeneities (Futamase 1989)). Second, to define a *Newtonian* cosmology, it is necessary to fix boundary-conditions. We shall assume periodic boundary conditions, so we consider a "windowing" of the universe, which divides the universe into cells of size $(\lambda_{max})^3$, which are self-similar to each other. Besides the mathematical simplification implied by this requirement to confine the inhomogeneities to a compact space, it implies further that in each "window" the statistical properties are identical. This third point is heuristic, since there might be no scale in the universe, on which statistical properties become stable. In other words, there might be no scale, on which the universe can be represented totally in terms of measurable quantities.

2.2. The "scale of coherence of structure":

This is the second intrinsic length-scale which will be built into the model. Large-scale structures in the universe appear to be coherent on scales around λ_{min} = 25Mpc/h (this is about the mean size of regions which are found to be devoid of bright galaxies (the so-called "Voids") in complete surveys of the universe. Such a scale can be explained on physical grounds: A coherence-length arises naturally in (1): A model dominated by massive neutrinos with masses of several tens of eV/c^2, HDM ("Hot Dark Matter")-model as a result of the so-called "free-streaming"-effect; (this effect is similar to the so-called 'Landau-damping mechanism' in plasmas); (2): A baryon-dominated model, here a viscous damping mechanism in the radiation-dominated epoch (the so-called 'Silk-damping mechanism') reduces fluctuations on small scales significantly (calculated to linear approximation). Both mechanisms imply that, roughly on the same scale $\approx \lambda_{min}$, (at the time when initial conditions for the model are set up, i.e. the time of decoupling of matter and radiation), perturbations on scales $\lambda < \lambda_{min}$ are strongly damped. In the popular interpretation of the HDM-picture, the assumption to "cut-off" perturbations at the scale λ_{min} implies that the formation of structure begins on large scales, galaxies are then thought to form via fragmentation of such large-scale structures later on. This "top-down"-scenario does not satisfy observational constraints mainly as a result of the assumption of late formation of galaxies for initial amplitudes, which obey the constraints of the microwave background; (for details of this picture see the review by Shandarin & Zel'dovich 1989). In contrast, the model to be constructed below implies formation of galaxies even before the formation of large-scale structures, which turns the traditional 'Pancake-picture' into a "bottom-up"-scenario and brings it into the vicinity of the competing CDM ("Cold Dark Matter")-model, which assumes perturbations also on smaller scales in the initial conditions and is physically based on more exotic Dark matter consisting of particles with larger masses.

In summary, we have introduced the following scales of inhomogeneity: The horizon-scale (12000 Mpc/h), the scale on which inhomogeneities are smoothed into a FL-background (λ_{max}= 300 Mpc/h) and the scale of smallest perturbations (λ_{min}= 25 Mpc/h). The wave-lengths λ_{max} and λ_{min} enter directly into the *initial conditions* given as Fourier-sums with Gaussian-distributed amplitudes for gravitational and velocity-potentials. The k-space ($k = 2\pi/\lambda$) is filled with modes ranging from a minimal to a maximal wavenumber related to the two scales. Since this range is large for large "windows", the spectrum is highly generic, it contains a huge number of harmonics, here 1248. The spectrum is assumed to be 'flat', i.e. k-independent for the density-contrast, (see Buchert 1989a).

We refer to the scale of galactic halos to be resolved according to assumptions discussed later. We shall have to resolve large-scale structures down to the scale of about 300 Kpc/h or less, such that the model covers at least three orders of magnitude in scale. The latter scale can also be identified with the limit-scale where the hydrodynamical approximation of "dust-matter" breaks down. This could be due to the velocity-dispersion of neutrinos, which "wipes out" structures on Kpc-scales, or, in the baryonic component, due to pressure-forces, which affect the picture e.g. as a result of collisions of galaxies. One model has been proposed which is based on the idea of "sticking" matter together after collision (see Shandarin & Zel'dovich 1989, Buchert 1988b).

2.3. Generic motions

The concept of 'generic motions' has been defined in the context of Lagrange-singularity theory for inertial motions (Arnol'd et al. 1982a). It can be stated in the form of a cosmological principle: We expect that structures in the universe are the result of a generic process, which means e.g. that 'almost all' initial conditions yield similar structures. This strong statement must be weakened in practice, since we have to confine initial conditions as well as evolution-models to certain restricted classes. Nevertheless, the statement applies e.g. to rule out shear-free motions in a class of shear-flows, if the physical assumptions made allow for shear. Such a discrimination is possible, since the shear-free case is a set of measure zero in that class. Also related to this concept is the question, if structural properties of the universe can be described and measured in terms of quantities, which are independent of the specific shape of the initial power-spectrum (e.g. the Void-size in the distribution of bright galaxies can be related to the "cut-off"-length of the perturbations).

The concept of generic caustic-patterns in the density-field has been applied to Zel'dovich's approximation (Arnol'd et al. 1982b). It applies also to self-gravitating flows, at least for some large classes of solutions (Buchert 1988a). The particular example of shear-flows mentioned above has been illustrated in the talk by opposing spherical symmetric infall-motions (where wave-fronts degenerate to a single point) to generic anisotropic motions (where the density of the d-dimensional medium degenerates to $(d-1)$-dimensional caustics). These caustics can be classified in terms of Legendrian singularities of the wave-fronts in the framework of Arnol'd's theory. Generic initial conditions can be characterized by using the concept of structural stability of Legendrian submanifolds with respect to canonical deformations; e.g. a deformation of the spherical front will render the motion aspherical in general. In three spatial dimensions, volume-elements of the fluid degenerate generically to surface-elements of two-dimensional sheets, which bifurcate into two sheet-boundaries enclosing a three-stream-system of the flow, which we call a Pancake. If the initial conditions are sufficiently smooth, then their continuous transformation according to the evolution-model will also result in smoothly varying surface-elements forming large sheetlike structures in space.

In addition to this property, numerical simulations of gravitational collapse suggest that the principal kinematical feature of any model should be a tendency to 'locally one-dimensional' motion. Perturbations of sphericity, i.e. anisotropies are (according to numerical and analytical results) strongly amplified in the non-linear regime of structure formation. One of the eigenvalues of the deformation-tensor of the fluid will dominate after some time causing the volume-elements to degenerate into surface-elements (spanned by the other two eigenvalues) at a finite time. This time determines the onset of formation of large-scale structures and is approximately the time, when high-density objects such as galaxies form.

In order to construct an inhomogeneous *evolution-model* which describes a *generic* collapse of structure, a class of exact 3D-solutions of the Euler-Poisson-system without assuming any symmetry has been derived. This class forms according to its properties the basis of an approximation (Buchert 1989a). The class provides an exact description for globally and, which is important in this context, for locally one-dimensional flows superimposed on 3D-FL-cosmologies ('local' means: in each volume-element of the fluid). Linearization of the solutions yields the 'linear theory of gravitational instability', which has been in use until the advent of non-linear models. The "Zel'dovich-approximation", which so far was the only model which formally extrapolates the 'linear theory' into the non-linear regime, is recovered as a subclass of solution-curves of this class. Using the solution-curves as approximate trajectories implies, however, ignoring the constraints on initial conditions. In comparison with numerical simulations, the approximate theory has proved to describe well strongly anisotropic motions, which are idealized by the solution. This idealization, however, gives rise to strong restrictions on initial data. The mathematical result concerning the constraints demands a generalization of the solution, which reflects well kinematical features, but fails to catch e.g. tidal forces of a self-gravitating fluid. Work is in progress, which generalizes the solution-class; (first results are given in Buchert 1988a).

2.4. Density-fields in 'Pancake-models'

The following pictures show density-fields as predicted by the model defined above: To demonstrate the principal *kinematical* features of the model, a highly symmetric (and therefore non-generic) spectrum has been evolved into the regime of caustic-formation, (a corresponding point-process of this model has been discussed as 'toy-model' in Mo & Buchert 1989). This calculation shows the successive stages in the kinematics of the density-field: First, the density becomes infinite at knots, separated roughly by a coherence-length, which immediately afterwards break to form a connected network-structure. The next stage is an infall-motion inside the network towards the knots, which now increase in mass drastically to form proto-structures of what we shall call later a "rich cluster of galaxies". A similar calculation for a generic spectrum with "cut-off" shows similar kinematical features, (figures are published in Buchert 1988b, Buchert 1989a, see Fig.1). However, Pancakes have the form of a sickle initially, as predicted by Arnol'd's theory. The boundary of such sickles breaks (according to a higher codimensional singularity) and this yields to an out-burst of mass-fingers with high speed connecting up other Pancakes as a result of continuity and smoothness of the density-distribution on the scale of the "cut-off". The kinematical mechanism of Pancake-formation, the break of Pancake-boundaries and the formation of a network has been explained in the talk on the basis of overlapping wave-fronts. Examples of caustic-metamorphoses calculated from the two-dimensional model at high-resolution as well as examples of three-dimensional caustics (Arnol'd 1982) have been shown, (some caustic-metamorphoses are published in Buchert 1989c).

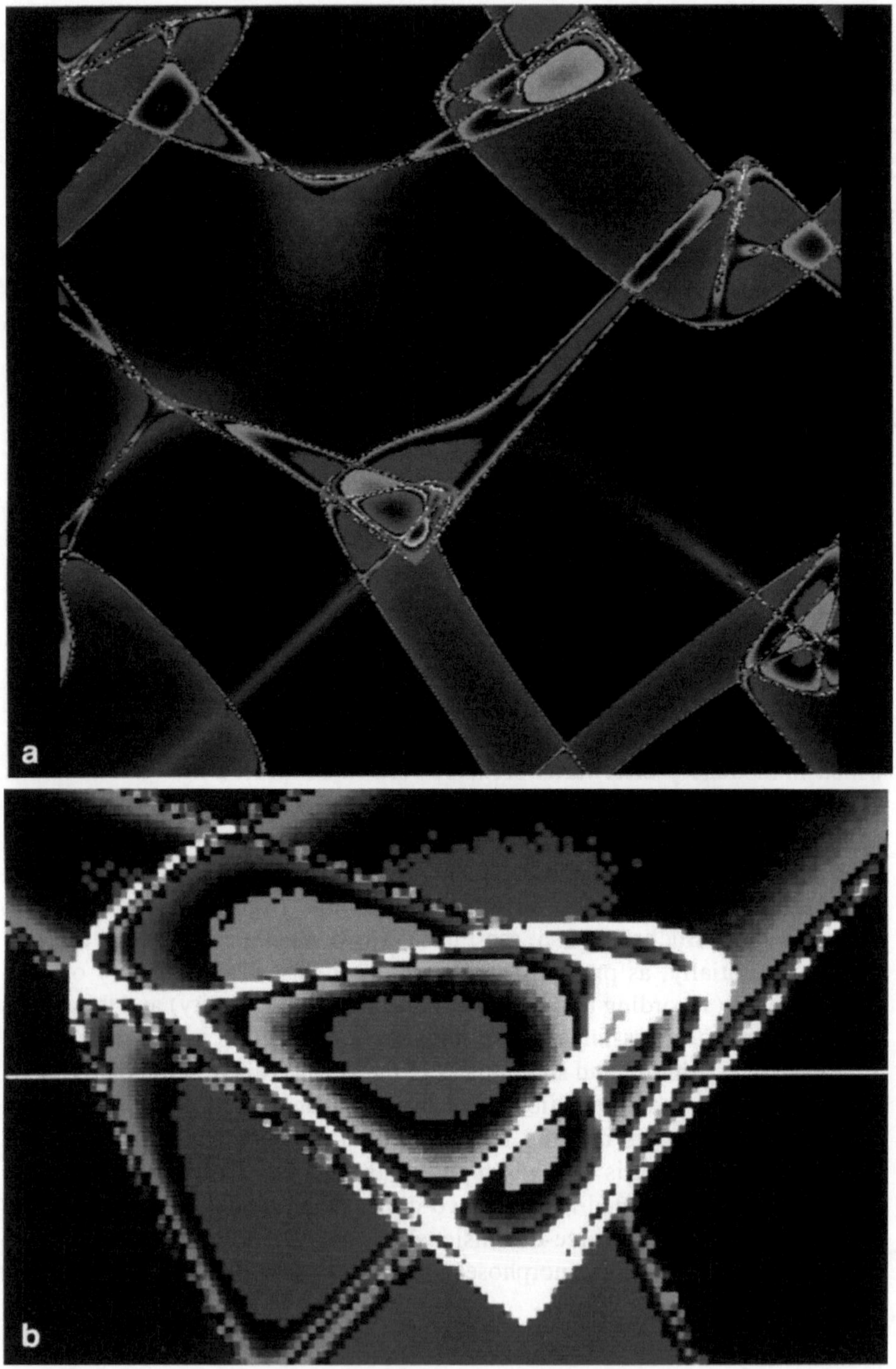
a
b

3. Lighting up Pancakes

Up to now, we considered and discussed patterns of a *continuous* density-field. The galaxy-distribution as we map it from the sky consists of a *discrete* point-set. So, either we can pursue the way of smoothing observed galaxy-samples into a continuous density-field, or we can extract a point-set from the continuous model, in order to compare observation and theory. Both ways have advantages, the former is helpful to distinguish topological and/or geometrical properties of the galaxy-distribution (Mo & Buchert 1989), the latter requires a fundamental understanding of the physical mechanisms to build galaxies in a theoretical model. Here, I present a way of the latter type and give extraction-principles to derive a large-scale distribution of luminous matter from the 'Pancake-models'. I propose the following three concepts to which the model should be subjected to obtain point-samples, which are ready to be compared to observation, (the concepts are more explicitly stated in Buchert 1989b).

3.1. "Fair" samples

The concept of a statistically 'fair' sample introduced by Peebles (1973) states that the space-region surveyed should be large enough to guarantee equal statistics in different regions of the sky, and, statistical measures should scale properly with distance according to Hubble's law. This concept is yet unfulfilled in complete surveys of the universe reaching only to about 100 Mpc/h. Such regions contain e.g. too few "Voids" to do good statistics. Also most numerical simulations do not respect this concept (due to computation-time and storage-capacity limits). In analytical models, as the one presented here, we are able to calculate a 'fair' slice (although we perform high resolution calculations). The concept of 'fairness' will be assumed to hold in slices of linear scale 300 Mpc/h. Peebles' concept should be adapted and altered as far as it concerns the redshift-direction of space-surveys; effects of evolution of galactic distributions in nearby samples are of no importance, but they are important e.g. in studies of quasar-clustering and are also important in models, where galaxies form early (like in the model presented). This first concept applies to the sample-size to be calculated.

Fig.1a,b: The density-distribution in a $(100\,\text{Mpc/h})^2$ box with periodic boundaries as predicted by the constructed 'Pancake-model' (section 2). The contemporary stage ($Z = 0$) is shown. (**b**): Enlargement of (a): Shown are two Pancakes with broken boundary (as a result of the degeneracy of two eigenvalues in the 2D-calculation). The colour-code assigns different colours to different density-levels $\chi = 1 + \delta$: Blue corresponds to levels around the mean density, $\chi = 1 \pm 1$; ; objects with density-contrast within a pixel larger than 9, $\chi > 10$, include caustics and are pictured White; inbetween, the colour-scale is linear and is divided into the colour-fields Cerulean Blue ($2 \le \chi \le 4$), Green ($4 \le \chi \le 6$), Yellow ($6 \le \chi \le 8$) and Red ($8 \le \chi \le 10$); (the colour-fields itself are devided into 10 intensity-steps).

3.2. High-spatial resolution

As mentioned in section 2, the resolution of large-scale structures down to scales of galactic halos (refering to the Dark matter constituent) is necessary to relate model-assumptions to the predictions of locations of galaxies (supposed to lie at the center of resolved halo-cubes of size $(\lambda_{halo})^3$). This concept together with the first determines the resolution of the calculation. To give numbers, a calculation of the sort shown previously resolves structures of 30 Kpc/h distributed as 10000^2 elementary cells on the whole $(300\text{Mpc/h})^2$ slice. Such halos correspond to dwarf objects. A resolution of giant objects with halos of 300 Kpc/h in diameter will demand a grid of 1000^2 cells, a resolution which is yet above most currently performed numerical simulations. The masses of those objects range from $10^9 M_\odot$ to $10^{12} M_\odot$. To map those masses into a correct Eulerian density-profile, a collection-procedure collecting small masses inside larger cells is necessary to account for the multiplicity of streams inside Pancakes; for the continuous model this has been demonstrated in (Buchert 1989c), the figures mentioned in section 2 and shown in the talk are based on this collection-algorithm, (see Fig.1).

3.3. Dynamical thresholding

One method which is commonplace in numerical simulations of CDM-scenarios is the so-called 'biasing'. The idea of 'biasing' states that galaxies form preferentially at points in 'Eulerian space' where the density exceeds some threshold-value. Galaxies are, as a result of this, more clustered than Dark matter. This is done either in the initial density-distribution or in the distribution assigned to the contemporary stage. The first method has the disadvantage that the distribution at the non-linear stage is determined by thresholding in the linear regime, the system is not allowed to forget initial conditions or to redistribute masses in 'Eulerian space', (we know that accretion of matter is a consequence of gravitational interaction) and, the second method depends on the normalization of the calculation, the galaxy-evolution cannot be traced dynamically. With the help of 'biasing' the CDM-picture is able to reconstruct the contemporary galaxy-distribution, not however, the observed peculiar-velocities for identical bias-parameters. To overcome the disadvantages of 'Eulerian thresholds', a method which follows the trajectories of elementary volume-elements of the fluid is useful to introduce a 'dynamical thresholding' of density inside such elementary cubes. This method is Lagrangian and determines the distribution of thresholded matter in 'Eulerian space' (mapped through the evolution-model after thresholding in 'Lagrangian space') dynamically. It is, however, assumed that flow-lines of Dark matter and baryonic matter are congruent. The result of a first calculation of 'fair' slices at low resolution (here 64^2 particles) and with 'dynamical thresholding' are shown in Fig.2. Different evolution-stages are shown: The panel to the

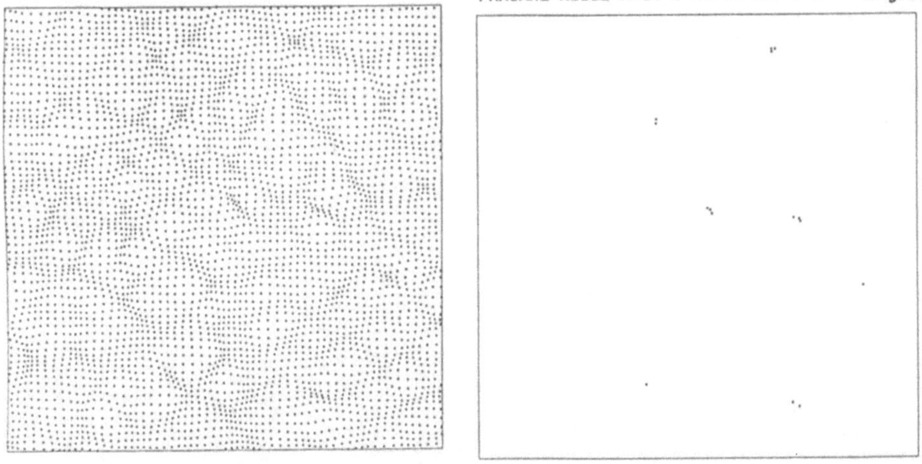

Fig.2a

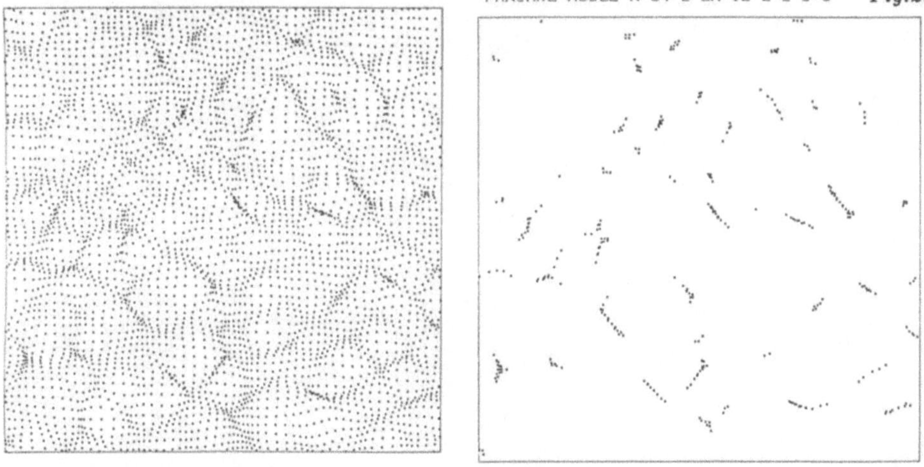

Fig.2b

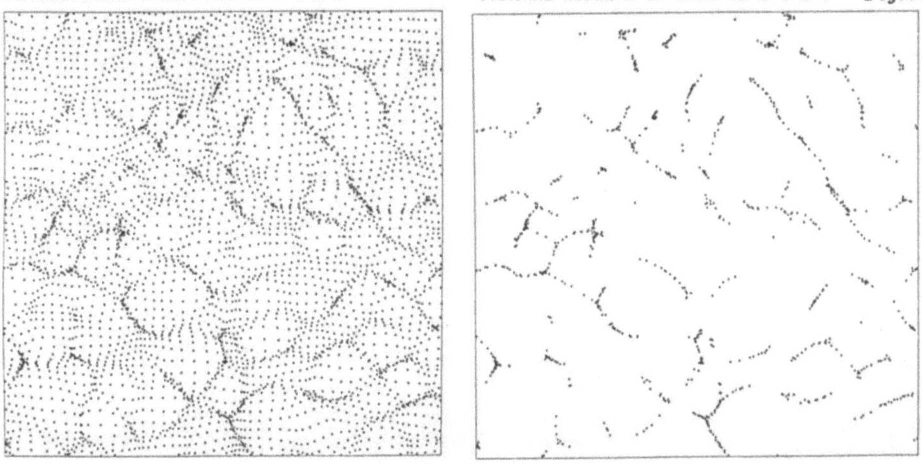

Fig.2c

PANCAKE-MODEL N-64^2 LM-12 Z-0.5 C-0

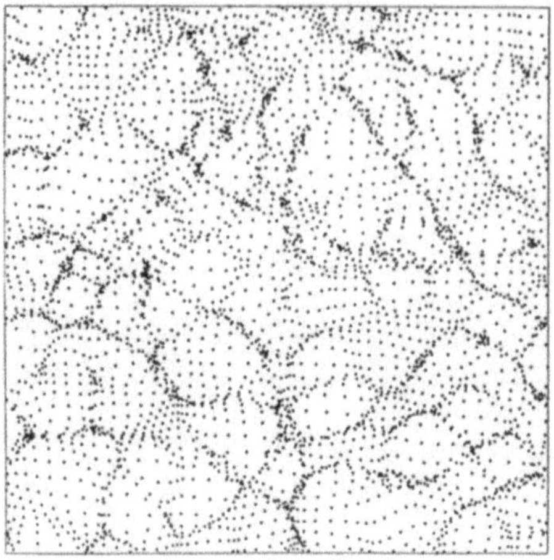

PANCAKE-MODEL N-64^2 LM-12 Z-0.5 C-3 ***Fig.2d***

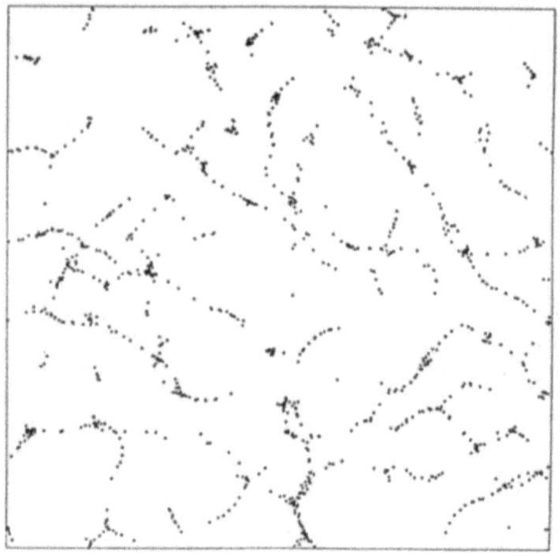

PANCAKE-MODEL N-64^2 LM-12 Z-0 C-0

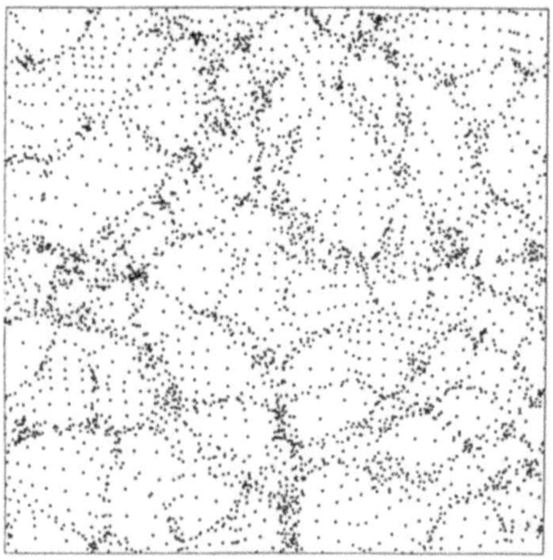

PANCAKE-MODEL N-64^2 LM-12 Z-0 CHIC-3 ***Fig.2e1***

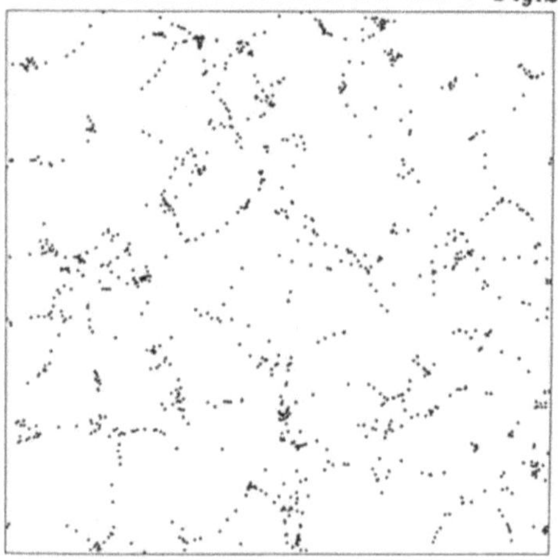

PANCAKE-MODEL N-64^2 LM-12 Z-0 CHIC-6 ***Fig.2e2***

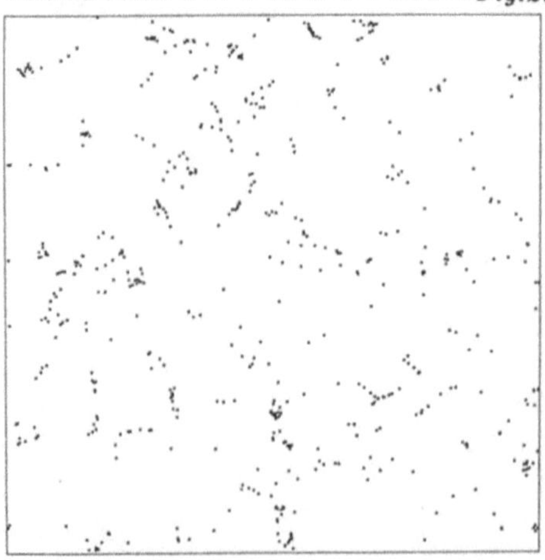

STICKMAN N=128 2 Z=0 C=1 LM=12 ***Fig.2f***

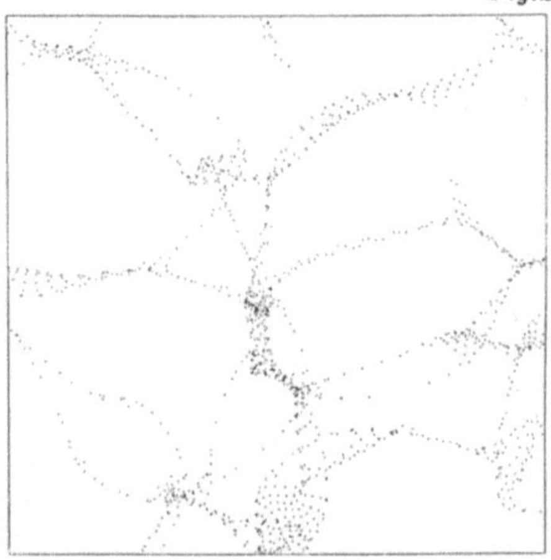

right shows "luminous points", the panel to the left shows a point-distribution at low resolution which traces the density-field. From the latter the "galaxies" have been extracted. We infer that (at an amplitude of 10^{-3} for the mean density-contrast on the "cut-off" scale at $Z = 1400$) first galaxies show up at redshift 4; at redshift 2, flattened, elongated structures ressembling "pearl-strings" form which connect up to a network at redshift 1; the infall-motion on "rich clusters" starts at a redshift of 0.5, all according to the kinematical properties of the model discussed in section 2.

Slices of high-resolution calculations (1024^2 particles; see Buchert 1989b, here not displayed) show the distribution of bright objects (density-contrast is 1000) down to the distribution of dwarf objects (density-contrast is 100). In the calculations mentioned, an implicit selection of "galaxies" has been applied, since after formation, only points with densities $|\rho| > \rho_c$ (ρ_c: threshold) are kept, imposing artificially an efficiency- or stability-criterium on galaxies. In a resolution-limited calculation, the resolution also determines a selection implicitly. The selection-procedure and the resolution-dependence of galactic densities, selection- and luminosity-functions as well as a proper normalization of the model are subjects of current studies. The dynamical determination of the model-parameters according to galaxy-evolution models is also a basic topic of further work.

4. Observational Evidence of Pancake-structures and their Statistical Measurement

Most of the support for discussions of Pancake-structures in observed slices of the universe comes from recent results of systematic complete surveys. The universe is sliced either in the direction of the declination coordinate, (see Fig.3a and the short review by Geller & Huchra 1988), or in the direction of the redshift-coordinate (Houjun Mo, priv. comm., Fig.3b). In terms of spatial distances, such slices should not be thicker than 10 Mpc/h, since the slice-thickness has to be considerably smaller than e.g. the Void-size to avoid projection-effects. Also a slicing of the southern sky shows similar Pancake-structures. A pattern of the calculations shown previously even suggests a direct comparison with 'CfA'-slices (a subslice of linear scale 100 Mpc/h is displayed in Fig.2f). Similar subslices can be produced to provide a statistical

Fig.2a–f: Evolution stages of galaxy-formation in the constructed 'Pancake-model': A (300 Mpc/h)^2 box with periodic boundaries is shown for five subsequent redshifts (**a–e**). The panel to the left in each figure shows point-processes which trace the density-field, the panels to the right show the extracted spatial distribution of "luminous objects". (**e**) is displayed for two different density-thresholds representing samples of 'typical' (*e1*) and bright (*e2*) objects.(**f**) shows a (100 Mpc/h)^2 subslice of Fig.2e which should be compared qualitatively with Fig.3a.

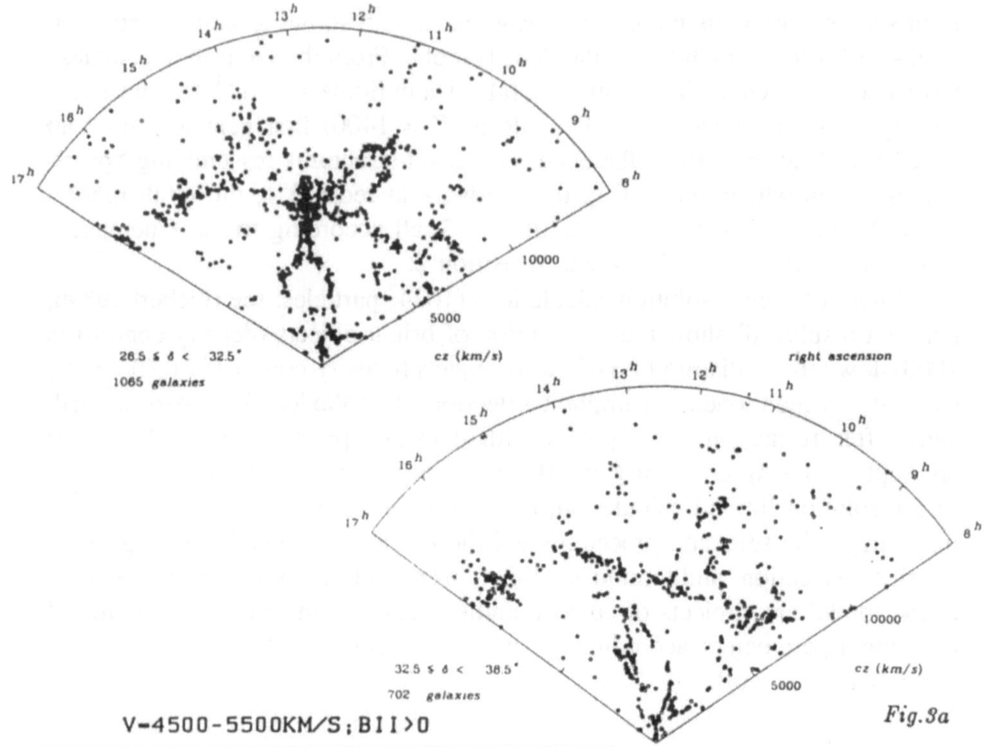

Fig.3a

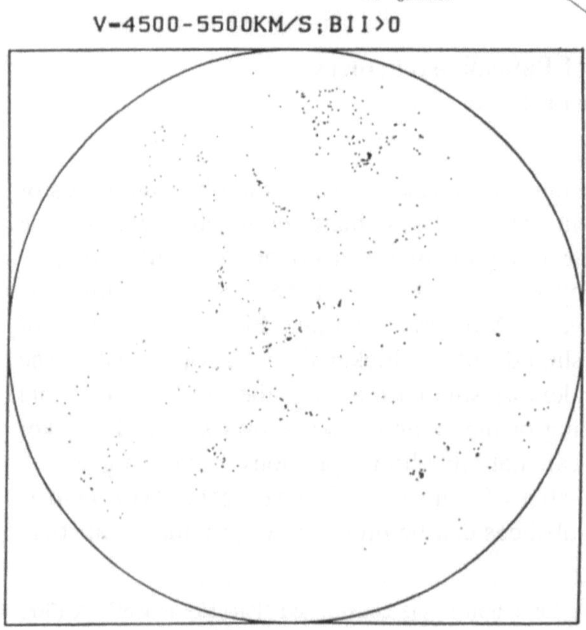

Fig.3b

Fig.3a,b: Some examples of observational evidence of Pancake-structures: (**a**) represent a slicing of the universe in the declination-direction from the 'CfA'-survey, (figures taken from Geller & Huchra 1988). (**b**) represents a slicing in Z-direction from the ZCAT: Shown is the whole sky ($0 \leq \alpha \leq 24$h, $-90° \leq \delta \leq +90°$); (Houjun Mo, priv. comm.).

sample of calculations, which form the basis of a "geometrical normalization" of patterns like the Void-size.

The comparison of observed samples and theoretical slices must be based on statistical measures. Most statistical methods complement each other with respect to their advantages. I mention here work which is in progress together with Alain Blanchard (Meudon), Bernard Jones (Copenhagen), Vicent Martinez (Valencia), Houjun Mo (MPA) and Rien van de Weygaert (Leiden) on the statistical comparison between 'Pancake-models' and observations and the 'fair sample hypothesis'. After these tests have been performed, both in observed samples and in point-processes extracted from the presented model, we are able to decide, whether or not this theory provides a consistent approach to the galaxy-formation problem. The study of the peculiar-velocity-field complements the study of the point-distribution in space and will test the consistency with respect to observed velocity-distributions. An open problem are the amplitude-limitations imposed by the microwave background. One suggestion to look for a solution of this problem is to relax the requirement that the Hubble-law describes the evolution of the mean quantities.

Remark: A preprint of this talk appears in (Börner et al. 1989) including references to other talks of the workshop.

Acknowledgements: I am thankful to G. Börner and J. Ehlers for useful comments on this talk and to my collaborators B.J.T. Jones, V. Martinez, H.J. Mo and P. Schiller for fruitful conversations on this subject. This work was supported by DFG (Deutsche Forschungsgemeinschaft).

References

Arnol'd V.I. (1982): Trudy Sem. Petrovskogo 8, p.21 (in Russian).

Arnol'd V.I., Gusein-Zade S.M., Varchenko A.N. (1982a): 'Singularities of Differentiable Maps', Nauka Moskau (1982); Birkhäuser Boston, Vol. 1 (1985).

Arnol'd V.I., Shandarin S.F., Zel'dovich Ya.B. (1982b): Geophys. Astrophys. Fluid Dyn. 20., p.111.

Börner G., Buchert T., Schneider P. (eds.) (1989): 'Progress Report on Cosmology and Gravitational Lensing', Workshop at Ringberg-castle, June, 22nd and 23rd, MPA-report No....

Buchert T. (1988a): Ph.D.-thesis, LMU Munich (in German).

Buchert T. (1988b): in: Proc. 'The World of Galaxies', Paris, April, 12th–14th, Springer.

Buchert T. (1989a): Astron. & Astrophys., in press.

Buchert T. (1989b,c): to appear in: Proc. ERAM, 'New Windows to the Universe', La Laguna, July, 3rd–8th.

Futamase T. (1989): M.N.R.A.S. 237, p.187.

Geller M.J., Huchra J.P. (1988): in: 'Large-scale Motions in the Universe', ed. by Rubin & Coyne, Princeton Series in Physics, Princeton Univ. Press.

Mo H.J., Buchert T. (1989): Astron. & Astrophys., submitted.

Peebles P.J.E. (1973): ApJ. 185, p.413.

Shandarin S.F., Zel'dovich Ya.B. (1989): Rev. Mod. Phys. 61, p.185.

The Simulation of Hydrodynamic Processes with Large Computers

H.W. Yorke

Institut für Astronomie und Astrophysik, Universität Würzburg,
Am Hubland, D-8700 Würzburg, Fed. Rep. of Germany

Summary: An overview of various methods used by computational physicists to simulate hydrodynamic phenomena is presented. Emphasis has been placed on readability by the interested non-expert. Thus, the presentation is necessarily mathematically non-rigorous and incomplete in the sense that for no problem has an attempt been made to give all the equations required for its numerical solution. Four discretization methods are discussed in general terms: finite difference methods, finite element methods, spectral methods, and particle methods. A few astrophysical applications for each of these discretization procedures are discussed.

1.0 Introduction

Due to advances in computer hardware technology and the development of efficient numerical algorithms computer simulations have attained a high degree of sophistication and have become a valuable investigative tool in all branches of science. There are a number of obvious advantages of numerical simulations as an investigative tool. In the following I would like to illuminate a few of these advantages – in a slightly critical light.

1) Computer simulations are often cheaper than the corresponding experiments. However, computer hardware development, production and maintainance, system software development, documentation and maintainance, and application software development all require the investment of many man-years of effort, before the first production runs can be started. The numerical simulations themselves require interpretive monitoring and perhaps some "fine tuning" of parameters while they run, and much effort is spent bringing the results into a digestable (understandable) form.
2) One is able to simplify the problem in order to understand the essentials. Numerical experimentation is possible; certain "contaminating" effects – always present in reality – can be switched off. However, because the numerical simulations deal with an idealized and simplified problem, there is a danger that important physical effects are neglected and the numerical results are overinterpreted.
3) The results of simulations are sometimes more accurate and more reliable than the corresponding experiment. Then again, numerical simulations sometimes supply incorrect answers to the posed physical problem.
4) Simulations do not pollute the environment. Experiments frequently do. However, use of simulations (rather the experiment) may encourage the development of environmentally risky technologies or give one a false sense of security.
5) Numerical simulations sometimes offer the only possibility to study physical phenomena. This is often the case in astrophysics. Astronomical densities, temperatures, distances, evolutionary time scales, etc. are generally prohibitably large or prohibitably small and are thus inaccessible to direct experimentation. Observations give us limited information on what is happening. Either we see only a very thin shell of an astronomical object – say, the photosphere of the sun – or only a "snapshot" during its very slow evolution, or

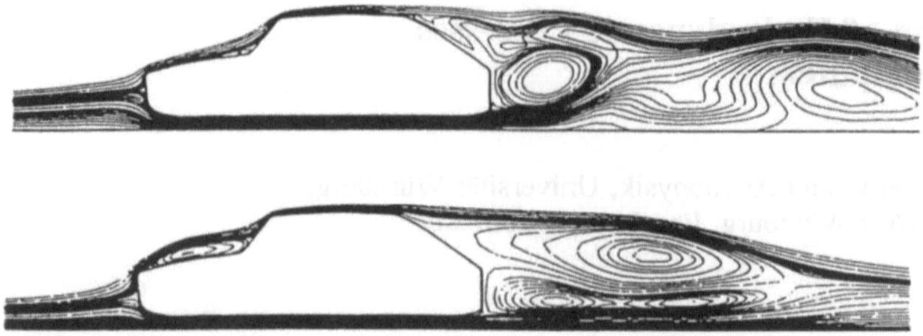

Figure 1. *Numerical simulation of the airflow around a stationary vehicle (top) and a moving vehicle (bottom). Calculations performed by Dornier GmbH (cf. Trottenberg 1988).*

both. The study of early epochs (*e.g.* the big bang or galaxy formation) would be seriously hampered without the use of simulations. However, the inaccessibility of many astrophysical phenomena to direct experimentation or observation makes "quality control" problematical. The numerical investigator must exercise extreme caution; internal consistency checks must be made during the calculation. It would be desirable if some observable predictions can be made.

As a single terrestrial example consider the optimization of the aerodynamical properties of an automobile. The goal of such an investigation is to reduce fuel usage, noise and emissions. There are significant differences in the structure of the airflow around a stationary vehicle and a moving one (see Fig. 1). Classical wind tunnel experiments measure the former. The wind tunnel experiment to accurately measure airflow around a moving vehicle would be very difficult. Numerical simulations allow one to do this experiment. The vehicle's shape can be easily altered, and the time between conception and production can be significantly shortened.

What steps are necessary for a hydrodynamical simulation of the evolution of an astrophysical object? First of all, the physical problem must be simplified. Assumptions must be made about the geometry, for example, either spherical or plane symmetry (one dimension or 1D), axial symmetry (2D) or no special symmetry (3D) is imposed upon the problem. The question of which physical effects can be neglected or at least highly simplified must be answered. Here, one is strongly influenced by hardware considerations (CPU time and memory requirements).

The next step is the mathematical formulation of the problem. The physical equations for gas dynamics, radiation transport, thermodynamics, electrodynamics, gravitation ..., together with their corresponding boundary conditions represent a system of coupled partial differential equations which must be supplemented by equations describing the state of the medium, its optical properties, nuclear or chemical reaction networks, population densities of excited states, etc.

These physical equations describe the evolution of a continuum, whereas computers manipulate numbers. Thus, the system of (in general, non-linear) partial differential equations must be transformed into a system of numerical equations which can be handled by computers. I shall call this transformation "discretization" and in the following section briefly introduce the four procedures commonly used by numerical astrophysicists: finite difference methods, finite element methods, spectral methods and particle methods.

The system of discretized equations must then be solved numerically. The choice of the method of numerical solution depends on the choice of discretization procedure, the structure of the matrices involved and memory and CPU time constraints. This is problem-dependent and a detailed discussion is beyond the scope of this brief overview. I refer the interested reader to standard monographs and the original publications describing the individual discretization methods.

The result of the numerical solution of the discretized equations is a huge set of numbers. This set of numbers has to be transformed into a series of interpretable and publishable numbers and/or pictures: *e.g.* tables, graphs, x-y plots, contour plots, 2D grey-scale representations, 2D false color representations, which describe what is happening during the simulated evolution. It is not sufficient, however, to simply describe the simulated evolution of an object. One would also like to learn why a particular path of evolution results. How sensitive are the final results to the assumptions made? Often further numerical experimentation is necessary. The investigator also acts as a "filter". What part of the results should one accept quantitatively, what should be considered a qualitative representation of reality, and what should be ignored and rejected as a spurious numerical peculiarity? There are no patent recipes for this and the interpretation and presentation of the numerical results thus has a subjective component.

2.0 Discretization Procedures

A wide variety of discretization procedures has been developed in order to transform systems of partial differential equations into systems of algebraic, computer-solvable equations. It would be impossible to make an exhaustive treatment of this complex subject in such a short overview. Instead, I shall discuss in general terms four methods of discretization, ***finite differences***, ***spectral methods***, ***finite elements***, and ***particle methods***, and present a few illustrative examples. Except for particle methods, these discretization procedures can be further classified into ***Eulerian***, ***Lagrangian***, and ***mixed*** or ***adaptive mesh*** schemes, depending on the motion of the grid. In Eulerian schemes the mesh (or elements) are assumed to be fixed in space. Quantities such as energy, mass, and momentum have to be advected across grid boundarys, a fact which causes some problems with accuracy. Lagrangian schemes accurately treat advection, but for multidimensional problems run into difficulties with mesh shearing. Adaptive grid schemes are a compromise between the two aimed at optimizating the grid movement (see *e.g.* Dorfi & Drury 1987). These schemes are the most complex and the most difficult to implement.

2.1 Finite difference methods

The derivative df/dx of a function $f(x)$ is commonly defined as the limit of a finite difference

$$\frac{df}{dx} = \lim_{h \to \infty} \frac{f(x+h) - f(x)}{h} \quad .$$

The discretization of a system of partial differential equations by finite differencing is a more or less straightforward application of this definition, except that the parameter h is finite and does not tend to zero. More specifically, the volume under consideration is divided into *cells* and the internal structure of a dependent variable within a cell, say density, temperature, or velocity are defined by one or a few parameters. The quotient obtained by dividing the difference of a variable in neighboring cells by the cell size is a finite difference representation of a partial derivative.

Consider, for example, the equation of continuity in slab (1D) geometry:

$$\frac{\partial n}{\partial t} + \frac{\partial}{\partial z}(nu) = 0 \quad , \tag{1}$$

where $n(z,t)$ is the gas particle density at the position z within an interval $z_a \leq z \leq z_b$ and at the time t; $u(z,t)$ is the gas velocity. That equation (1) is an equation for the conservation of

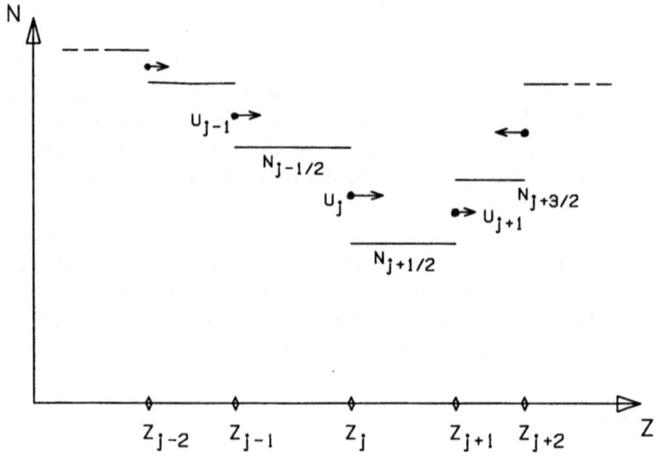

Figure 2. *Grid in the vicinity of Z_j showing the discretization of velocity and density.*

the number of particles in the volume considered (here, "volume" is really an interval multiplied by a unit area) can be shown by spatially integrating it from one boundary to the other:

$$\frac{\partial}{\partial t}\left(\int_{z_a}^{z_b} n\,dz\right) = n_a u_a - n_b u_b \quad , \tag{2}$$

where $n_i u_i$ is the flux of particles through the boundary 'i' ($i = a$ or b). In a closed volume, $n_a u_a = n_b u_b = 0$, the total number of particles per unit area in the interval, $\int_a^b n\,dz$, is constant in time.

For the sake of discussion, I discretize equation (1) by the following procedure. The interval $[z_a, z_b]$ is divided into J subintervals $[Z_j, Z_{j+1}]$, where $0 \le j \le J-1$ and $Z_0 = z_a$, $Z_J = z_b$ (see Fig. 2). Discretized variables will be denoted by capital letters. The parameter U_j is to be interpreted as the discretized approximation of the gas velocity u at the grid point $z = Z_j$ and the parameter $N_{j+1/2}$ as the approximation of the average gas density n in the subinterval $[Z_j, Z_{j+1}]$. Furthermore, the partial derivative $x = \partial(nu)/\partial z$ at a cell's midpoint, $z = (Z_j + Z_{j+1})/2$, will be approximated by

$$X_{j+1/2} = \frac{U_{j+1}\tilde{N}_{j+1} - U_j\tilde{N}_j}{Z_{j+1} - Z_j} \quad , \tag{3}$$

where $\tilde{N}_j$ is an appropriate approximation of the gas density defined at Z_j. $\tilde{N}_j$ is a (yet to be specified) function of the parameters Z_i, U_i, and $N_{i+1/2}$ (i in the neighborhood of j). Note that equation (3) is not the only possibility of discretizing x. For instance, x could be written

$$x = \frac{\partial nu}{\partial z} = u\frac{\partial n}{\partial z} + n\frac{\partial u}{\partial z} \tag{4}$$

and the partial derivatives $\partial n/\partial z$ and $\partial u/\partial z$ each be individually discretized. However, equation (3) has the distinct advantage that it is *conservative* in the sense that the number of particles is globally conserved (to the limit of rounding errors, *i.e.* 17 decimal place accuracy for double precision variables). This means the sum $\sum X_{j+1/2} \cdot (Z_{j+1} - Z_j)$, which is the approximation of $\int(\partial nu/\partial z)dz$ on the right-hand side of equation (2), will attain the numerical value

$$\sum_{j=0}^{J-1} X_{j+1/2}(Z_{j+1} - Z_j) = U_J \tilde{N}_J - U_0 \tilde{N}_0 \quad . \tag{5}$$

For any straightforward discretization of equation (4) the relationship (5) will not hold.

2.1.1 Three spatial finite difference schemes

How can one express $\tilde{N}_j$ as a function of the parameters Z_i, U_i, and $N_{i+1/2}$? I will consider three cases. First of all, one could use the arithmetic mean of the average densities $N_{j\pm1/2}$ in the neighboring cells

$$\tilde{N}_j = \frac{N_{j+1/2} + N_{j-1/2}}{2} \quad . \tag{6a}$$

This form has the advantage that the discretization error ϵ_D expressed as $\epsilon_D = |x - X_{j+1/2}|$, where x is evaluated at $z = (Z_{j+1} + Z_j)/2$, tends towards zero as $|Z_{j+1} - Z_j|^2$ (second order difference scheme). However, (6a) has the disadvantage that it is numerically unstable, which means that the numerical solution of the discretized problem, U_j and $N_{j+1/2}$, can deviate wildly from the solution of the mathematical problem posed after only a few time steps (see example in Fig. 4). Of course, the question of stability can only be answered after considering the entire system of discretized equations, including boundary conditions. The equation of motion at least would have to be included. Following a naive *cell-centering* philosophy such as in (6a) for all partial differential equations will most likely lead to numerical instabilities.

A second possibility for $\tilde{N}_j$ is first order donor-cell differencing, *i.e.*

$$\tilde{N}_j = \begin{cases} N_{j+1/2} & \text{for } U_j < 0 \\ N_{j-1/2} & \text{otherwise} \end{cases} \tag{6b}$$

Thus, depending on the direction of gas flow, one uses the average density of the cell out of which material is flowing (the *donor* cell). The discretization error ϵ_D converges to zero only as the first power of $|Z_{j+1} - Z_j|$, but the scheme developed by following this philosophy for all comparable terms is likely to be numerically stable. (I have to be very careful how I say this, because *every* individual term in a discretized system of equations is a potential source of numerical instability!) Form (6b) introduces numerical viscosity into the system of equations, damping instabilities, both numerical and physical. This effect may or may not be desirable.

The amount of numerical viscosity introduced by donor-cell differencing can be reduced by considering higher-order schemes, which require use of information over a larger interval than simply the neighboring cells to the grid point Z_j. For illustrative purposes I will consider one possibility (*cf.* van Leer 1977), describing it more in terms of pictures and words than in equations. For this example $\tilde{N}_j$ is a specified function of densities in the nearest four cells:

$$\tilde{N}_j = f(N_{j-3/2}\, , \, N_{j-1/2}\, , \, N_{j+1/2}\, , \, N_{j+3/2}\, , \, U_j\Delta t) \quad , \tag{6c}$$

where Δt is the size of the time step. In the following discussion I will implicitly assume that $\Delta t < (Z_{j+1}-Z_j)/(|U_{j+1}|+|U_j|)$ for all j. First, one determines the arithmetic means of densities of the cells bounding Z_{j-1}, Z_j and Z_{j+1}, namely $(N_{j-3/2} + N_{j-1/2})/2$, $(N_{j-1/2} + N_{j+1/2})/2$ and $(N_{j+1/2} + N_{j+3/2})/2$, respectively. Next one approximates the internal density structure in the two neighboring intervals to Z_j by linear relationships of the form $\hat{N}(z) = a + bz$, which fulfills the following requirements (see Fig. 3): i) the average density in each interval is preserved, and ii) at each endpoint of the interval $\hat{N}$ deviates from the average density no more than the arithmetic mean does. Condition ii) guarantees monotonicity in the vicinity of Z_j. Finally, the average of $\hat{N}(z)$ in the interval $[Z_*, Z_j]$ is determined, where $Z_* = Z_j - U_j\Delta t$. This average is

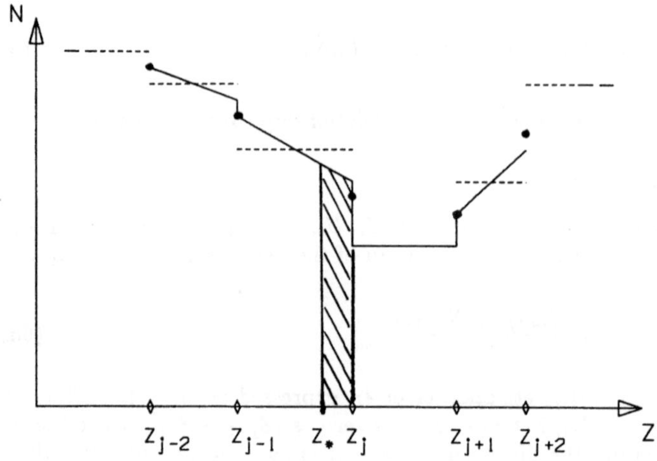

Figure 3. *Grid in the vicinity of Z_j showing the second-order internal density structure within a cell.*

$\tilde{N}_j$. In Fig. 3 I show an example of this procedure, assuming U_j to be positive. The average of $\hat{N}(z)$ in the shaded area to the left of Z_j is $\tilde{N}_j$. Its value is less than the value $N_{j-1/2}$ that first order donor-cell differencing would have provided.

The scheme described above automatically changes to first order accuracy close to extrema, because condition ii) results in $\hat{N}(z) = N_{j+1/2}$ in the interval $[Z_j, Z_{j+1}]$, if $N_{j+1/2}$ is either a maximum or minimum (see Fig. 3 for an example of a minimum). This is sometimes a useful feature, for instance close to shock fronts of moderate strength (see Fig. 4).

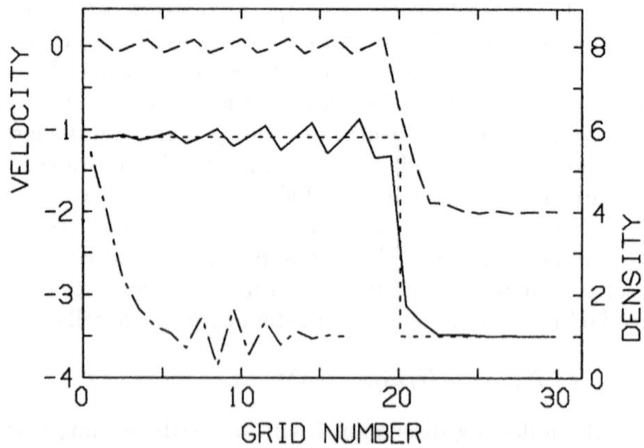

Figure 4. *Numerical solution of the shock front problem discussed in the text. The density (solid line) and velocity (dashed line) are shown after 970 time steps with $\Delta t = 0.05\,\Delta Z$. The analytic solution for density (dotted line) is shown for comparison. The numerical calculations were performed using first order donor-cell finite differencing. Significant improvements are possible by including artificial viscosity, using higher order difference schemes and/or adaptive grid techniques (Dorfi & Drury 1987). The density distribution of a numerically unstable calculation after 97 time steps is also shown (dashed-dotted line).*

2.1.2 Time differencing

Consider the following discretization of density $n(z,t)$ and spatial partial derivative $x(z,t) = \partial(nu)/\partial z$ with respect to time

$$n(z,t) \rightarrow N^k_{j+1/2} \qquad \text{and} \qquad x(z,t) \rightarrow X^k_{j+1/2}$$

where the superscript 'k' is used to denote the values of $N_{j+1/2}$ and $X_{j+1/2}$ at the k-th time step $t = t^k$. A popular method of time differencing equations such as (1) is the θ-method (see *e.g.* Richtmyer & Morton 1967):

$$\frac{N^{k+1}_{j+1/2} - N^k_{j+1/2}}{\Delta t} = \theta X^{k+1}_{j+1/2} + (1-\theta) X^k_{j+1/2} \quad , \tag{7}$$

where $0 \leq \theta \leq 1$ and $\Delta t = t^{k+1} - t^k$ is the size of the time step.

The choice of θ is critical. For $\theta = 0$ one speaks of an explicit calculation, because the values for the density at the new time t^{k+1} can be given explicitly algebraically as a function of known values at the previous time t^k, *i.e.* $N^{k+1} = N^k + X^k \Delta t$. For $\theta > 0$ one speaks of an implicit calculation, because in the general, non-linear case (including the equation of motion) N^{k+1} and U^{k+1} can only be determined iteratively. It is much easier to write an explicit hydrodynamics code and explicit computations require much less memory and CPU time per time step than comparable implicit computations. However, for $\theta \leq 1/2$ the time discretization is only conditionally stable. The time step Δt must be smaller than a critical value which is of the order of the grid spacing divided by a characteristic velocity. For magneto-hydrodynamics (MHD) this characteristic velocity is the sum of the absolute values of the gas velocity, the local sound speed and the local Alfvén velocity (Courant-Friedrichs-Lewy stability condition: see *e.g.* Potter 1977).

For $\theta = 1/2$ the time discretization is second order accurate, *i.e.* the discretization error converges to zero as Δt^2. However, this procedure is neither unconditionally stable nor explicit. For $\theta > 1/2$ the time discretization is unconditionally stable, so that the only restriction on the size of the time step is dictated by the nature of the problem one is solving. In the sun, for example, the pulsation time scale is ~ 1 hr., whereas the quasi-hydrostatic evolution time scale is $\sim 10^{10}$ yr. Thus, the choice of time step in a code modelling the sun depends on whether one is interested in the sun's pulsations or its slow evolution.

2.1.3 A simple numerical example

I now consider the evolution of a moderate strength isothermal shock front in slab geometry. Without loss of generality I set the ratio of pressure to particle density p/n equal to unity, the left boundary to $z_a = 0$, and the right boundary to $z_b = 1$. Using the same notation as before, the equation of momentum conservation can be put into the form

$$\frac{\partial(nu)}{\partial t} + \frac{\partial}{\partial z}(nu^2 + n) = 0 \quad . \tag{8}$$

This equation and equation (1), supplemented by the boundary conditions $u(0,t) = 0$ and $n(1,t)u(1,t) = -2$ for $t > 0$ and the initial conditions $n(z,0) = 1$, $u(z,0) = -2$ for $0 \leq z \leq 1$, can be solved analytically. At $t = 0$ a discontinuity develops at $z = 0$ and subsequently moves to the right with a velocity $u_S = \sqrt{2} - 1 \simeq 0.414$. Here, the gas velocity jumps from $u_1 = -2$ to $u_2 = 0$ and the density from $n_1 = 1$ to $n_2 = 3 + 2\sqrt{2} \simeq 5.828$, where I use the subscripts '1' and '2' to denote quantities to the right of and left of the shock front, respectively.

First order explicit donor-cell differencing could be done as follows. I approximate the particle flux nu at $z = Z_j$ and $t = t^{k-1/2}$ by $F_j^{k-1/2}$. The velocity U_j as it appears in equations

(3), (6b) and (6c) is calculated from $U_j = 2F_j^{k+1/2}/(N_{j-1/2}^k + N_{j+1/2}^k)$. The particle flux is updated in time by an explicit time step

$$F_j^{k+1/2} = F_j^{k-1/2} - \frac{2\Delta t}{Z_{j+1} - Z_{j-1}}(Y_{j+1/2} - Y_{j-1/2}) \quad , \tag{9}$$

where $Y_{j+1/2} = \tilde{U}_{j+1/2}\tilde{F}_{j+1/2} + N_{j+1/2}^k$. Depending on the sign of the cell-centered velocity $\tilde{U}_{j+1/2} = (F_j^{k-1/2} + F_{j+1}^{k-1/2})/2N_{j+1/2}^k$, the cell-centered particle flux $\tilde{F}_{j+1/2}$ is determined by

$$\tilde{F}_{j+1/2} = \begin{cases} F_{j+1}^{k-1/2} & \text{for } \tilde{U}_{j+1/2} < 0 \\ F_j^{k-1/2} & \text{otherwise} \end{cases} \quad . \tag{10}$$

Initial conditions can be expressed by $N_{j+1/2}^0 = 1$ and $F_j^{-1/2} = -2$ for $0 \le j \le J$. The boundary conditions are $F_0^{k+1/2} = 0$ and $F_J^{k+1/2} = -2$ for $0 \le k$. By alternatively solving equation (9) for $F_j^{k+1/2}$ and equation (3) for $N_{j+1/2}^{k+1}$ for $0 \le j \le J-1$, the evolution of the shock front can be calculated numerically.

The numerical solution is shown in Fig. 4 together with the analytic solution. Also shown in the lower left corner are the results of a numerically unstable simulation after 97 time steps. Here, the spatially centered form (6a) was used for $\tilde{N}_j$ rather than first order donor-cell differencing (6b). An attempt was made to stabilize the calculation by including artificial viscosity (*cf.* Tscharnuter & Winkler 1979), because without it the density became negative after 18 time steps. Otherwise the calculations were identical. Merely changing one term led to a numerical instability – both the velocity and density oscillated wildly and the calculations eventually had to be stopped.

2.2 Spectral methods

Rather than give a general discussion of the use of spectral methods in computational hydrodynamics, I will consider a single 3D example, the Poisson equation

$$\nabla^2 u = n \quad . \tag{11}$$

Multiplying equation (11) by the spherical harmonic function

$$Y_{l\,m}(\Omega) = \left[\frac{2l+1}{2\pi}\frac{(l-|m|)!}{(l+|m|)!}\right]^{1/2} P_l^{|m|}(\cos\theta) \times \begin{cases} \cos m\phi & \text{if } m \ge 0 \\ \sin m\phi & \text{if } m < 0 \end{cases}$$

and integrating it over all solid angles Ω yields

$$\frac{1}{r^2}\frac{\partial}{\partial r}\left(r^2\frac{\partial U_{l\,m}}{\partial r}\right) + \frac{l(l+1) - m^2}{r^2}U_{l\,m} = N_{l\,m} \quad , \tag{12}$$

where the coefficients $U_{l\,m}(r)$ and $N_{l\,m}(r)$ are defined by

$$n(r,\Omega) = \sum_{l=0}^{\infty}\sum_{m=-l}^{+l} N_{l\,m}(r)\, Y_{l\,m}(\Omega) \tag{13a}$$

$$u(r,\Omega) = \sum_{l=0}^{\infty}\sum_{m=-l}^{+l} U_{l\,m}(r)\, Y_{l\,m}(\Omega) \quad . \tag{13b}$$

The above spectral decomposition of the Poisson equation has reduced the 3D problem (equation 11) to an infinite number of ordinary differential equations (equation 12) for the coefficients of u. In order to handle this problem in the computer, the expansions (13a and b) are truncated at $l = L$. We can easily accomodate certain symmetries in the problem. For symmetry with respect to the plane $\theta = \pi/2$, for example, we only use even values of l in the expansion. For axial symmetry only terms with $m = 0$ are necessary. If the function $n(r, \Omega)$ is known, rather than its expansion $N_{l\,m}(r)$, one utilizes that fact that the functions $Y_{l\,m}$ are orthonormal. Thus, the coefficients for the expansion (13a) can be calculated

$$N_{l\,m}(r) = \int_{4\pi} Y_{l\,m}(\Omega)\, n(r, \Omega)\, d\Omega \quad . \tag{14}$$

The integral in equation (14) is done numerically; *e.g.* the integral over $d\mu = d(\cos\theta)$ can be performed quite efficiently for this problem by Gauss quadrature. There are also efficient methods available for performing the ϕ integration (*e.g.* fast fourier transforms).

Generalizing this procedure to other partial differential equations is straightforward. Partial derivatives of a dependent variable $f(r, \theta, \phi, t)$ with respect to the angle θ or ϕ can be expressed

$$\frac{\partial f(r, \theta, \phi, t)}{\partial \theta} = \sum_{l=0}^{L} \sum_{m=-l}^{+l} F_{l\,m}(r, t)\, \frac{\partial Y_{l\,m}}{\partial \theta} \tag{15a}$$

$$\frac{\partial f(r, \theta, \phi, t)}{\partial \phi} = \sum_{l=0}^{L} \sum_{m=-l}^{+l} F_{l\,m}(r, t)\, \frac{\partial Y_{l\,m}}{\partial \phi} \tag{15b}$$

where $F_{l\,m}(r, t)$ are the corresponding expansion coefficients. Non-linear terms in a partial differential equation may have to be decomposed according to a formula similar to (14). Note that use of the same expansion for all dependent variables may not be advisable. Consider, for example, the radial and θ components of the velocity, v_r and v_θ, in a hydrodynamic problem for which symmetry with respect to a plane is imposed. Whereas v_r is symmetric with respect to the plane $\theta = \pi/2$, v_θ is anti-symmetric. Thus, different orthonormal expansion functions should be used for $v_\theta(r, \theta, \phi)$, such as $dY_{l\,m}/d\theta$ (*c.f.* Różyczka *et al.* 1980).

The associated partial differential equations for the coefficients $F_{l\,m}(r, t)$ can be further decomposed by a series expansion for the radial dependence. However, it may be advantageous to utilize spectral decomposition for only a few independent variables and finite difference methods or finite element methods to solve for the coefficients. Examples of the latter are given in section 3.4.

Note that the discretization error of spectral methods depends solely on the spectral coefficients that have been discarded in the truncation. By a careful choice of the expansion functions, it is possible to fulfill the boundary conditions, the symmetry conditions and still have good series expansion convergence properties. The hallmark of a good spectral method is a discretization error that decreases *exponentially* with increasing resolution, whereas in a second-order accurate finite difference method, the error decreases only as the square of the spatial resolution (see *e.g.* Marcus 1986).

2.3 Finite element methods

In discussing finite element methods (FEMs) I will restrict myself to general aspects of the Galerkin formulation. The underlying mathematical theory of FEMs is actually better developed than for finite difference methods. Still, finite difference methods are more popular amoung numerical astrophysicists, perhaps because they appear to be intuitively simple. Also, being non-dissipative, the Galerkin FEMs are not suitable for solving the hydrodynamic equations except for a short period of time. A modification of this method is necessary (Petrov-Galerkin

FEMs), which introduces some numerical dissipation. Since FEMs have been applied to time dependent fluid problems only rather recently, one can expect further developments to take place to improve their overall efficiency.

Let $\{\phi_1, \phi_2, \ldots, \phi_N\}$ denote a set of *trial* functions defined over the volume V considered for the Poisson equation (11). I will furthermore assume that u and the trial functions vanish on the boundary of V. The solution u is approximated by the function

$$U(\mathbf{x}) = \sum_{j=1}^{N} a_j \, \phi_j(\mathbf{x}) \quad , \tag{16}$$

where the coefficients a_j are computed from the condition that the residual $\nabla^2 U - n$ be orthogonal to each of the *test* functions ϕ_i (minimal energy condition):

$$\int_V \phi_i \, (\nabla^2 U - n) \, dV = 0 \, , \qquad i = 1, 2, \ldots, N \, . \tag{17}$$

If the trial functions are linearly independent and sufficiently smooth, equations (17) yield an $N \times N$ system of linearly independent equations

$$- \, \mathbf{S} \cdot \mathbf{a} = \mathbf{F} \quad , \tag{18}$$

where the stiffness matrix $\mathbf{S}$ and the source term $\mathbf{F}$ are defined by

$$F_i = \int_V \phi_i n dV \quad \text{and} \quad S_{i\,j} = - \int_V \phi_i \nabla^2 \phi_j dV \quad .$$

Because U vanishes on the boundary of V, we may employ Green's Theorem, so that $S_{i\,j} = \int_V \nabla\phi_i \nabla\phi_j dV$ only utilizes the first derivatives of the trial functions. In this case it is sufficient that the trial functions be continuous in V.

The crucial step in transforming the above definitions into a FEM is to subdivide the volume V into a large number of non-overlapping subregions – called elements – and each trial function is defined in such a manner as to vanish on all but a small number of elements. Since the matrix entry $S_{i\,j}$ is zero, except when ϕ_i and ϕ_j are both non-zero on at least one element, the matrix $\mathbf{S}$ will contain a large proportion of zeros and equation (18) can be solved by use of sparse matrix techniques.

The extension of these ideas to time dependent problems is straightforward. For the heat equation $\partial u / \partial t = \nabla^2 u + n$ the coefficients a_j in (16) are allowed to vary in time. The residual is now $\dot{U} - \nabla^2 U - n$ and this leads to the system of ordinary differential equations

$$\mathbf{M} \cdot \dot{\mathbf{a}} + \mathbf{S} \cdot \mathbf{a} = \mathbf{F} \quad , \tag{19}$$

where the dot denotes differentiation with respect to t and the mass matrix $\mathbf{M}$ is defined by

$$M_{i\,j} = \int_V \phi_i \phi_j dV \quad . \tag{20}$$

Time discretization can be by the θ-method described in section 2.1.2. This leads to the equation

$$[\mathbf{M} + \theta \Delta t \; \mathbf{S}] \cdot \mathbf{U}^{k+1} = [\mathbf{M} - (1 - \theta) \Delta t \; \mathbf{S}] \cdot \mathbf{U}^k + \mathbf{F}^{k+\theta} \quad , \tag{21}$$

where $\mathbf{F}^{k+\theta}$ is an appropriate approximation of $\mathbf{F}$ at the time $t^k + \theta \Delta t$. This scheme is clearly implicit for all values of θ and is the price to be paid for the inclusion of the mass matrix. There may still be some advantage in setting $\theta = 0$, since the matrix $[\ \mathbf{M} + \theta \Delta t \ \mathbf{S}\ ]$ is then symmetric.

2.3.1 FEM for 1D problems

Trial functions and elements must be defined in such a manner that $U \to u$ as $N \to \infty$. Generally the trial functions are constructed so that the coefficients a_j have a physical interpretation. For instance, for the piecewise linear functions (see Fig. 5) a_j would refer to the value of U at a particular point (node). These are the lowest degree polynomials consistent with the admissibility conditions.

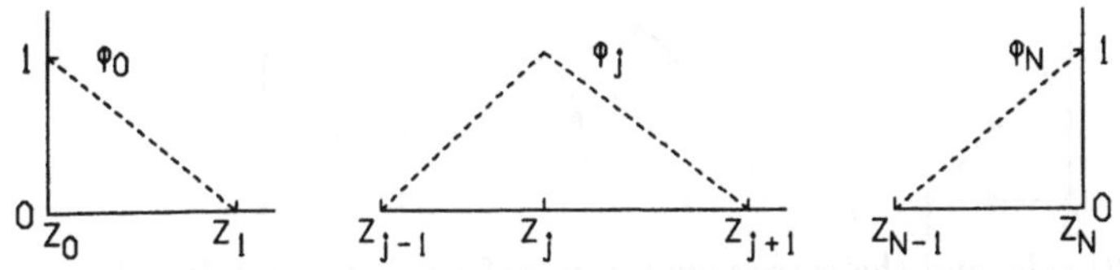

Figure 5. *Piecewise linear trial functions ϕ_j defined over grid $Z_0, Z_1, \ldots, Z_N$.*

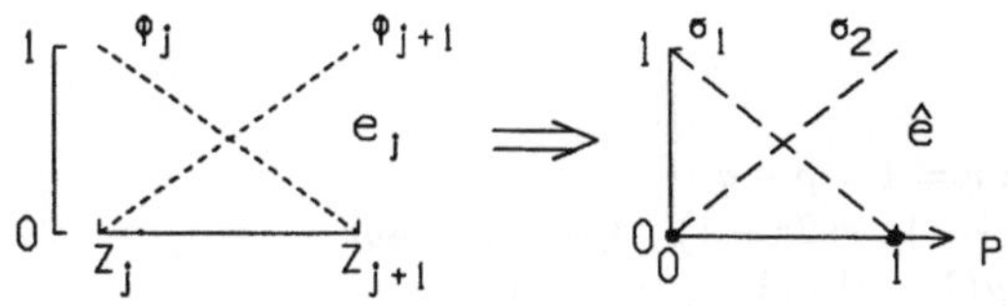

Figure 6. *Transformation of piecewise linear trial functions in the element e_j to the unit cell $\hat{e}$ with its shape functions σ_1 and σ_2.*

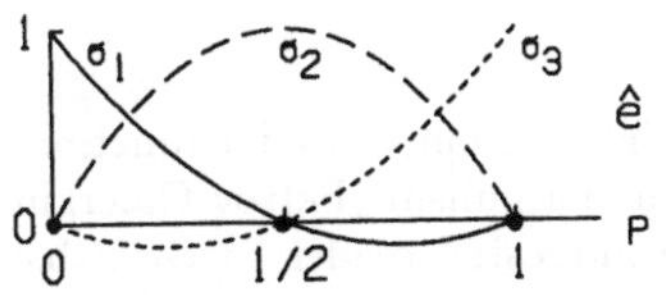

Figure 7. *Unit cell showing the three quadratic shape functions and the three nodes.*

Rather than define trial functions ϕ_j, it is more convenient to consider individual elements e_j and transform them to the unit element $\hat{e}$ (Fig. 6). The two pieces of overlapping trial functions ϕ_j and ϕ_{j+1} in e_j transform to the linear shape functions $\sigma_1 = 1 - p$ and $\sigma_2 = p$, where p varies from 0 to 1 between adjacent nodes. The generalization to second order polynomial expansions is straightforward. Since three points are necessary to uniquely define a quadratic function, the unit element will have three nodes (Fig. 7) and three shape functions, σ_1, σ_2 and σ_3, defined so that at each node one shape function is unity and the other two vanish, *i.e.* $\sigma_1 = (1-p)(1-2p)$, $\sigma_2 = p(1-p)$, $\sigma_3 = p(2p-1)$. Similarly, one can define four cubic shape functions on a unit element with four nodes. Alternatively, one can insist that the trial functions and their derivatives be continuous, so that the global approximation is continuously differentiable. This so-called Hermite cubic approximation has the shape functions: $\sigma_1 = (1+2p)(1-p)^2$, $\sigma_2 = p(1-p)^2$, $\sigma_3 = (3-2p)p^2$, $\sigma_4 = -p^2(1-p)$.

2.3.2 FEM for 2D problems

Most finite element approximations in 2D begin by dividing the domain V into non-overlapping elements, usually triangles, quadrilaterals or a combination of the two. The transformations of these elements to a unit element are shown schematically in Fig. 8, where I only show the orientation of the nodes and do not attempt to include a represenation of the shape functions. The role of the variable p, describing the inner structure of a 1D element, is adopted by the variable pair (p, q) in 2D. The shape functions for each case depicted in Fig. 8 are as follows:

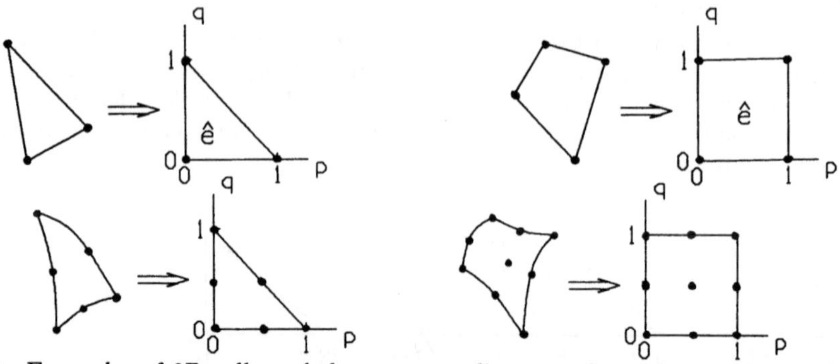

Figure 8. *Examples of 2D cells and the corresponding transformations to unit cells showing the location of nodes for the linear triangle (upper left), the 4-node bilinear element (upper right), the quadratic triangle (lower left), and the 9-node biquadratic element (lower right).*

The linear triangle:	$p\ ,\ q\ ,\ r\ ,$ where $r = 1 - p - q$
The quadratic triangle:	$p(2p-1)\ ,\ q(2q-1)\ ,\ r(2r-1)\ ,\ 4pq\ ,\ 4pr\ ,\ 4qr$
4-node bilinear element:	$p\,q\ ,\ (1-p)\,q\ ,\ p\,(1-q)\ ,\ (1-p)\,(1-q)$
9-node biquadratic element:	$\sigma_j^*(p)\sigma_k^*(q)$, where $j,k \leq 3$ and σ_i^* are the 1D quadratic shape functions.

By allowing the test function in equation (17) to differ slightly from the trial function, a controllable amount of dissipation can be introduced into this formulism (Petrov-Galerkin FEMs). A number of variations are in use today. I refer the interested reader to Griffiths' (1986) introduction to this subject for a list of references.

2.4 Particle methods

The common distinguishing feature of particle methods is the use of a set of particles, carrying with them certain characteristics (eliminating the problem with advection) and moving according to some equation of motion. The most obvious application is the study of N-body systems, where there is a one to one correspondence between physical particle and simulation particle. However, each simulation particle can correspond to an arbitrary number of physical particles and thus simulate magnetofluids not ordinarily considered in terms of their microparticulate nature.

The equations of motion to be satisfied by each particle can be written:

$$\frac{d\mathbf{x}}{dt} = \mathbf{v} \qquad \text{and} \qquad \frac{d\mathbf{v}}{dt} = -\frac{1}{\rho}\nabla P - \nabla\Phi + \mathbf{S} \quad ,$$

where $\mathbf{S}$ represents additional acceleration (from *e.g.* artificial viscosity or magnetic forces) to be considered. The pressure acceleration $\nabla P/\rho$ and gravitational acceleration $\nabla\Phi$ terms are calculated either by considering individual particle-particle forces (here the computational expense increases as N^2, where N is the number of particles) or by use of some smoothing algorithm to obtain averaged quantities from which these terms can be calculated by more efficient means (*cf.* Hockney & Eastwood 1988). Sometimes meshes are employed to facilitate smoothing. Efficient N-body techniques exist to advance particle positions in time.

Rather than attempting to describe the wide range of particle methods in use, I will discuss the smoothed particle hydrodynamics (SPH) scheme introduced in an astrophysical context by Lucy (1977) and since considerably improved (*cf.* Gingold & Monaghan 1983, Monaghan 1985). We first note that any physical quantity may be written in the form

$$A(\mathbf{x}) = \sum_{i=1}^{N} m_i \left[\frac{A(\mathbf{x}_i)}{\rho_i}\right] W(|\mathbf{x} - \mathbf{x}_i|, h) \quad ,$$

where $W(r,h)$ is the smoothing kernel, h the smoothing length, and m_i the particle "mass", a weighting factor. Using this formulism, the equation of motion can be put into the form

$$\frac{d\mathbf{v}_i}{dt} = -\sum_{j=1}^{N}\left[m_j\left(\frac{P_j}{\rho_j^2} + \frac{P_i}{\rho_i^2}\right)\nabla_i W(r_{ij}, h) - \frac{G\, M_{ij}}{r_{ij}^3}(\mathbf{x}_i - \mathbf{x}_j)\right] + \mathbf{S}_i \quad ,$$

where $r_{ij} = |\mathbf{x}_i - \mathbf{x}_j|$ and M_{ij} is the mass of particle j within a sphere of radius r_{ij}:

$$M_{ij} = 4\pi\, m_j \int_0^{r_{ij}} r^2 W(r,h)\, dr \quad .$$

The unusual form of the pressure term arises from the identity $\nabla P/\rho = \nabla(P/\rho) + P/\rho^2 \nabla\rho$. Its symmetric form ensures exact conservation of linear and angular momentum. A similar SPH equation can be formulated for the conservation of energy.

To complete the description of the method we need to specify the kernel. The only constraints to be imposed on $W(r,h)$ are that it be bounded, differentiable and normalized so that $M_{ij} \to m_j$ as $r_{ij} \to \infty$. Usually a simple analytic form is chosen. In the example discussed in section 3.2 the kernel $W(r,h) = e^{-r/h}/8\pi h^3$ was used.

3.0 Numerical Examples

In this chapter I briefly discuss the results of selected numerical calculations. They are more or less typical astrophysical applications, which means that they were calculated under the constraints of producing a reasonable number of scientific papers of reasonable quality within a reasonable time using computer facilities currently available to the astrophysical community. Since I will be more concerned with numerical aspects of the calculations than with details of the results, I refer the interested reader to the original publications.

3.1 Collision of a white dwarf with a main sequence star

Although collisions between stars are very rare under normal galactic conditions, they may significantly influence the evolution of dense stellar systems such as the central regions of galaxies and globular cluster cores. The percentage of old stars in these regions is high, so that many of these collisions will occur between a compact stellar remnant and low-mass main sequence stars. Most of these events are in fact grazing or off-center collisions. The main expected effects are production of rejuvenated red giants (mass transfer to white dwarfs), formation of thick, massive accretion disks around compact objects, ejection of gas from the stars, enrichment of the interstellar medium with products of nucleosynthesis, production of X-rays, coalescense of stars and the formation of more massive objects.

In Fig. 9 the results of a numerical simulation of a head-on collision between a 0.5 $M_\odot$ compact object and a 0.5 $M_\odot$ main sequence star of radius 6×10^{10} cm are shown (Różyczka *et al.* 1989). The initial relative velocity between the two stars was –472 km s^{-1}, which except for sign is the escape velocity at their assumed initial center of mass separation (1.2×10^{11} cm).

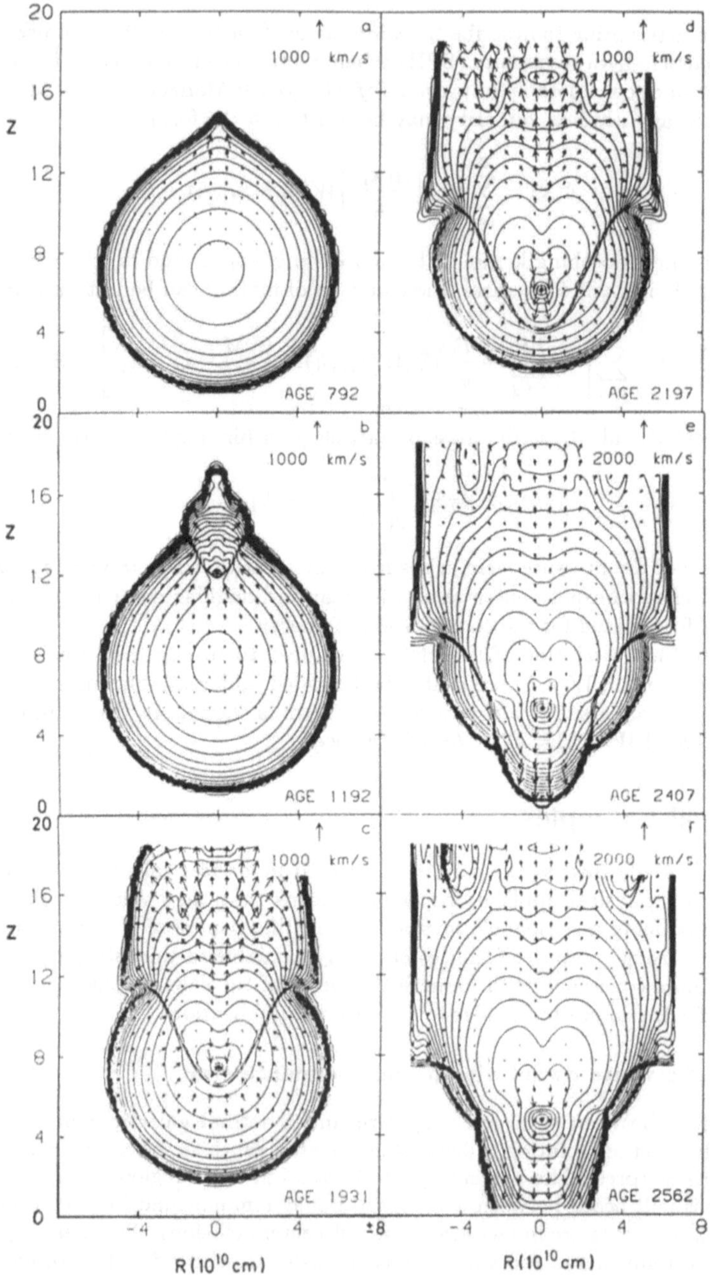

Figure 9. *Isodensity contours and velocity vectors in the meridional plane are shown for selected times during the course of a white dwarf – main sequence star collision. The age in seconds is given in the lower right corner of each frame, the velocity scale in the upper right. Contour intervals are 0.25 dex. The highest values for* $\log \rho$ *are: 0.75* (**a**)*, 1.25* (**b**)*, 1.75* (**c d e** *and* **f**)*. The position of the white dwarf is marked with an asterisk.*

An adiabatic energy equation was assumed. After completion of the hydrodynamic calculation, a detailed nuclear reaction network was solved, using the temperature and density distribution obtained. It was shown that nuclear energy generation was negligible during the course of the evolution ($\approx$ 2000 s). About 50% of the mass of the main sequence star remained gravitationally bound to the white dwarf; the remainder was expelled from the system.

The 2D axisymmetric numerical code employed finite differences of second order spatial accuracy (corresponding to the third scheme discussed in section 2.1.1) on an equidistant 190×70 Eulerian grid. Due to operator-splitting it was quasi-second order in time. Artificial viscosity was used to treat shock fronts. A pseudo-Lagrangian surface was used to treat the outer boundary of the star. The Poisson equation for the stellar contribution to gravity and the energy equation were solved implicitly; otherwise the calculation was explicit. Approximately 2 seconds of CRAY-1 time were required for each time step; about 5000 time steps were calculated.

3.2 Off-center collisions of two main sequence stars

Most of the stars in a globular cluster or a galactic nucleus will be low mass ($M \lesssim 1\ M_\odot$) main sequence stars. Thus, collisions between two main sequence stars are more common than between a white dwarf and a main sequence star. Because such collisions can be expected to significantly influence the overall evolution of compact stellar systems, it is important to understand this basic process. In particular, the possibility of building up more massive stars by coalescence, the complete remixing which occurs during collisions and the partial (or complete) disruption of colliding stars are interesting questions to be answered by numerical simulations.

Assuming a simplified equation of state ($P \propto \rho^{1.5}$), Benz & Hills (1987) investigated off-center collisions of two identical main sequence stars, varying the impact parameter and the collision velocity over a wide range. A 3D SPH method with 1024 particles was employed. In Fig. 10 I show their results for the case of zero relative velocity at infinity $V_\infty = 0$ and closest approach $R_{min}/(R_1 + R_2) = 0.953$ (R_1 and R_2 are the stellar radii). The stars become a gravitationally bound system during the first periastron passage (due to mutual tidal interactions) and coalesce into a single star during the second periastron passage. This two-stage process in which a binary first forms after a grazing collision, and then coalesces into a single star during a subsequent periastron passage occurred quite often in their simulations, even for a relatively high velocity ($V_\infty/V_{esc} = 1.67$, where V_{esc} is the escape velocity from the surface of the stars) and low impact parameter $R_{min}/(R_1 + R_2) = 0.1$. Mass loss from the system, carrying away excess angular momentum, was an important part of this process.

3.3 Non-Central Collisions of Galaxies

From the mean-square velocities, sizes, and number densities of galaxies in clusters it can be shown that interactions and collisions between entire galaxies are not seldom. Such interactions are believed to be an important mechanism to produce the hot dilute intra- cluster gas observed in X-rays and to trigger episodes of high star formation activity.

The hydrodynamical and thermal evolution of the interstellar gas during non-central collisions of two spherical galaxies was calculated with a 3D explicit finite difference code using a non-equidistant Eulerian grid of $\approx 10^6$ grid points by Müller *et al.* (1989; see Fig. 11). In this so-called second-order alternating direction implicit/explicit code (SADIE for short), the fluxes in the advection terms were calculated using the flux-vector splitting method of van Leer (1982). The gravitational potential was assumed to be given by the distributions of stars in each system and, except for their relative movement, each stellar system was held fixed in time. This assumption is valid for the low gas densities and high collision velocities considered. At lower velocities the mutual tidal distortion of the stellar systems would have to be considered, and

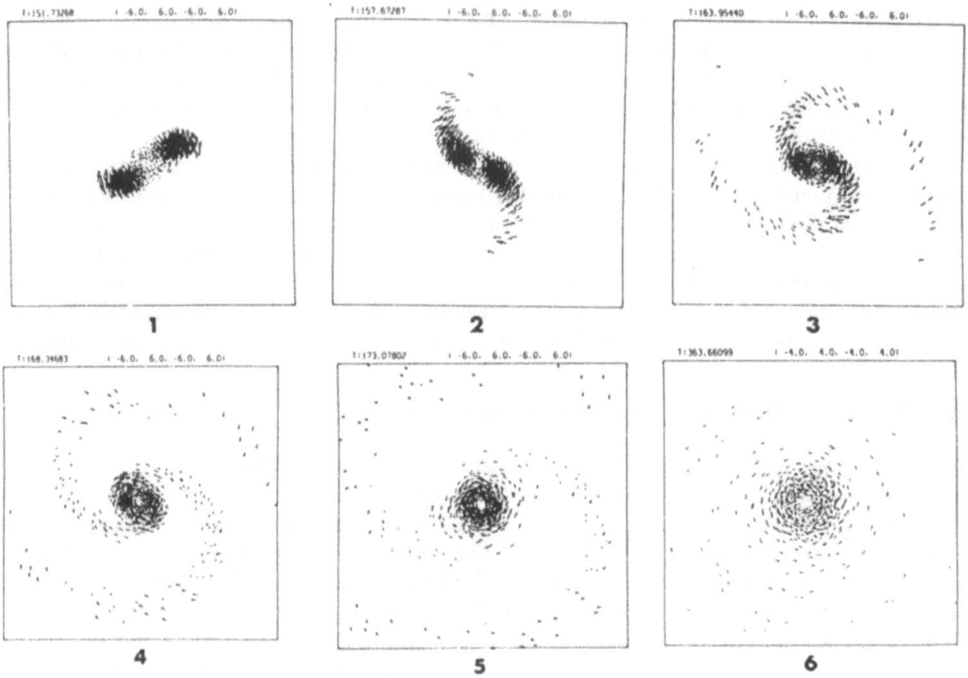

Figure 10. *Evolution of two colliding main sequence stars as simulated with a 3D SPH code (from Benz & Hills 1987). The frames show the velocity vectors of the fluid particles projected onto the plane containing the stars' centers of mass. The first five frames occurred during second periastron passage; the last is a magnified view near the end of the computer simulation. The time unit is the pre-collision oscillation time of an undisturbed star.*

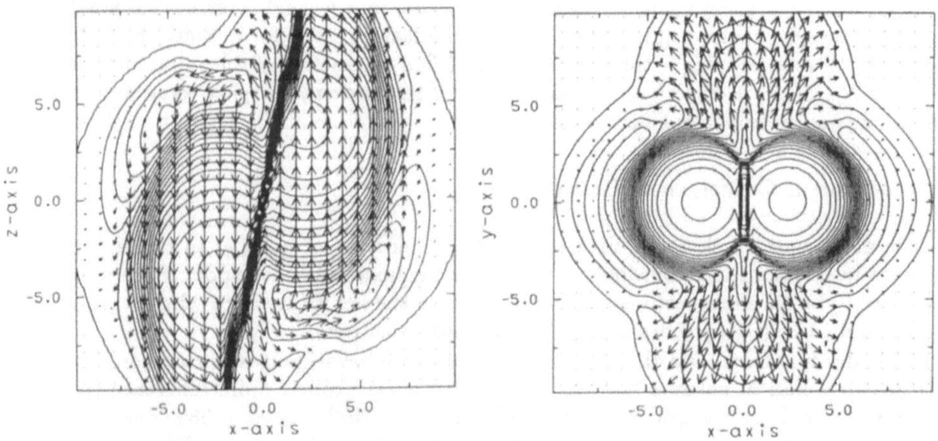

Figure 11. *Isodensity contours and velocity vectors in the x-z (left) and x-y (right) planes are shown 40 million years after contact of two colliding galaxies. The centers of mass of the two galaxies move along trajectories $x = \pm 2.26 \times 10^{22}$ cm, $y = 0$ at constant velocities $v_z = \pm 500$ km s^{-1}. The linear scale shown is $\pm 10^{23}$ cm. Typical velocities are ≈ 500 km s^{-1}. The 21 contour levels of $\log \rho$ are equally spaced in the range [min,max] = [–31.5,–24.3] (left) and [–30.4,–25.3] (right).*

solution of the Poisson equation cannot be avoided. Note that the stellar systems pass through one another with a very low probability for a single direct star-star collision – due to the low mean density of stars.

In Fig. 11 the distribution of the gas and its velocity are shown for two planar cuts at a time about 40 million years after initial contact between the two systems. The impact parameter was chosen to be $4/3R_g$, where R_g is the radius of the two equilibrium gas spheres, and the relative velocity was set at 1000 km s^{-1}. The calculations show that the shocked gas is heated to several 10^6 K, but quickly cools, forming a thin sheet-like feature in the central shock zone. This is likely to be torn apart by the separating galaxy potentials and will subsequently fragment. Some of the cooled gas remains behind and – after a period of star formation – results in dwarf galaxies as are commonly associated with giant galaxies.

This simulation was performed on a CRAY-2. Approximately 400 MByte of central memory was required, because SADIE extensively uses scratch arrays of size $N_x \times N_y \times N_z$, in order to enhance code performance. Each time step required about 30 seconds of CPU time using the CFT77 (version 1.2) compiler, corresponding to about 53 MFLOPS (million floating point operations per second).

3.4 Fragmentation of self-gravitating rotating disks

In an explicit 3D hybrid spectral/finite difference calculation Różyczka *et al.* (1980) examined the stability of a differentially rotating disk with respect to non-axisymmetric perturbations. First, a hydrostatic disk was constructed by following the collapse of an initially homogeneous, isothermal sphere rotating as a solid body (see Fig. 12), using an implicit 2D version of the code with viscosity (but without angular momentum transport). After about 100 initial free-fall times a hydrostatic disk resulted, all motions except differential rotation having been damped out. At this time the ratio of thermal energy to gravitational binding energy was 0.25; the ratio of rotational energy to gravitational was 0.33.

The disk thus obtained was perturbed by multiplying the density by a factor $(1+0.1 \cos 2\phi)$, and the subsequent evolution was followed with the 3D code (see Fig. 13). The result of this calculation was gravitational fragmentation of the disk and formation of a binary system. Thus,

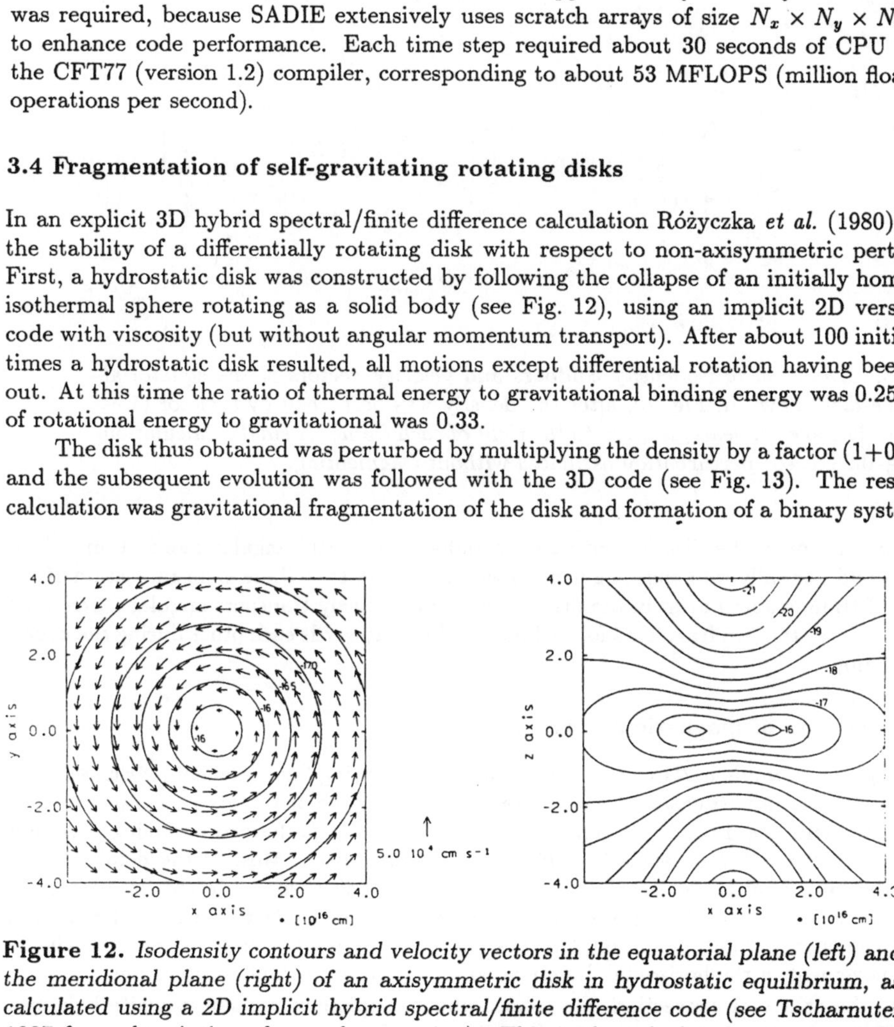

Figure 12. *Isodensity contours and velocity vectors in the equatorial plane (left) and the meridional plane (right) of an axisymmetric disk in hydrostatic equilibrium, as calculated using a 2D implicit hybrid spectral/finite difference code (see Tscharnuter 1987 for a description of a modern version). This is the calculated outcome of the gravitational collapse of a homogeneous gas sphere, assuming initial solid body rotation but no subsequent angular momentum transport.*

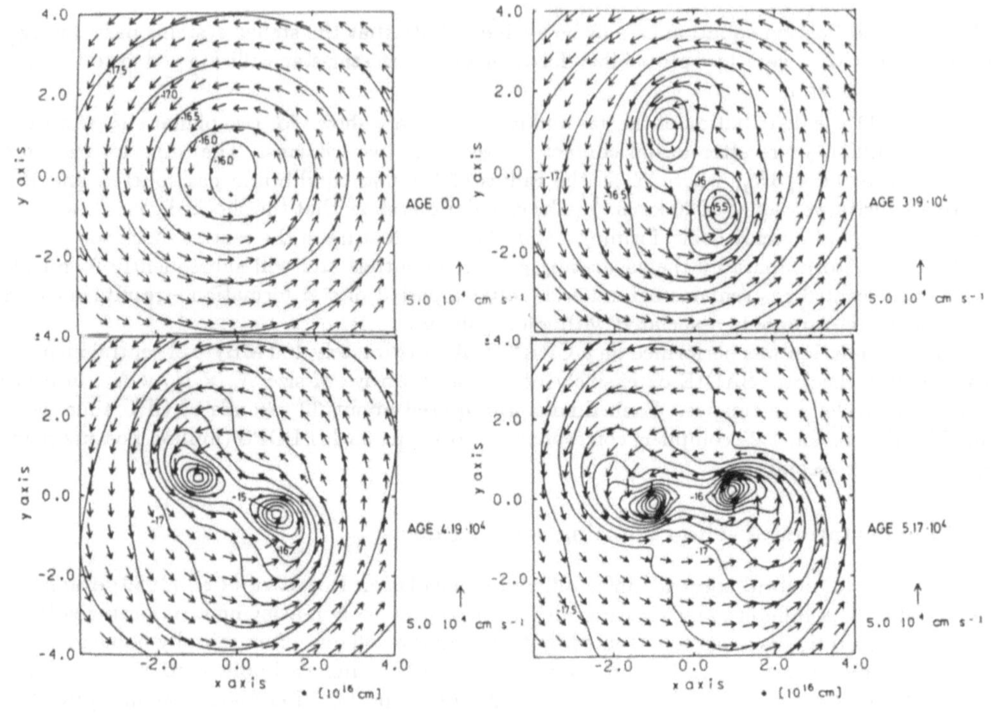

Figure 13. *Evolution of isodensity contours and velocity vectors in the equatorial plane of the disk shown in Fig. 12, after the density was perturbed by a factor* $(1 + 0.1 \cos 2\phi)$. *The age (in years) is given to the right of each frame. A similar calculation with a* $(1 + 0.2 \cos 4\phi)$ *perturbation oscillated without fragmenting.*

angular momentum in the disk is converted partially into orbital angular momentum (of the fragments) and partially into spin angular momentum. Note that there is an indication of the formation of trailing spiral arms in the disk which can effectively transport angular momentum outwards. Indeed, the two fragments formed moved slowly towards each other during the course of the evolution.

3.5 The evolution of magnetic flux tubes

As an example of an implicit 2D-MHD simulation using a moving finite element (MFE) scheme, I discuss briefly the final (quasi-stationary) state of evolving magnetic flux sheets in the photosphere of the sun (see Fig. 14). Such calculations have been performed assuming a variety of conditions (see Knölker & Schüssler 1989 for a reference list). Most of the details of the code and the numerical model can be found in Deinzer *et al.* (1984). Recent improvements are replacement of the diffusion approximation by full angle-dependent (grey, LTE) radiation transfer in the energy equation and implementation of a fully implicit time integration scheme with grid movement in the horizontal direction, solved by Newton-Raphson iteration.

Due to the effects of convection, the (photospheric) surface of the sun in white light has a mottled (granular) appearance. Slowly ascending and diverging hot gas releases its heat in a thin layer, cools and subsequently descends quickly, converging into finger-like downdrafts. The resulting temperature contrast manifests itself in continuum light as a network of extended,

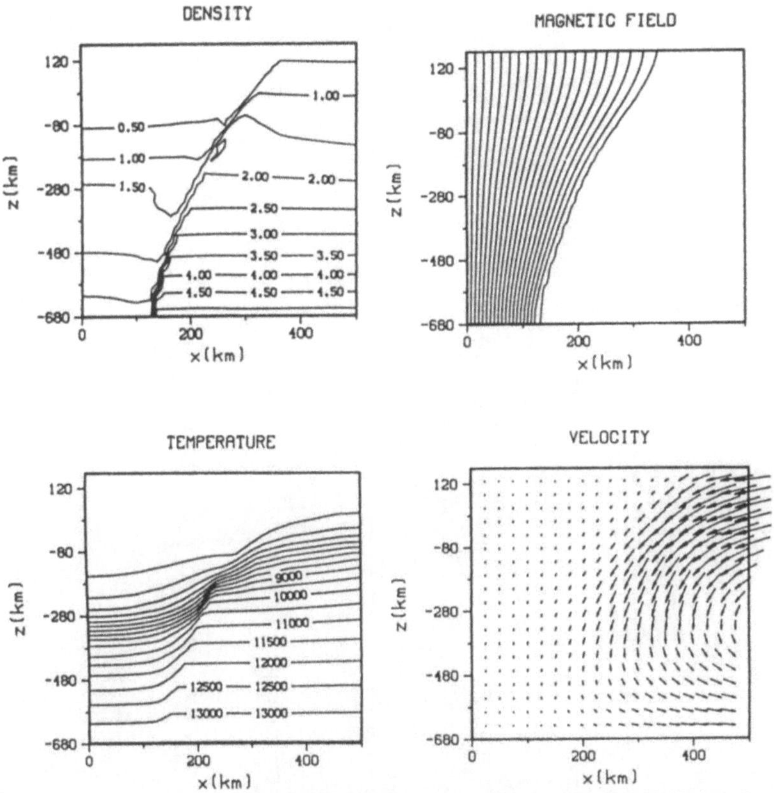

Figure 14. *Structure of a solar magnetic flux sheet (from Knölker & Schüssler 1989). Symmetry with respect to the $x = 0$ plane is assumed. The horizontal coordinate x has been stretched for better visibility; only one third of the region of numerical integration is shown. Densities are in units of 3×10^{-7} g cm^{-3}, temperatures in Kelvin. The vertical position $z = 0$ corresponds to unit continuum optical depth in the undisturbed photosphere. The longest velocity arrows correspond to approximately 300* m s^{-1}.

slightly brighter regions (granules) and more concentrated, slightly darker regions (intergranular lanes). Granules grow readily in regions of upflow and move towards locations of downdrafts, where granule growth is suppressed.

Most of the magnetic flux throughout the photosphere outside sunspots occurs in the form of concentrated filamentary structures embedded in a virtually field-free medium in the intergranular lanes. The formation of these features is probably due to advection of magnetic flux by convective flows, cooling by suppression of convective energy flux within highly magnetized regions, and a magneto-convective instability. After formation of such a magnetic flux concentration, it presumably evolves towards a quasi-stationary or oscillatory state. The evolution of flux concentrations has been followed by MHD simulations to the point of a stationary flow pattern (*e.g.* Fig. 14), *i.e.* a static, cool flux sheet is surrounded by a stationary circulation cell.

3.6 The simulation of convection in the atmosphere of the sun

As a second example of explicit finite difference hydrodynamics in 3D, I display the results of simulated convection for a thin layer (2.5 Mm), at the solar photospheric surface (*cf.* Stein *et al.*

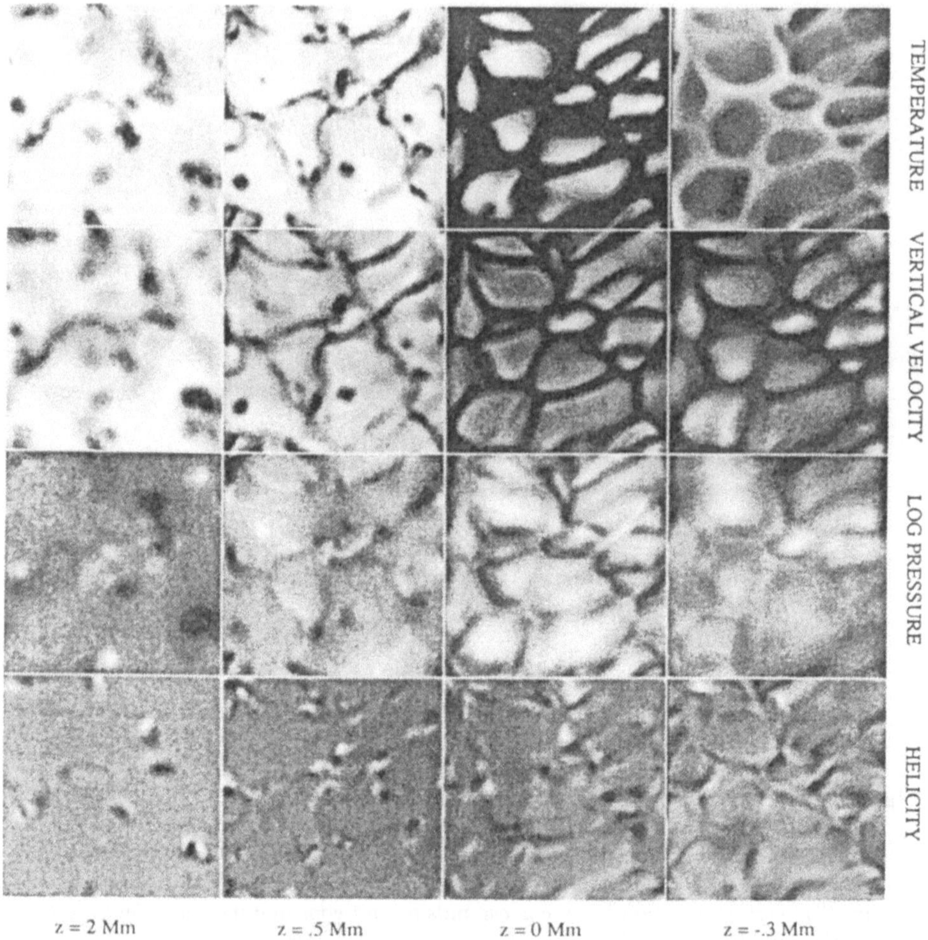

Figure 15. *Structure of simulated solar convection on horizontal slices at four depths as indicated.* $z = 0$ *corresponds to unit continuum optical depth (from Stein et al. 1988).*

1988, 1989), extending 6 by 6 Mm horizontally. Due to effects of stratification, the density and pressure can vary over many orders of magnitude over a relatively small vertical distance. For the simulation shown in Fig. 15 over five orders of magnitude variation had to be incorporated in the computational volume. For this reason, the authors used $\log \rho$ and $\log P$ as dependent variables, rather than ρ and P, and the fluid equations had to be solved in a non-conservative manner. The grid resolution was $63 \times 63 \times 63$.

4.0 A Few Concluding Remarks

It is difficult to express the performance of a computer with just a few numbers. It would be misleading to simply run a standard FORTRAN program (a so-called benchmark program) on several machines and compare the results. The reason for this is that there is a wide variety of memory and CPU architectures and the execution speed of a program can depend critically on how it is written and how it is compiled/optimized.

Table 1. Computer hardware parameters (very approximate).

	# Processors	MFLOPS (maximum)	MFLOPS (application)	PRICE Mega–$
SUN 4/110	1	1	1	0.05
CRAY-1	1	100	50	
CRAY-2	1	150	80	
	4	600	250	5
CRAY-X/MP	1	200	100	
	4	800	350	10
CRAY-Y/MP	8	2000	800	25
SUPRENUM	256	4000	?	25
NEC SX-3	4*16	22000	?	25

Consider, for example, the FORTRAN segment:

```
      DO 10 J=1,200
      DO 10 I=1,200
      A(I,J)= ...
10    CONTINUE
```

Switching the order of the first two statements could have catastrophic results on a computer which relies on page swapping (transfering a section of central memory to mass storage, for example a disk) to be able to handle large arrays. The performance can decrease by many orders of magnitude if extensive page swapping is required. To a lesser extent this is also true for CPUs which utilize extremely fast cache memory. Depending on the 'hit rate' (how often the next number or next instruction is located in cache), the performance can vary over a wide range. Another important aspect is the degree to which a program can utilize pipelining features and parallel processing via multiple CPUs. Generally speaking, the numerical discretization methods discussed in this overview allow a high degree of *vectorization*. In the following I will assume that a reasonable effort has been made to optimize the high level language programs for a particular machine.

I consider a simple 2D hydrodynamic application. The grid is 100×100; each grid point requires 20 numbers (dependent variables + scratch arrays) of type DOUBLE PRECISION or equivalent. The memory requirements for this application are therefore ≈ 2 Mbyte. The number of floating point operations (or equivalent) required for each number is about 100 per time step (this number can be many orders of magnitude larger if iterations are necessary). I therefore estimate that on the order of 2×10^7 operations per time step are required. On a 2 MFLOPS (million floating point operations per second) computer a total of 10 seconds per time step would be necessary. A typical application requires on the order of 10^4 time steps, or a total of 10^5 s $\simeq 1$ day CPU time on a 2 MFLOPS machine. On a 50 MFLOPS supercomputer the same application requires approximately 1 hour CPU time.

Consider now a typical 3D application with a $100 \times 100 \times 100$ grid. The total memory requirements are ≈ 200 Mbyte, which is generally beyond the range of a 2 MFLOPS minicomputer (here, ≈ 10 Mbyte is typical). The total CPU time required on a 50 MFLOPS machine is about 120 hours (≈ 40 s per time step). For the 3D calculation discussed in section 3.3 the corresponding numbers were: 400 Mbyte central memory, 30 s per time step, 1.5×10^4

time steps, 125 hours CPU on one processor of a CRAY-2 (53 MFLOPS effective usage). Note that this application did not solve the Poisson equation. Solution of the Poisson equation would probably have doubled the total CPU time requirements.

In Table 1 I give a crude summary of the performance of several computers and their prices. The numbers have been pulled out of a hat but are guaranteed to be correct within an order of magnitude. From the above discussion it appears clear that a present-day mini-computer is sufficient for simple (explicit) 2D problems, but that for a 3D problem, a supercomputer is necessary. As soon as the problem to be solved involves more detailed physics or an iterative scheme, these simple-minded estimates can be orders of magnitude too low.

What is the bottom line to all of this? This I would like to summarize in a series of theses:

1.) Numerical simulations are an important aspect of astrophysical research, supplementing and complementing the types of experiments discussed at this meeting.
2.) The degree of sophistication of hydrodynamic simulations is hardware-limited.
3.) A large class of problems can only be solved with the biggest available machines.
4.) There is a certain degree of frustration among computational (astro-)physicists, because access to big machines is limited.
5.) In order that the quality of numerical simulations keep pace with a) space experiments, b) new technology ground-based observations, and c) overseas competition, a large, expensive (30 Mega-$ + 10 Mega-$ yearly) computing center in Germany is needed, which is dedicated to big astrophysical computations.
6.) As remarked during the discussion at this meeting, all of the above (except hopefully item 5.) are time-independent.

Acknowledgements: I am grateful to U. Trottenberg, E. Müller, W. Hillebrandt, M. Knölker for providing some of the figures and to R. Kunze for carefully reading and correcting the manuscript.

References

Benz, W., Hills, J.G.: 1987, *Astrophys. J.* **323:** 614.

Deinzer, W., Hensler, G., Schüssler, M., Weisshaar, E.: 1984, *Astron. Astrophys.* **139:** 426.

Dorfi, E.A., Drury, L.O'C.: 1987, *J. Comp. Phys.* **69:** 175.

Gingold, R.A., Monaghan, J.J.: 1983, *J. Comp. Phys.* **52:** 374.

Griffiths, D.F.: 1986, in *Astrophysical Radiation Hydrodynamics*, ed. K.-H. Winkler and M.L. Norman (Dordrecht: Reidel), p. 327.

Hockney, R.W., Eastwood, J.W.: 1988, *Computer Simulation using Particles*, (Adam Hilger).

Knölker, M., Schüssler, M.: 1989, *Astron. Astrophys.* **202:** 275.

Leer, B. van: 1977, *J. Comp. Phys.* **23:** 276.

Leer, B. van: 1982, *Lecture Notes in Physics* **170:** 507.

Lucy, L.B.: 1977, *Astron. J.* **83:** 1013.

Marcus, P.S.: 1986, in *Astrophysical Radiation Hydrodynamics*, ed. K.-H. Winkler and M.L. Norman (Dordrecht: Reidel), p. 359.

Monaghan, J.J.: 1985, *Comp. Phys. Rep.* **3:** 71.

Müller, E., Mair, G., Hillebrandt, W.: 1989, *Astron. Astrophys.* **216:** 19.

Potter, D.: 1977, *Computational Physics*, (New York: John Wiley).

Richtmyer, R.D., Morton, K.W.: 1967, *Difference Methods for Initial-Value Problems*, 2nd ed. (New York: Interscience).

Różyczka, M., Tscharnuter, W.M., Winkler, K.-H., Yorke, H.W.: 1980, *Astron. Astrophys.* **83:** 118.

Różyczka, M., Yorke, H.W., Bodenheimer, P., Müller, E., Hashimoto, M.: 1989, *Astron. Astrophys.* **208:** 69.

Stein, R.F., Nordlund, Å., Kuhn, J.R.: 1988 in *Proc. Symp. Seismology of the Sun and Sun-like Stars*, held in Tenerife, Spain, 26–30 Sept. 1988, ESA SP–286, p. 529.
Stein, R.F., Nordlund, Å., Kuhn, J.R.: 1989 in *Solar and Stellar Granulation*, eds. R.J. Rutten and G. Severino (Kluwer Acad. Publ.), p. 381.
Trottenberg, U.: 1988, *c't Magazin für Computertechnik* No. 8, (Heise Verlag), p. 150.
Tscharnuter, W.: 1987, *Astron. Astrophys.* **188:** 55.
Tscharnuter, W., Winkler, K.-H.: 1979, *Comp. Phys. Comm.* **18:** 171.

Evolution of Massive Stars

N. Langer

Universitäts-Sternwarte Göttingen, Geismarlandstr. 11,
D-3400 Göttingen, Fed. Rep. of Germany

1. Introduction

The term *massive star* is not very precise, and in the following we shall use it for stars with zero age main sequence (ZAMS) masses M_{ZAMS} above approximately $15\,M_\odot$. Note, however, that these abjects may achieve actual masses well below $15\,M_\odot$ during their evolution, as a consequence of mass loss due to stellar winds, pulsations, or other processes. $M_{ZAMS} \simeq 15\,M_\odot$ is a critical initial mass concerning several features. Only stars initially more massive develop superadiabatic layers above the convective hydrogen burning core, which turns already the core hydrogen burning phase to be difficult and uncertain. Only for stars above $M_{ZAMS} \simeq 15\,M_\odot$, mass loss due to stellar wind is important on the main sequence. Stars of lower initial mass always turn to the Hayashi-line after core hydrogen exhaustion, while for massive stars the post main sequence behaviour depends sensitively on physical details. This critical mass of $\sim 15\,M_\odot$ is not to be thought of as a strict limit. It depends especially on the stellar metallicity, and $15\,M_\odot$ may be a good number for Population I stars. It should be sufficiently high, anyway, in order to guarantee a nondegenerate carbon ignition, which ensures that the final evolutionary state of those objects is not a White Dwarf.

Massive stars, though they are relatively rare, deserve much attention due to their extreme properties. They are the most luminous individual objects and thereby rather easy to identify even in external galaxies. Their lifetime is very short ($\lesssim 10^7\,yr$), and so they are excellent tracers of active star formation. Their surface temperature is very high during most of their evolution, turning them into very powerful sources of high energy photons, which may consequently ionize the gas in a large system around them. They are also powerful sources of kinetic energy due to both, high velocity stellar winds ($\sim 10^{51}\,erg$) and the final supernova explosion ($\sim 10^{51}\,erg$). And finally they are the major site of cosmic nucleosynthesis, due to hydrostatic and explosive burning processes. Thereby, they are mainly responsible for the chemical evolution of galaxies.

It is beyond the scope of this paper to discuss all the items mentioned above in detail, rather we concentrate our discussion on structure and evolution of massive stars in phases, which are directly accessible to observations. More specifically, this concerns the hydrostatic evolution up to central helium ex-

Reviews in Modern Astronomy 2 Editor: G. Klare

haustion (which accounts for $\sim$ 99% of the stellar lifetime), and the supernova explosion, which is a very short phase, but so luminous as to improve the observational statistic sufficiently.

The aim of the present work is to report on the current status of theoretical evolutionary computations for massive stars in the mentioned phases, and to discuss our knowledge and ignorance concerning abserved luminous stars which may (or may not) be identified with those computer models. Furthermore, we restrict ourselves to the case of spherically symmetric, non-rotating, non-magnetic massive single stars, which — as we show below — is not necessarily a simple case. However, there is some evidence that such stars exist in nature, i.e. that the above restrictions do not necessarily imply oversimplifications (cf. discussion on SN 1987a in Sections 6 and 7).

2. Main sequence evolution

Previously we already mentioned that the modelling of the main sequence evolution of massive stars encounters two main problems: the first is stellar mass loss, the second convection theory or, more specifically, internal mixing processes. Both processes are very difficult theoretical problems involved in stellar evolution calculations in general. However, it is just that for massive stars they are important from the beginning of the evolution, which renders even the main sequence phase as problematic.

Massive main sequence stars suffer mass loss due to a stellar wind which is driven by radiation pressure in absorption lines. Corresponding theories (see Castor et al., 1975) are meanwhile in a state which allows quantitative predictions of mass loss rates and terminal wind velocities within an accuracy of 10-20% (Kudritzki, 1988; Pauldrach et al., 1989). However, all published evolutionary calculations for massive main sequence stars used mass loss rates from fits to observational data (e.g. Lamers, 1981; de Jager et al., 1986). Those empirical mass loss rates have an accuracy of a factor of 2 in general. Especially for very massive stars, where observational data is rare, the error may be even larger. Furthermore, those formulae do not account for the dependence of the mass loss rate on the stellar metallicity, which is predicted from the wind theory and observationally confirmed in Local Group galaxies (cf. Kudritzki et al., 1987). Recently (Langer and El Eid, in prep.) the first stellar evolution calculations using theoretical mass loss rates have been performed by using the analytical solutions for wind models of Kudritzki et al. (1989), which approximate full non-LTE hydrodynamic wind models to high precision. Sequences for stars in the mass range $20\,M_{\odot} - 200\,M_{\odot}$ and for four different metallicities ($Z = 0.03,\ 0.02,\ 0.05,\ 0.002$) have been performed. The amount of mass lost during central hydrogen burning was found to be considerably less compared to computations which used empirical mass loss relations. E.g. a $40\,M_{\odot}$ star with 2% metallicity looses only $\Delta M = 2.5\,M_{\odot}$ according to our new calculations, while Maeder and Meynet (1987) found a total amount of $\Delta M = 7.5\,M_{\odot}$ to be lost during hydrogen burning using the mass loss formula of Jager et al. (1986).

This difference reflects mainly differences in the mass loss rates, since the duration of the hydrogen burning phase was comparable in both cases ($4.52\,10^6\,yr$ and $4.80\,10^6\,yr$). For a $100\,M_\odot$ sequence, Langer and El Eid (1986) found, by using the Lamers (1981) formula, $\Delta M = 24.2\,M_\odot$, while the new calculations yield $\Delta M = 11.5\,M_\odot$. These results demonstrate that theoretical mass loss rates are distinctively smaller than those obtained by current empirical mass loss relations. Since Kudritzki and co-worker compared the theoretical wind models with observations in great detail (e.g. not only the mass loss rate and wind velocity, but also spectral line profiles) and found very good agreement in most cases (cf. e.g. Kudritzki, 1988) we tend to favour the theoretical mass loss rates from observational fits, which, anyway, yield quite different results depending on which one is used (cf. discussion in Chiosi and Maeder, 1986).

The effect of mass loss during core hydrogen burning on the evolution of massive stars has been widely discussed in the literature (e.g. de Loore, 1980; Chiosi and Maeder, 1986) and will not be repeated here. We want to point out the aspect that low mass loss rates favour the occurrence of semiconvection in intermediate layers during core hydrogen burning which may have a large influence on the hydrogen profile and thereby on the post main sequence evolution (cf. discussion of Langer et al., 1985).

Massive main sequence stars are basically composed of a convective core and a radiative envelope. Both parts, however, are not unproblematic. For the convective core, there has been a long debate in the literature, whether its extension can be determined by the Schwarzschild criterion, i.e. $\nabla > \nabla_{ad}$ (where $\nabla = d\ln T/d\ln P$, and $\nabla_{ad} = (d\ln T/d\ln P)_{ad}$), or whether it is larger due to the so called convective overshooting. Current convection theories don't give a unique answer to this problem (cf. e.g. Langer, 1986; Renzini, 1987), and also a comparison of stellar tracks with observations does not yield really strict constraints (cf. Doom, 1985; Mermilliod and Maeder, 1986; Langer and El Eid, 1986). The consequences of an extended convective core for the evolution of a massive star are considerable, not only for the main sequence phase itself, which is prolongated and leads the star to cooler effective temperatures (see e.g. Stothers and Chin, 1985), but also for the post main sequence evolution (cf. Sect. 3), which makes a solution of the overshooting problem very desirable. In this context we refer to Sect. 6, where we include a discussion of the progenitor evolution of SN 1987a, which may bring us somewhat closer to such a solution. The difficulty which may occur in the radiative envelope of massive main sequence stars is the semiconvection. Semiconvection is a process which develops in layers where the temperature gradient is superadiabatic (i.e. the Schwarzschild criterion is fulfilled) but where also a molecular weight gradient is present, which prevents the onset of convection according to the Ledoux criterion (Ledoux, 1941). Instead of convection, which would act on a dynamic timescale, an overstability develops (Kato, 1966), which leads to a mixing process on a thermal time scale (Langer et al., 1983). In massive main sequence stars, the convective core mass decreases with time, leaving behind an extended, chemically inhomogeneous zone. The high contribution of radiation pressure to the total pressure favours the occurrence of superadiabaticity in these layers, which then leads to the onset of semiconvection.

In the main sequence phase, the evolutionary timescale (and therefore the timestep in numerical calculations) is large compared to the thermal timescale. However, one may not assume a complete mixing (i.e. homogenisation) in semiconvective layers during this phase, since the radiative diffusion coefficient is linked to the hydrogen abundance via the opacity (electron scattering dominates). Complete mixing may therefore lead to subadiabatic temperature gradients (cf. Chiosi and Summa, 1970), i.e. to a completely stable situation, which is a contradiction to the assumption that mixing may occur. Iben (1974) realized that a very fine spatial zoning avoids this contradiction. However, in post main sequence phases it is no longer guaranteed that the numerical timestep exceeds the thermal timescale, and the semiconvective mixing has to be treated as timedependent, e.g. by solution of a diffusion equation (Langer et al., 1983). Both mass loss and convective core overshooting tend to choke off the occurrence of semiconvection in massive main sequence stars. This is visible in Fig 1, where the internal convective structure of a star of initially $30\,M_\odot$ is plotted as a function of time. Note that in the computations, the mass loss rate of Lamers (1981) has been used for the main sequence phase, which drastically reduces semiconvection compared to the case of no mass loss (cf. Fig. 10 of Langer et al., 1985). In computations including mass loss and convective overshooting, practically no superadiabaticities occur outside the convective core of the $30\,M_\odot$ sequence during central hydrogen burning.

We argued above that the mass loss rates of massive main sequence stars may have been somewhat overestimated in the past. In Sect. 6 we will see that probably efficient overshooting would have prevented the SN 1987a progenitor to evolve towards high effective temperture before exploding. Therefore, it seems likely that semiconvection actually occurs in the main sequence phase of massive stars. Since energy transport due to semiconvection is negligible (Langer et al., 1985), its effect is just a modification of the intermediate chemical profile. However, this has a large influence on the post main sequence evolution of massive stars (see below, and cf. Fig. 1).

3. The supergiant phase

A supergiant is usually classified as such from spectroscopic criteria, which indicate an extended atmosphere, i.e. a large stellar radius. The ignition of the hydrogen burning shell after core hydrogen exhaustion leads to a strong expansion of the overlying stellar envelope in massive stars, increasing thereby considerably their radius and reducing their surface temperature. Therefore, models of massive post main sequence stars may certainly be identified as supergiants (as long as they still possess a sufficiently massive envelope; cf. Sect. 5). Whether the reverse is true is not certain: towards the end of core hydrogen burning, the radius of a massive star may be much larger compared to its radius on the zero age main sequence (ZAMS). E.g. a $100\,M_\odot$ star increases

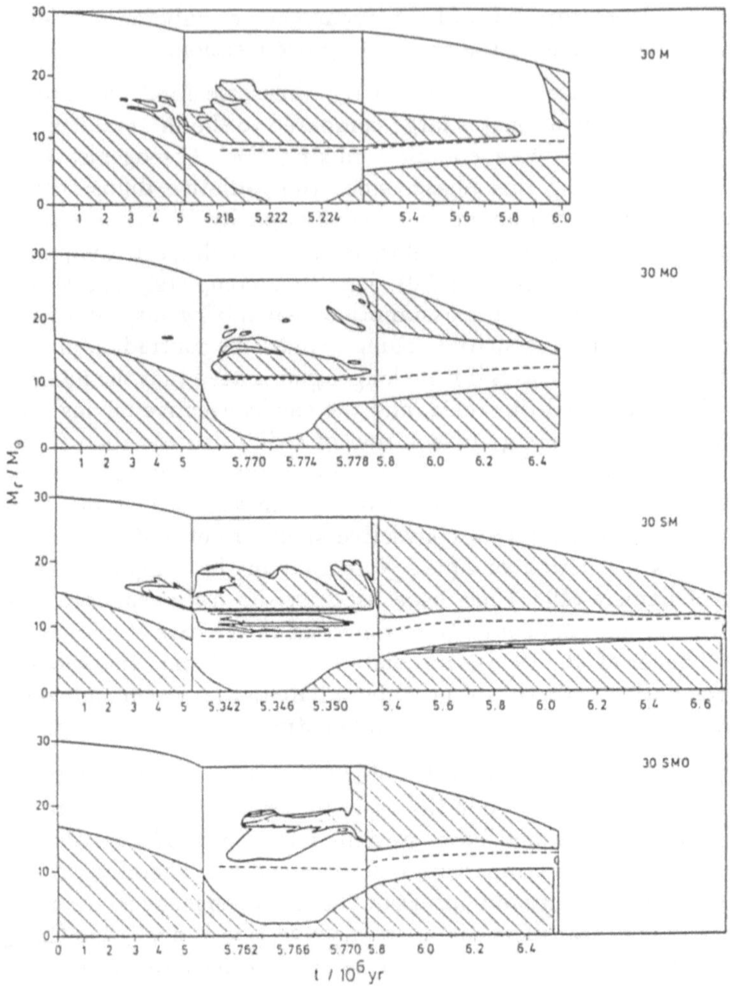

Fig. 1: Internal structure as function of time for a star of 30 $M_\odot$ initial mass with 4 different assumptions on convection: 1. Schwarzschild criterion for convection (30M); 2. Schwarzschild criterion with overshooting of 0.2 pressure scale heights (30MO); 3. Semiconvection (efficiency parameter $\alpha = 0.1$) and without overshooting (30SM); and 4. Semiconvection ($\alpha = 0.1$) and overshooting (30SMO). Hatched areas indicate convection, surrounded white areas semiconvection. The time scale breaks twice, separating the three phases core hydrogen burning, contraction phase towards helium ignition, and core helium burning. The dashed line indicates the position of the H-burning shell. The upper borderline of the figures shows the total mass as a function of time. All sequences have been computed with Lamers (1981) mass loss rate for effective temperatures above 6 500 K and Reimers (1975) for lower values. For central He-burning, a convective envelope means that the star is in the Red Supergiant stage, while otherwise it is a Blue Supergiant. A detailed description of these sequences can be found in Langer (1986a).

its radius from 13 $R_\odot$ at H-ignition to 53 $R_\odot$ at core hydrogen exhaustion. In the following, anyway, we refer to post main sequence phases when we use the expression "supergiant".

The effective temperture, at which models of massive stars perform core helium burning depends sensitively on the input physics used in the model computation, and has been subject to many studies in recent years. Mass loss and convective overshooting, which both diminish the envelope mass of the star, tend to push the star towards the Hayashi line, i.e. towards the Red Supergiant stage (cf. Stothers and Chin, 1979; Maeder, 1981; Maeder and Meynet, 1987). Semiconvection also favours core helium ignition as a Red Supergiant due to a limitation of the hydrogen content at the position of the H-burning shell (cf. Fig. 1). However, when the H-shell propagates outwards with time it may eventually reach layers with high hydrogen concentration and reactivate, leading to a "blue loop" in the Hertzsprung-Russell diagram, i.e. to hotter effective temperatures (cf. Langer et al., 1985). However since the efficiency of all three processes is somewhat uncertain, reliable post main sequence tracks are hard to predict. In this context the supernova 1987a is a very useful event, since it helps constrain these uncertainties considerably (cf. Sect. 6). From the computations of Maeder and Meynet (1987) we can say that moderate overshooting and current standard mass loss rates yield almost no Blue Supergiants at all for "massive" stars (i.e. $M_{ZAMS} \gtrsim 15\,M_\odot$). Therefore, one or both of these processes should have a reduced efficiency compared to that adopted by Maeder and Meynet.

New computations for 20 and 40 $M_\odot$ stars with galactic metallicity performed with the Schwarzschild criterion for convection (Langer and El Eid, in prep.) indicate a movement on the nuclear timescale from hot effective temperatures to the Red Supergiant stage in the HR diagram during central helium burning. This appears to be more consistent with the distribution of Supergiants in the HR diagram (see Humphreys and McElroy, 1984).

Stars more luminous than $\sim 10^{5.7}\,L_\odot$ (corresponding to $M_{ZAMS} \gtrsim 50\,M_\odot$), are known to have no observed Red Supergiant counterparts in the HR diagram (Humphreys and McElroy, 1984; Humphreys, 1984). In order to avoid that a very massive star evolves to low effective temperatures after core hydrogen exhaustion, its envelope mass has to be drastically reduced, which can be achieved either by adopting an extremely high mass loss rate after core hydrogen exhaustion, or by assuming convective overshooting to be very efficient during core hydrogen burning (cf. Langer and El Eid, 1986; Maeder and Meynet, 1987), or a combination of both. The question which of this two processes is the dominant one is enlightened by an analysis of the so called Luminous Blue Variables (LBVs), which is the subject of the next Section.

4. Luminous Blue Variables

As mentioned above, there are no observed stars in the upper right corner of the HR diagram (see Humphreys and Davidson, 1979; Humphreys and McElroy, 1984). The borderline between this empty region and the rest of the HR diagram, the so called Humphreys-Davidson (HD) limit, depends only weakly on the effective temperature for values lower than $\sim 15\,000\,K$, lies at $\log L/L_\odot \simeq 5.7$, and corresponds to the absence of very luminous Red Supergiants mentioned in the previous Section. For surface temperatures in excess of $15\,000\,K$, the luminosity at the HD-limit increases with increasing temperature.

From observations it is known that the mass loss rate due to stellar wind increases quite drastically for stars closer to the HD-limit (de Jager et al., 1986). Furthermore, many — if not all — stars close to the HD-limit show irregular variability, and on timescales of decades some (e.g. P Cygni, η Carina) show real outbursts, which have to be interpreted as shell ejections (Lamers, 1989). Humphreys (1989) has tried to give a definition of the Luminous Blue Variables (LBVs), which is a collective designation for the variable stars close to the HD-limit, and which comprises the P Cygni type stars, the S Doradus variables, and the Hubble-Sandage variables. According to Humphreys, besides their variability, LBVs have the following properties: they show emission- and/or P Cygni-lines in their spectra, show a surface temperature variation in the range $25\,000\,K - 8\,000K$, have high luminosities, very high mass loss rates, and show evidence for circumstellar ejecta by e.g. an infrared excess. Quantitative analysis of the ejecta, such as in η Car, and the atmospheres and circumstellar envelopes of these stars show that they are nitrogen and helium rich (Davidson et al., 1982, 1986; Allen et al., 1985). This supports the idea of a reduced envelope mass due to mass loss and/or mixing in post H-burning very massive stars (cf. Sect. 3), which leads to H-burning products at the stellar surface.

Lamers (1989) has analysed the mass loss rate of several LBVs in different galaxies. It ranges from values, which can possibly explained by the radiation driven wind theory, i.e. $\sim 10^{-5}\,M_\odot\,yr^{-1}$, corresponding to quiescent phases, up to values of $10^{-2} - 10^{0}\,M_\odot$ in one violent outburst. Lamers estimates the time averaged mass loss rate to be of the order of $10^{-4}\,M_\odot\,yr^{-1}$, but the uncertainty is large. Depending on the recurrence time of violent outbursts and on the average ejected mass it could also be higher. Adopting a value of $10\,M_\odot$ as the order of magnitude of the mass which a massive star has to loose in the LBV stage in order to avoid an evolution to cool effective temperatures, a time average mass loss rate of $10^{-4}\,M_\odot\,yr^{-1}$ would result in a duration of the LBV phase of $\sim 10^5\,yr$. This value seems too large regarding the small number of LBVs as compared to the number of Wolf-Rayet stars, the probable LBV descendants with lifetimes of some $10^5\,yr$ (cf. Sect. 5).

Another way of estimating the duration of the LBV phase for massive stars comes from evolutionary considerations. Consider a star close to the HD-limit, moving towards cool surface temperatures. In a certain (small) time interval

Δt the star increases its radius by ΔR. In order to stop the redwards motion, the star has to loose an amount of mass ΔM which leads to a radius decrease of ΔR (see Heisler and Alcock, 1986). Thus, the condition $R(t) = const.$ yields a critical mass loss rate $\Delta M/\Delta t$. In stellar evolution calculations, this critical rate can be found by adopting a mass loss formula of the form $\dot{M} = \dot{M}_0(T_{eff,0}/T_{eff})^{\alpha}$ with large α (say $\alpha = 5$). Using standard mass loss rates for $\dot{M}_0$, e.g. those predicted by the radiation driven wind theory, and for $T_{eff,0}$ a value close to the HD-limit, ensures a continuous but drastic increase of the mass loss rate at the HD limit, and the star will establish the condition $R(t) = const.$ Langer and El Eid (in prep.) performed computations of this type and found $\dot{M}(\dot{R} = 0) \simeq 5\,10^{-3}\,M_{\odot}\,yr^{-1}$ for a sequence with $M_{ZAMS} = 100\,M_{\odot}$, which leads to a duration of the LBV phase of $8\,10^3\,yr$.

However, we note that the radius increase per time at the HD-limit, i.e. the speed of the redwards evolution close to the HD-limit, may depend much on the previous evolution of the star, i.e. on the adopted input physics of the computations, especially concerning convection/semiconvection and main sequence mass loss rate. E.g. very massive stars without mass loss and overshooting evolve redwards only on a nuclear timescale rather than on a thermal timescale (cf. Stothers and Chin, 1976). In this case, the LBV phase would last considerably longer and would proceed at much smaller mass loss rates.

In this context it is especially interesting to point out that a lower main sequence mass loss rate may prolongate the LBV phase for two reasons: it leaves more mass to get rid of in this phase, and it reduces the speed of redwards evolution after central H-exhaustion, thereby reducing the LBV mass loss rate. Therefore, since the main sequence mass loss rates in the Magellanic Clouds are known to be smaller compared to those in the Milky Way (Kudritzki et al, 1987), this may explain the relatively large number of LBVs in the LMC. Preliminary computations for a $100\,M_{\odot}$ star of LMC composition and main sequence mass loss according to Kudritzki (1987, 1989) indicate $\dot{M}(\dot{R} = 0) \simeq 10^{-3}\,M_{\odot}\,yr^{-1}$.

We should note here, that the LBV phase does not necessarily proceed always in the same way for all stars, rather that qualitative differences may occur for stars of different ZAMS mass or actual mass. Langer (1989a) proposed that the behaviour of stars in the LBV phase may be determined by the internal hydrogen profile, which (for a fixed convection theory) is a function of the stellar mass. He argues that LBVs of very high mass will produce late WN stars as descendants, while those of smaller mass will produce early WN stars (cf. Sect. 5). Furthermore, he discusses the possibility of a post-Red Supergiant LBV phase as well as a post-LBV Red Supergiant phase. Neither are prohibited by observations for a narrow mass range corresponding to the luminosities of the most luminous Red Supergiants, i.e. there exist LBVs of similar bolometric brightnes.

Nothing has been said up to now about the physical mechanism of the LBV mass loss, which in reality is not continuosly varying but rather erratic. At a recent IAU meeting (Davidson and Moffat, 1989) several mechanisms for the diversity of unstable behaviours of LBVs have been proposed (cf. review of Walborn, 1989), e.g. radiation pressure (cf. Lamers and Fitzpatrick, 1988; but also Pauldrach and Puls, 1989), internal density inversion and related super-Eddington luminosity (Maeder, 1989) or convectively initiated turbulent pressure (Kiriakidis, 1987). Which of these proposed mechanisms really act in LBVs is yet unclear. It is worth pointing out, however, that they are all related to the recombination of hydrogen at temperatures of $\sim 15\,000\,K$. This changes drastically the opacity and the mean molecular wheight, which has effects on radiation transport, convection and equation of state.

5. Wolf-Rayet stars

Wolf-Rayet (WR) stars are supposed to be the descendants of massive stars which have lost all or the main part of their hydrogen rich envelope (cf. Chiosi and Maeder, 1986), and therefore the WR stage is probably succeeding the LBV phase. However, not all single WR stars are necessarily formed via the LBV scenario. Some may be direct descendants of Red Supergiants, while others could perhaps be formed already during the main sequence evolution., e.g. due to rotationally induced mixing (cf. Maeder, 1987). We should mention also that some WR stars might have formed through mass exchange in a close binary system, but that this is not the dominant WR formation scenario (cf. Chiosi and Maeder, 1986, and references therein).

WR stars exist, and independently of how they might have been formed one may consider their internal structure and evolution. For that we want to divide the WR stars into two classes, those which still contain some hydrogen in their envelope — let us call them late WN or WNL stars here — and those which do not contain any hydrogen, which we will designate as (true) WR stars. Our definition of WNL stars is roughly consistent with that derived from spectroscopic criteria (cf. Willis, 1982). We regard the WNL stars as a separate class, since there is evidence from both observation and theory that they have very different properties compared to the other WR stars. WNL stars are known to be much more massive (Niemela, 1983; Moffat, 1982) and luminous (cf. e.g. Smith and Maeder, 1989) than other WR subtypes. Furthermore, they are much less compact, i.e. have much larger radii, which may be provoked by the presence of hydrogen: hydrogen raises the opacity, lowers the mean molecular wheight, and — probably most important — allows the presence of a second nuclear burning region besides the He-burning center, i.e. sets up a hydrogen burning shell. These structural differences may result in the occurrence of absorbtion lines in WNL spectra (van der Hucht et al., 1988; see below), and may be related to spectroscopic similarities of WNL and Of stars (cf. Bahannan and Walborn, 1989).

Hydrogenless WR stars, in contrast, which appear either as early WN stars or as WC/WO stars, are compact, much dimmer and less massive than WNL stars. Since they do not contain hydrogen, their only nuclear energy source is the helium burning center. This makes their internal structure relatively simple, and results e.g. in a narrow mass luminosity relation (cf. Maeder, 1983). Langer (1989b) constructed a grid of WR models and investigated the dependence of observable quantities as function of the WR mass, composition, and convective structure. He found that due to the dominance of radiation pressure all over the star except in a tiny mass fraction close to the stellar surface, WR mass and surface chemical composition alone determine all observable quantities. He could derive e.g. the theoretically allowed areas for WR stars in the L-M, R-M, and L-T diagrams (see Figs. 2 and 3), where R and T correspond to radius and surface temperature of the hydrostatic WR core. He also derived estimates for the optical thickness τ of the WR wind zone, which may hide

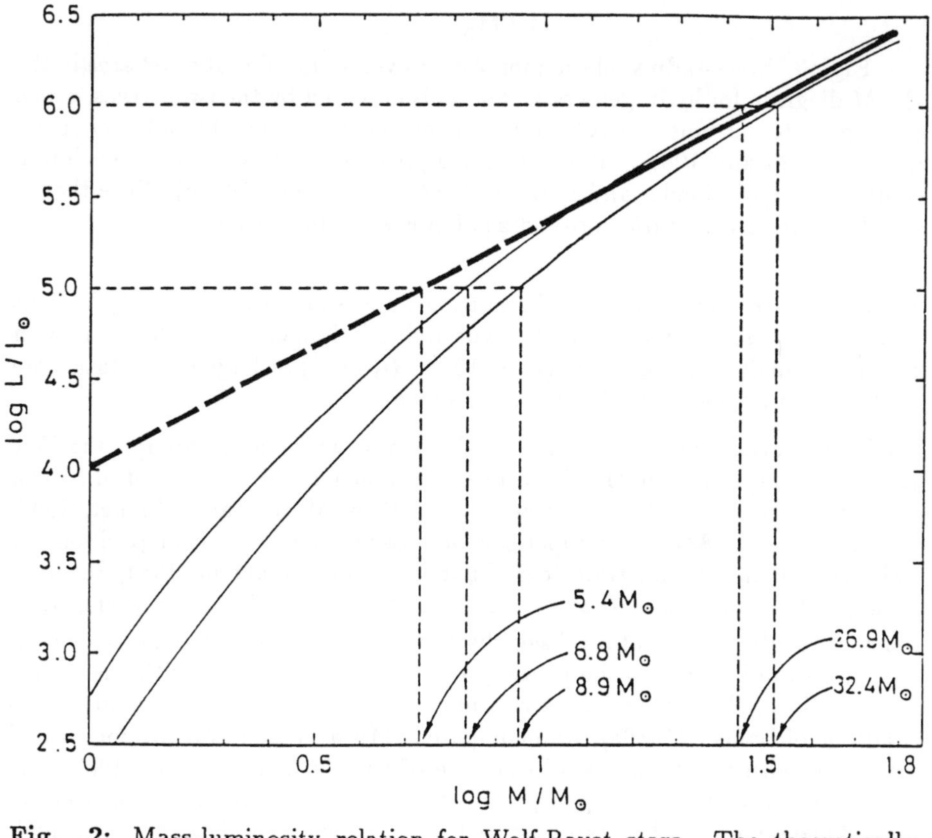

Fig. 2: Mass-luminosity relation for Wolf-Rayet stars. The theoretically allowed area in the $\log M - \log L$ diagram for hydrogenless WR stars is the area in between the two continous lines. The thick continuous line corresponds to the $M-L$ relation of Maeder and Meynet (1987) and is extrapolated towards lower masses (dashed part). Two examples of a mass determination for a given luminosity are indicated.

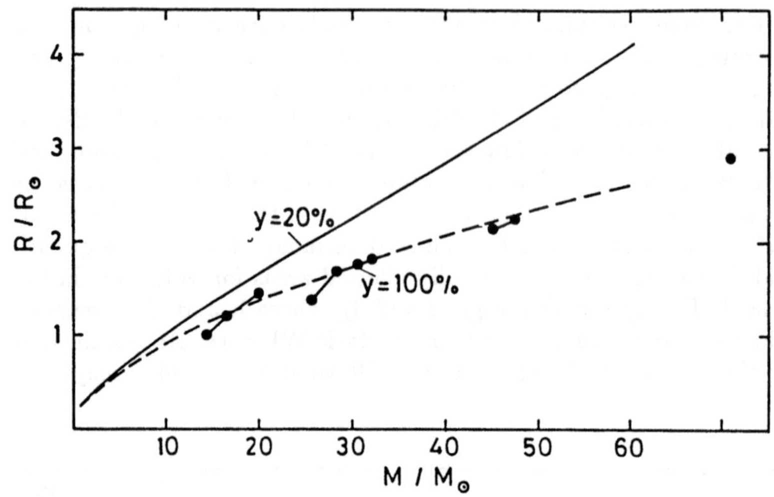

Fig. 3: Mass-radius relation for Wolf-Rayet stars. The allowed area in the $R-M$ diagram for hydrogenless WR stars is bounded by the two curves, which correspond to extreme cases of surface chemical composition, i.e. a helium mass fraction of 20 and 100%. The dots correspond to models from the evolution computations of Maeder and Meynet (1987); cf. Langer (1989b). Note that R stands for the hydrostatic core radius of WR stars (see text).

the hydrostatic WR core from observations, and showed that $\tau \sim \dot{M}/R$. This means that absorbtion lines can be expected in WR spectra only for stars with relatively small mass loss rates ($\dot{M} < 10^{-5}\ M_\odot\ yr^{-1}$) or large hydrostatic core radii (which applies to WNL stars; see above).

Stellar evolution calculations for massive stars, which lead through the WR phase, are rather rare in the literature (e.g. Maeder, 1983; Doom et al., 1986; Langer and El Eid, 1986; Prantzos et al., 1986; Maeder and Meynet, 1987; Langer et al., 1988). A comparison of observationally derived positions of WR stars in the HR diagram (e.g. Lundström and Stenholm, 1984; Schmutz et al., 1988) with those tracks — though it is very difficult to perform; cf Langer (1989b) — indicates a basic discrepancy, which is that the observed luminosities (and therefore the masses) are too low to compare with theoretical computations. This discrepancy is confirmed by simplified WR evolution calculations of Langer (1989c), who computed IMF-averaged WR properties and compared with the complete WR sample of van der Hucht et al. (1988). He also showed that this discrepancy can be removed by adopting the concept of mass dependent mass loss rates for WR stars, which is motivated by the models of Langer (1989b), and which has already been suggested by Abbott et al. (1986) from observational data. Langer finds that a mass loss law of the form $\dot{M}_{WR} \simeq (0.6-1.0)10^{-7}(M_{WR}/M_\odot)^{2.5}\ [\ M_\odot\ yr^{-1}]$ yields good agreement of theory and observations in many respects. This concerns IMF-averages of WR masses and luminosities, WR progenitor ZAMS masses, WR subtype ratios, and WR lifetimes. Also the final WR masses, i.e. the masses of the supernova configurations resulting from massive stars which evolve into WRs, is much smaller adopting mass dependent WR mass loss rates rather than keeping the WR mass loss rate constant with time (as usually done), which may have important implications for the explanation of the newly identified class of type Ib supernovae, as will be discussed in the next Section.

6. Supernovae from massive stars

Massive stars up to $M_{ZAMS} \simeq 100\, M_\odot$ are supposed to proceed hydrostatic nucleosynthesis up to silicon burning, thereby forming a degenerate Fe/Ni-core. The ensuing core collapse is thought to give rise to the standard scenario of supernovae (SNe) from massive stars (Woosley and Weaver, 1986). Note that stars initially more massive than $\sim 100\, M_\odot$ explode through the $e^\pm$-pair creation scenario by explosive oxygen burning (cf. again Woosley and Weaver, 1986), potentially as very massive WR stars of type WNL (El Eid and Langer, 1986; Langer, 1987a,b). Possible observational counterparts of such pair creation SNe could be the peculiar SN 1961v (Langer,1987c), or the SN remnant Cas A (El Eid and Langer, 1986). The optical display of such supernovae depends greatly on the presence of hydrogen in the envelope and the amount of radioactive nickel produced by the explosive burning (Woosley and Weaver, 1982; Herzig, 1988; Ensman and Woosley, 1989; Herzig et al., in prep.).

The presupernova configuration of a star with initial mass in the range $15\, M_\odot \lesssim M_{ZAMS} \lesssim 100\, M_\odot$ is either a Supergiant or a Wolf-Rayet star. Since most of the WR stars do not contain hydrogen in their envelope, such objects would be classified as type I SNe when exploding. The usual type I SNe, also designated as type Ia, are probably related to low mass binary stars (cf. Woosley and Weaver, 1986). The recently identified class of type Ib SNe (cf. Branch, 1986), however, shows a distinct preference to occur in regions of active star formation and should therefore be related to more massive stars. Ensman and Woosley (1988) performed lightcurve computations for WR stars exploding due to the core collapse mechanism and concluded that most type Ibs could not originate from WR stars, since the theoretical lightcurves are only consistent with observed ones when very low WR masses ($M_{WR} \lesssim 8\, M_\odot$) are adopted for the computations, which they found to be inconsistent with existing stellar evolution calculations (cf. Sect. 5). However, as mentioned above, the concept of mass dependent mass loss rates for WR stars naturally leads to such small final WR masses (Langer, 1989c) for all WR stars (except WNLs), independent of their initial mass. Therefore, the idea of WR stars as type Ib SN progenitors may be realistic.

The usual type II SN explosions are thought of as being relatively well understood in general. The standard model is that of an exploding Red Supergiant which blows up due to the core collapse mechanism (cf. Woosley and Weaver, 1986), and the SN 1987a in the Large Magellanic Cloud (LMC) confirmed the general picture. An unexpected peculiarity of SN 1987a was, however, that its progenitor star was a Blue rather than a Red Supergiant.

Several reasons have been suggested in order to explain why a Blue Supergiant may explode and why nobody though about that before it happened. First, the SN 1987a is the only observed SN in an irregular dwarf galaxy, which are generally low metallicity systems. The metallicity of the LMC is $\sim 1/4$ of that in the Milky Way, and this circumstance was first thought of as explaining the Blue Supergiant explosion (cf. e.g. Hillebrandt et al.,1987). However, soon it became evident that the SN 1987a progenitor must have been a Red Supergiant several $10^4\,yr$ before the explosion (Fransson et al., 1989). Woo6ley (1988) found that massive low metallicity stars may return from the Hayashi track to hotter surface temperatures during the contraction phase towards central carbon burning (i.e. at the right time), when the Ledoux criterion for convection is used instead of the usual Schwarzschild criterion. Also Weiss (1989) performed evolutionary computations and got a Blue Supergiant SN progenitor with the Ledoux criterion, but in his case the so called blue loop occurred during central helium burning (i.e. too early). Recent evolutionary computations of Langer , El Eid, and Baraffe (1990), who used the semiconvection theory suggested by Langer et al. (1983), indicate the following results for a $20\,M_\odot$ star with LMC composition (cf. also Fig. 4): Blue Supergiant explosions are obtained for values of the semiconvective efficiency parameter α of $0.05 \leq \alpha \leq 0.008$, while red explosions occur for larger or smaller values. Note that $\alpha = 0$ corresponds to the Ledoux criterion for convection, while $\alpha = \infty$ recovers the Schwarzschild criterion. The value recommended by Langer et al. (1983, 1985) for theoretical reasons was $\alpha \simeq 0.1$. For the blue explosions, the effective temperature at the time of explosion was found to be in the range $21\,000\,K \gtrsim T_{eff,expl} \gtrsim 13\,000\,K$, depending on α. Some sequences performed a blue loop during central helium burning, most of them returning to the Hayashi line at central helium exhaustion. These results turned out to be insensitive to changes in the adopted mass loss rates. Note that for $\alpha = 0.01$ but galactic metallicity ($Z = 2\%$) we obtained a Red Supergiant SN progenitor structure.

These results indicate that probably molecular weight barriers in the stellar interior cannot be overcome by convection instantaneously, which is implicitly assumed by adopting the Schwarzschild criterion for convection, rather that mixing in chemically inhomogeneous and superadiabatic regions is performed on a thermal timescale. Also convective overshooting should have a limited efficiency, at least as long as the molecular weight within the convective zone and the adjacent radiative layers is different. If this picture will be confirmed, it would mean that most stellar evolution calculations performed yet for massive stars contain a basic oversimplification. Note that e.g. the mass of the C/O-core at central He-exhaustion obtained for a $20\,M_\odot$ sequence with semiconvection and an efficiency parameter in the range which yields a blue explosion is of the order of $2\,M_\odot$, while adopting the Schwarzschild criterion results in a C/O-core mass of $\sim 4\,M_\odot$. Also the central carbon mass fraction at core helium exhaus-

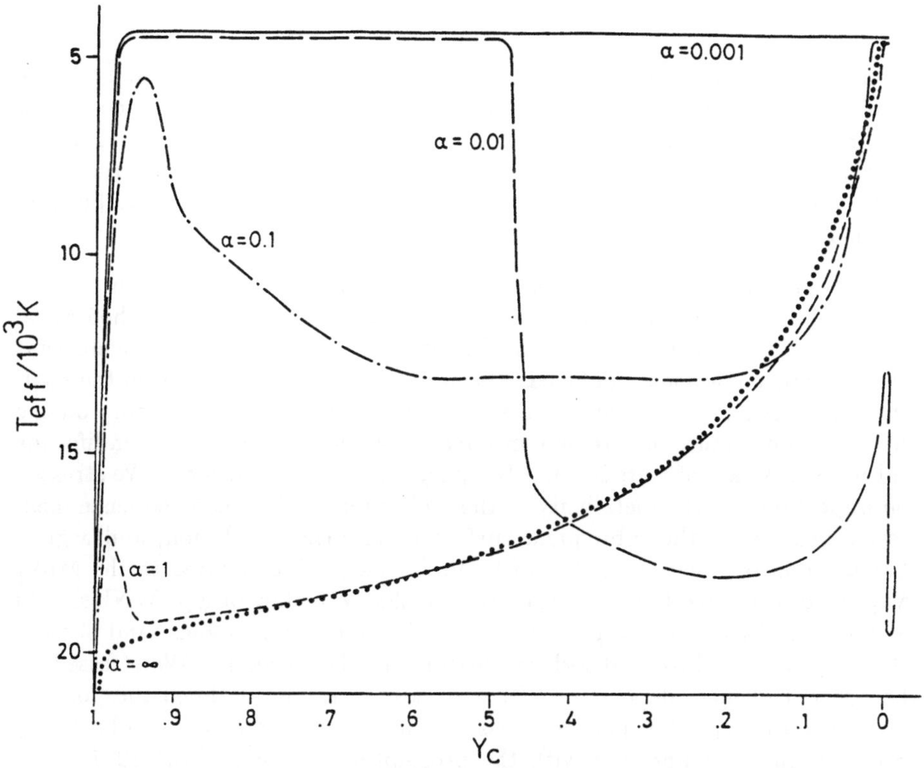

Fig. 4: Surface temperature during central helium burning as function of the central helium content for $20\,M_\odot$ sequences of LMC composition computed with mass loss and semiconvection for varying semiconvective efficiency parameter α. Tracks for $\alpha = 0.001, 0.01, 0.1$, and 1.0 are shown, as well as that of a sequence computed with the Schwarzschild criterion, which corresponds to $\alpha = \infty$. The sequences with $\alpha = 0.001$, 0.1, 1.0, and ∞ yield Red Supergiant explosions, while $\alpha = 0.01$ yields a blue explosion at $T_{eff} \simeq 17\,000\,K$. (Note that the computations have been performed up to carbon burning, which ensures that the surface temperature will not change any more.) For α slightly larger than 0.01 we got a track similar to that for $\alpha = 0.01$ but the surface temperature diminished towards central helium exhaustion up to the Hayashi line. For values slightly lower than 0.01 we got a track like that for $\alpha = 0.001$, but with a blue loop to $T_{eff} \simeq 15\,000\,K$ after central helium exhaustion. See Langer, El Eid, and Baraffe (1990) for more details.

tion is greatly influenced, values ranging from $X_C = 0.06$ for the Schwarzschild case up to $X_C = 0.20$ for inefficient semiconvective mixing. Hence it appears that internal mixing during He-burning will have a strong impact on the final stages of massive stars.

7. Summary

In the present work we intended to report on progress and problems in the simulation of the evolution of massive ($M_{ZAMS} \gtrsim 15\, M_\odot$) stars. It could not be our aim to be comprehensive in this respect, rather we selected topics which are controversial at present. We confined ourself to problems concerning the stellar structure and observable evolutionary phases, leaving out late burning stages and the whole item of stellar nucleosynthesis.

We argued that the main sequence phase already bears significant problems. New evolutionary calculations, which relied for the first time on theoretical mass loss rates according to the radiation driven wind theory indicate that main sequence mass loss might have been somewhat overestimated in the past. The consequences thereof and of assumptions on the convection theory on the effective temperature of core helium burning Supergiants as well as on lifetime and mass loss rate of stars in the LBV phase have been discussed. We stressed the importance of the metallicity of the stellar matter for main sequence mass loss, which affects the subsequent surface temperature evolution, and argued that the relatively large number of LBVs in the LMC compared to the Milky Way could be related to that. We motivated that WR stars of type WNL should be considered as a class separately from all other WR subtypes, and showed why they may well have absorbtion features in their spectra. We stressed on the concept of mass dependent WR mass loss rates and outlined the possible relation to the type Ib supernova class. Finally, we presented new evolutionary computations in connection with the progenitor evolution of SN 1987a, and showed that the observational constraint of the blue-red-blue evolution can be recovered by computations using input physics, which — if not standard — was propagated well before the supernova occurred.

Especially the last example demonstrates the strengths and weaknesses of current theoretical stellar models. Certainly, present day stellar evolution theory is yet unable to explain observations like — for example — the irregular and manifold behaviour of LBVs. But if really such a complicated evolution as that of the SN 1987a progenitor can be simulated by completely neglecting e.g. rotation, magnetic fields, and assuming spherical symmetry — which still has to be approved — it means that either stars are simple objects (at least some of them) or our evolution codes are quite good for their designed purpose. On the other hand, the fact that the Blue Supergiant explosion was so surprising shows clearly, that our codes may still be too simple to predict reliably how stars evolve and fade away.

Acknowledgement. The author is grateful to the members of the Stellar Evolution Group at Göttingen, i.e. I. Baraffe, M.F. El Eid, K.J. Fricke, K. Herzig, and M. Kiriakidis, for many stimulating discussions and coorporations. He also benefitted from discussions with many collegues, especially with M. Arnould, W. Hillebrandt, R.P. Kudritzki, A. Maeder, D. Sugimoto, and S.E. Woosley, which have been made possible by grants of the Astronomische Gesellschaft (AG), the Deutsche Forschungsgemeinschaft (DFG), and the Deutscher Akademischer Austauschdienst (DAAD).

References

Abbott, D.C., Bieging, J.H., Churchwell, E., Torres, A.V.: 1986, Astrophys. J. **303**, 239

Allen, D.A., Jones, T.J., Hyland, A.R.:1985, Astrophys. J. **291**, 280

Bohannan, B., Walborn, N.R.: 1989, P.A.S.P., in press

Branch, D,: 1986, Astrophys. J. *Letters* **300**, L51

Castor, J., Abbott, D.C., Klein, R.: 1975, Astrophys. J. **195**, 157

Chiosi, C., Summa, C,: 1970, Astrophys. Space Sci. **8**, 478

Chiosi, C., Maeder, A.: 1986, Ann. Rev. Astron. Astrophys. **24**, 329

Davidson, K., Walborn, N.R., Gull, T.R.: 1982, Astrophys. J. *Letters* **254**, L47

Davidson, K., Dufour,R.J., Walborn, N.R., Gull, T.R.: 1986, Astrophys. J. **305**, 867

Davidson, K., Moffat, A.F., eds.: 1989, IAU-Colloq. **113** on *Physics of Luminous Blue Variables*, in press

Doom, C.: 1985, Astron. Astrophys. **142**, 143

Doom, C., De Greve, J.-P., de Loore, C.: 1986, Astrophys. J. **303**, 136

El Eid, M.F., Langer, N.: 1986, Astron. Astrophys. **167**, 274

Ensman, L.M., Woosley, S.E.: 1988, Astrophys. J. **333**, 754

Ensman, L.M., Woosley, S.E.: 1989, priv. communication

Fransson, C., et al.: 1989, Astrophys. J. , in press

Heisler, J., Alcock, C.: 1986, Astrophys. J. **306**, 166

Herzig, K.: 1988, Diploma thesis, Göttingen University

Hillebrandt, W., Höflich, P., Truran, J.W., Weiss, A.: 1987, Nature **327**, 597

Humphreys, R.M.: 1984, in: IAU-Symposium No. **105** on *Observational Tests of the Stellar Evolution Theory*; A. Maeder, A. Renzini, eds., p. 279

Humphreys, R.M.: 1989, in: IAU-Colloquium No. **113** on *Physics of Luminous Blue Variables*, K. Davidson, A. Moffat, eds., , in press

Humphreys, R.M., Davidson, K.: 1979, Astrophys. J. **232**, 409

Humphreys, R.M., McElroy, D.B.: 1984, Astrophys. J. **284**, 565

van der Hucht, K.A., Hidayat, B., Admiranto, A.G., Supelli, K.R., Doom, C.: 1988, Astron. Astrophys. **199**, 217

Iben, I. Jr.: 1974, Ann. Rev. Astron. Astrophys. **12**, 215

de Jager, C., Nieuwenhuijzen, H., van der Hucht, K.A.: 1986, in: IAU-Symposium No. **116** on *Luminous Stars and Associations in Galaxies*, C. de Loore et al., eds., p. 109

Kato, S.: 1966, P.A.S.J. **18**, 374

Kiriakidis, M.: 1987, Diploma thesis, Göttingen University

Kudritzki, R.P.: 1988, in: 18th Saas Fee course, Swiss Society of Astron. Astrophys.

Kudritzki, R.P., Pauldrach, A., Puls, J.: 1987, Astron. Astrophys. **173**, 293

Kudritzki, R.P., Pauldrach, A., Puls, J., Abbott, D.C.: 1989, Astron. Astrophys. , in press

Lamers, H.J.G.L.M.: 1981, Astrophys. J. **245**, 593

Lamers, H.J.G.L.M.: 1989, in: IAU-Colloquium No. **113** on *Physics of Luminous Blue Variables*, K. Davidson, A. Moffat, eds., , in press

Lamers, H.J.G.L.M., Fitzpatrick, E.L.: 1988, Astrophys. J. **324**, 279

Langer, N.: 1986, Astron. Astrophys. **164**, 45

Langer, N.: 1986a, Ph.D. thesis, Göttingen University

Langer, N.: 1987a, Astron. Astrophys. *Letters* **171**, L1

Langer, N.: 1987b, Mitt. Astron. Ges. **70**, 202

Langer, N.: 1987c, in *Nuclear Astrophysics*, Lecture Notes in Physics **287**, W. Hillebrandt et al., eds., Springer, p. 180

Langer, N.: 1989a, in: IAU-Colloquium No. **113** on *Physics of Luminous Blue Variables*, K. Davidson, A. Moffat, eds., , in press

Langer, N.: 1989b, Astron. Astrophys. **210**, 93

Langer, N.: 1989c, Astron. Astrophys. , in press

Langer, N., Sugimoto, D., Fricke, K.J.: 1983, Astron. Astrophys. **126**, 207

Langer, N., El Eid, M.F., Fricke, K.J.: 1985, Astron. Astrophys. **145**, 179

Langer, N., El Eid, M.F.: 1986, Astron. Astrophys. **167**, 265

Langer, N., Kiriakidis, M., El Eid, M.F., Fricke, K.J., Weiss, A.: 1988, Astron. Astrophys. **192**, 177

Langer, N., El Eid, M.F., Baraffe, I.: 1990, Astron. Astrophys. *Letters*, in prep.

Ledoux, P.: 1941, Astrophys. J. **94**, 537

de Loore, C.: 1980, Space Sci. Rev. **26**, 113

Lundström, I., Stenholm, B.: 1984, Astron. Astrophys. Suppl. **58**, 163

Maeder, A.: 1981, Astron. Astrophys. **101**, 385

Maeder, A.: 1987, Astron. Astrophys. **178**, 159

Maeder, A.: 1989, in: IAU-Colloquium No. **113** on *Physics of Luminous Blue Variables*, K. Davidson, A. Moffat, eds., , in press

Maeder, A., Meynet, G.: 1987, Astron. Astrophys. **182**, 243

Mermilliod, J.C., Maeder, A.: 1986, Astron. Astrophys. **158**, 45

Moffat, A.F.: 1982, in: IAU-Symposium No. **99** on *Wolf-Rayet stars: Observations, Physics, Evolution*; C. de Loore, A.J. Willis, eds., p. 263

Niemela, V.S.: 1983, in: *Workshop on Wolf-Rayet stars*, Paris-Meudon, M.C. Lortet, A. Pitault, eds., p. III.3

Pauldrach, A., Puls, J.: 1989, Astron. Astrophys. , in press

Pauldrach, A., Kudritzki, R.P., Puls, J., Buttler, K.: 1989, Astron. Astrophys. , in press

Prantzos, N., Doom, C., Arnould, M., de Loore, C.: 1986, Astrophys. J. **304**, 695

Reimers, D., 1975: Mèm. Soc. Roy. Liège, 6th Ser. **8**, 369

Renzini, A.: 1987, Astron. Astrophys. **188**, 49

Schmutz, W., Hamann, W.-R., Wessolowski, K.: 1988, Astron. Astrophys. **210**, 236

Smith, L.F., Maeder, A.: 1989, Astron. Astrophys. **211**, 71

Stothers, R.B., Chin, C.-W.: 1976, Astrophys. J. **204**, 472

Stothers, R.B., Chin, C.-W.: 1979, Astrophys. J. **233**, 267

Stothers, R.B., Chin, C.-W.: 1985, Astrophys. J. **292**, 222

Walborn, N.R.: 1989, Comments on Astrophys., in press

Weiss, A.: 1989, Astrophys. J. **339**, 365

Willis, A.J.: 1982, in: IAU-Symposium No. **99** on *Wolf-Rayet stars: Observations, Physics, Evolution*; C. de Loore, A.J. Willis, eds., p. 87

Woosley, S.E.: 1988, Astrophys. J. **330**, 218

Woosley, S.E., Weaver, T.A.: 1982, in: *Supernovae: A Survey of Current Research*, M.J. Rees et al., eds., Reidel, Dordrecht, p. 79

Woosley, S.E., Weaver, T.A.: 1986, Ann. Rev. Astron. Astrophys. **24**, 205

Multi-dimensional Radiation Transfer in the Expanding Envelopes of Binary Systems

R. Baade

Hamburger Sternwarte, Gojenbergsweg 112,
D-2050 Hamburg 80, Fed. Rep. of Germany

Abstract. A generalized integral-operator method is developed to solve the radiative transfer problem in non-symmetric envelopes with arbitrary velocity fields. The major goal of this research is to design an improved radiation transfer code for the UV binary technique of studying mass-loss phenomena of late giants and supergiants. The application to the ζ Aur configuration requires a two-dimensional approach since the light source (B star) is offset from the centre of symmetry of the wind. First results for the system 32 Cyg yield a consistent fit of line profiles.

1. Introduction

In recent years it has been recognized that mass-loss phenomena in red giants are essential for the understanding of the final evolution of low and intermediate mass stars. IUE observations of ζ Aur systems and similar binaries offer an excellent possibility to obtain reliable mass-loss rates and other wind parameters (Reimers, 1987). In particular the eclipsing ζ Aur systems are the only stars (besides the Sun) where the structure of the chromosphere and the inner wind region can be observed with high spatial resolution.

The well-detached ζ Aur binaries consist of a late supergiant (G to M type) and an early main sequence star which moves around in the wind of the red giant. At IUE wavelengths one observes a pure B star spectrum (generally a smooth continuum) with superimposed P Cygni type lines formed by resonance scattering in the circumstellar envelope. In phases near eclipse the chromosphere of the giant star generates a rich absorption line spectrum.

The extended chromosphere - where the wind starts to expand - can be studied by means of a curve of growth analysis. Schröder (1985 and 1986) has shown that the resultant column densities suggest a density model of the form $\rho \sim r^{-2}[r/(r - R_*)]^{\beta}$ with $\beta = 2.5...3.5$. Assuming a steady wind and using the equation of continuity this density distribution implies a velocity law $v(r) = v_{\infty}\ (1 - R_*/r)^{\beta}$. Terminal velocities and mass-loss rates can be determined by theoretical modelling of wind line profiles and of their phase dependence. The basic procedure is quite different to the profile fit analysis of single stars since the source of photons is eccentric from the symmetry centre of the envelope. Computer codes that solve this problem have been developed in the framework of the generalized Sobolev Theory for the two-level approximation by Hempe (1982, 1984) and for the multi-level case by Baade (1986). Che et al. (1983) applied the

Reviews in Modern Astronomy 2 Editor: G. Klare

two-level code to several ζ Aur systems and were able to match the line profiles by chosing adequate mass-loss, wind velocity, and turbulence parameters. However, the strengths of the emission components of the P Cygni lines could not be reproduced exactly. The assumption of a considerable thermalization parameter must be refused a priori since the electron densities are much too low in the relevant part of the envelope.

Probably, the discrepancies can be explained by the inadequacy of the radiation transfer method. Actually, the condition for applying the Sobolev approximation is not fullfilled since the ratio of wind to stochastic velocity is small (≤ 4) in ζ Aur systems. For that reason an improved radiative transfer scheme will be the basic requirement for a better understanding of the outer envelope. I present a generalized integral-operator method which is applicable to the two-dimensional configuration of a ζ Aur system.

2. Transfer equation in operator form

The line transfer problem is defined by the simultaneous solution of the transfer equation and the rate equations for the ion under consideration. The equation of stationary transfer of unpolarized radiation may be written in the observer's frame (neglecting the background continuum):

$$(\mathbf{n}\cdot\nabla)\, I(\nu,\mathbf{r},\mathbf{n}) = k_l(\mathbf{r})\, \varphi\Big[\nu - \frac{\nu_0}{c}\,\mathbf{n}\cdot\mathbf{v}(\mathbf{r}),\mathbf{r}\Big]\Big[S_l(\mathbf{r}) - I(\nu,\mathbf{r},\mathbf{n})\Big]. \quad (1)$$

S_l denotes the frequency-independent line source function (complete redistribution is assumed) and φ the normalized profile function. The line opacity for the transition under consideration can be written in the Milne form:

$$k_l(\mathbf{r}) = \frac{h\,\nu_0}{4\,\pi}\Big(n_i\, B_{ij} - n_j\, B_{ji}\Big)\,. \quad (2)$$

For the equivalent two-level atom the equations of statistical equilibrium yield a relationship of the form (see e.g. Mihalas, 1978):

$$S_l(\mathbf{r}) = \frac{\bar{J}(\mathbf{r}) + (\varepsilon + \eta)\, B}{1 + \varepsilon + \sigma}\,, \quad (3)$$

where the quantities η and σ represent the influences of multi-level interlocking processes. The thermalization parameter ε is a measure for the fraction of photons removed by collisional de-excitation.

If the multiplicity of the transfer equation is entirely spatial an appropriate representation along rays or characteristics can be constructed. The specific intensity at a given field point on a given ray can be expressed in terms of the source function. The so-called integral-operator approach is extensively used in line and continuum transfer calculations for 1-D configurations. The general method was extended to multi-dimensional problems by Jones and Skumanich (1970 and 1973) and Jones (1973). However, the applications were restricted to velocity independent slab geometries. The present paper describes a generalization to arbitrary moving atmospheres with curvilinear geometries.

In the multi-dimensional case it is convenient to define transformation operators which relate functions of spatial position to optical depth. The optical path length along a characteristic is defined by

$$\tau(x,s) = \int_0^s k_1[\mathbf{r}(s')]\ \Phi[\mathbf{r}(s'), x - u_s(\mathbf{r}(s'))]\ ds' , \tag{4}$$

where $x = (\nu - \nu_0)/\Delta\nu_D$ denotes the dimensionless frequency variable and $u_s = \mathbf{n}\cdot\mathbf{v}/v_{sto}$ stands for the dimensionless velocity projected on the characteristic in the unit direction $\mathbf{n}$ (v_{sto} is the stochastic velocity containing all line - broadening contributions and $\Delta\nu_D$ some characteristic width of the profile). The corresponding profile function has to be modified according to the relation $\varphi(\nu) = \Delta\nu_D^{-1}\Phi(x)$. With a variable replacement operator $\mathbf{R_{xrn}}$ intensity and source function transform as

$$\tilde{I}(x,\tau) = \mathbf{R_{xrn}}\ I(x,\mathbf{r},\mathbf{n}) , \tag{5a}$$

$$\tilde{S}(x,\tau) = \mathbf{R_{xrn}}\ S(x,\mathbf{r}) . \tag{5b}$$

Using the relation (cf. Jones and Skumanich, 1973)

$$\mathbf{R_{xrn}} \left\{ \frac{\Delta\nu_D}{k(\mathbf{r})\ \Phi(x,\mathbf{r},\mathbf{n})}\ \mathbf{n}\cdot\nabla \right\} \mathbf{R_{xrn}^{-1}} = \frac{d}{d\tau} \tag{6}$$

the transfer equation can be written in its simplest form:

$$\frac{d\tilde{I}(x,\tau)}{d\tau} = \tilde{S}(x,\tau) - \tilde{I}(x,\tau) . \tag{7}$$

The well-known formal solution for a field point τ may be abbreviated by

$$\tilde{I}(x,\tau) = \mathbf{L}_\tau\ \tilde{S}(t) + \tilde{I}_b\ e^{-\tau} , \tag{8}$$

where $\tilde{I}_b$ represents the imposed boundary condition (i.e. the incident radiation field). Back transformation to the geometrical space yields

$$I(x,\mathbf{r},\mathbf{n}) = \left\{ \mathbf{R_{xrn}^{-1}}\ \mathbf{L}\ \mathbf{R_{xrn}} \right\} S(\mathbf{r}) + I_b\ e^{-\tau} . \tag{9}$$

Finally, the scattering integral (i.e., the profile-weighted mean intensity) can be written by means of a concise operator notation:

$$\bar{J}(\mathbf{r}) = \left\{ \boldsymbol{P}^{\Omega}\ \boldsymbol{P}^{x}_{\mathbf{rn}}\ \mathbf{R_{xrn}^{-1}}\ \mathbf{L}\ \mathbf{R_{xrn}} \right\} S(x,\mathbf{r}) + \bar{J}_c(\mathbf{r}) , \tag{10}$$

where the quantity $\bar{J}_c$ denotes the direct component of the radiation field, which can be evaluated by a direct calculus. Following Jones and Skumanich (1973) frequency and angular integration are represented by the so-called projection operators $\boldsymbol{P}^{\Omega}$ and $\boldsymbol{P}^{x}_{\mathbf{rn}}$.

The classical integral equation method is often accomplished in the escape operator (or flux-divergence operator) formulation, i.e., the statistical equilibrium equation is written in terms of $S - \bar{J}$. This approach generates an active transfer equation, which no longer contains passive components of the radiation field. Hence, the algorithm is well-conditioned even in the limit of large optical depths. A combination of the radiation operator formalism and the statistical equilibrium equation (3) yields the basic expression of the multi-dimensional integral-operator approach (***I*** is the identity operator):

$$\left\{ I + \frac{1}{\varepsilon + \sigma} \mathbf{P} \right\} S = \frac{\bar{J}_C + (\varepsilon + \eta) B}{\varepsilon + \sigma} , \qquad (11)$$

with the escape or flux-divergence operator

$$\mathbf{P} = P^{\Omega} P^{\times}_{\mathbf{rn}} \mathbf{R}^{-1}_{\mathbf{xrn}} (I - \mathbf{L}) \mathbf{R}_{\mathbf{xrn}} . \qquad (12)$$

A more general description including a continuum radiation field can be found by Baade (1988).

The discrete numerical representation of the operator equation can be achieved by introducing appropriate interpolation and weighting coefficients. For a given characteristic with a path grid $\{\tau_n | n = 1, \ldots, N\}$ the net transfer operator can be approximated by a quadrature sum:

$$(I - \mathbf{L})_{\tau} \; \tilde{S}(t) \approx \sum_{n=1}^{N} w_n(\tau) \, \tilde{S}(\tau_n) . \qquad (13)$$

With an adequate set of path interpolation functions for the source function the coefficients w_n can be evaluated by a simple algebraic procedure.

The representation of the transformation operator requires a multi-dimensional interpolation approach on a space grid $\{ (u,v,w)_i \mid i = 1,\ldots,I \}$. This problem may be solved by defining appropriate one-dimensional interpolation functions U, V, and W on the corresponding sub-grids $\{u_j\}$, $\{v_k\}$, and $\{w_l\}$. The source function at any point τ_n on the optical path grid may be related to the space grid by

$$S(\tau_n) \approx \mathbf{R}_{\mathbf{xrn}} \left\{ \sum_j \sum_k \sum_l U_j(u) \, V_k(v) \, W_l(w) \; S(u_j, v_k, w_l) \right\} . \qquad (14)$$

It is often sufficient to select a path grid based on the intersections of the particular characteristic with the space grid surfaces. Consequently, the resulting interpolation problem is reduced by one order.

The projection operators P^{Ω} and $P^{\times}_{\mathbf{rn}}$ are approximated by suitable algorithms for angular and frequency integration. At each field point the quadrature formulae prescribe discrete characteristics and frequencies for which the weighting functions and interpolation coefficients have to be evaluated. Finally, the discretized system of (11) can be written in the form:

$$\sum_{i'=1}^{I} \left[\delta_{ii'} + (\varepsilon_i + \sigma_i)^{-1} \rho_{ii'} \right] S_{i'} = Q_i \quad \text{for all } i = 1,\ldots,I . \qquad (15)$$

The quantities $\rho_{ii'}$ represent the numerical approximation of the escape operator **P**. Q_i replaces all terms on the right hand side of (11). In the two-level resonance line case the final system (15) can be solved directly by standard methods.

3. The binary problem

The requirements of the practical calculations suggest to chose a spherical coordinate system for defining the physical properties of the extended envelope. For the binary configuration it is convenient to fix the origin at the centre of the photon source (i.e. the B star). Due to the azimuthal

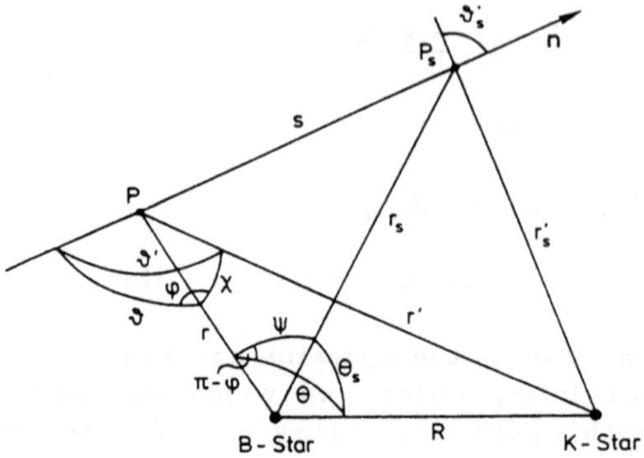

Fig. 1. Basic geometry of the binary problem. A characteristic is specified by the coordinates r and Θ of the reference point P and the local direction parameters μ (= cos ϑ) and φ

symmetry (i.e. rotational invariance with respect to the system axis) a field point P is determined by the coordinates r and Θ (see Fig. 1).

The projection operator $\boldsymbol{P}^{\Omega}$ can be written in terms of two local direction parameters μ (= cos ϑ) and φ. For each grid point P(r, Θ) the specific angular quadrature determines the set of characteristics which have to be considered. A source point P_s in the distance s from the reference point P is fixed by the coordinates

$$r_s = \sqrt{r^2 + s^2 + 2 r s \mu} \tag{16}$$

and

$$\Theta_s = \cos^{-1} (\cos\Theta \cos\Psi - \sin\Theta \sin\Psi \cos\varphi) , \tag{17}$$

where Ψ is given by the plane law of sines:

$$\frac{s}{\sin\Psi} = \frac{r_s}{\sin\vartheta} . \tag{18}$$

The wind velocity is spherically symmetric with respect to the late type star, i.e., the velocity field is completely specified by a simple function u = u (r'). The velocity component in the direction $\mathbf{n}$ (defined by $\mathbf{n} \cdot \mathbf{u} = \mu' u$) is specified by the relations

$$\mu' \equiv \cos\vartheta' = \frac{r - R\cos\Theta}{r'} \mu + \frac{R \sin\Theta}{r'} \sqrt{1-\mu^2} \cos\varphi \tag{19}$$

and

$$\mu_s' \equiv \cos\vartheta_s' = \frac{r_s'^2 + s^2 - r'^2}{2 r_s' s} . \tag{20}$$

4. Numerical details

Following Avrett and Loeser (1984) I introduce a local frequency transformation to include arbitrary velocity fields. Along the characteristic c($\mathbf{r}$,$\mathbf{n}$) the transformation relates the frequency x' at the reference point $\mathbf{r}$ to the frequency x_p' at the source point $\mathbf{r}_p$:

$$x'_p = x' + \mathbf{n}\cdot[\mathbf{u}(\mathbf{r}) - \mathbf{u}(\mathbf{r}_p)] \tag{21}$$

Thus it is possible to determine the optical depth from the reference point $\mathbf{r}$ to any other point on the characteristic, as seen in the local reference frame at $\mathbf{r}$. Now the scattering integral $\bar{J}(\mathbf{r})$ can be written in terms of intensities in the comoving frame. Consequently the normalized profile function and the resulting projection operator are isotropic. The improper integral with the weighting function $\exp(-x^2)$ can be approximated by a Gauss-Hermite quadrature (i.e. the division points are based on the zeros of the Hermite polynomial) without any normalizing problems.

The angular quadrature algorithm prescribes the rays for which the transfer operator calculus has to be performed. For a spherical configuration it is convenient to chose characteristics tangent to the system of concentric spheres generated by the radial grid. The basic set of optical depths is defined by the successive intersections with the surfaces of the radial grid. The optical depth is substantially affected by the presence of a velocity field. Consequently the line radiation processes can only be represented satisfactory if the path-grid resolves the spatial variation of the profile function. In regions with large velocity gradients intermediate grid points must be added to retain numerical accuracy. This procedure generates a fine mesh which contains the path-surface intersections plus a set of auxiliary points. Finally, it is necessary to find the discrete numerical representation of the transfer operator, i.e., to evaluate the weighting functions defined by (13). In the particular application given here the variation of S is assumed to be piecewise linear with the monochromatic optical depth. This procedure assures that the numerical solution is well behaved even in the case of very large changes in the atmospheric parameters from one depth point to the next. Test calculations for the spherical 1-D case are used for a suitable adjustment of numerical parameters. The reader interested in further details is referred to Baade (1988).

5. Theoretical line profiles

With known system parameters it is possible to solve the Kepler problem and to fix the orientation of the binary configuration in the plane of the sky. The emergent intensity and finally the flux profile can be determined by a three-dimensional ray-tracing technique. It is convenient to define a cylindrical coordinate system (R,φ,Z) with the core (i.e. the B star) at the origin and the Z-axis pointing towards the observer. The normalized emergent flux at a frequency x is given by

$$\frac{F(x)}{F_c(x)} = \frac{\int_0^{\pi}\int_0^{\infty} I^e(x,R,\varphi)\, R\, dR\, d\varphi}{\int_0^{\pi}\int_0^{R_c} I^c(x,R,\varphi)\, R\, dR\, d\varphi} \, . \tag{22}$$

The problem of calculating the emergent intensity from the formal solution reduces to the well-known integral

$$I^e(x,R,\varphi) = \int_0^{\tau_{max}} S(R,\varphi,Z)\, e^{-\tau(x,R,\varphi,Z)}\, d\tau + \begin{cases} I^c\, e^{-\tau_{max}(x,R,\varphi)} & R \le R_c \\ 0 & R > R_c \end{cases} . \tag{23}$$

In performing the ray quadrature in (23) there arises severe difficulties, since the source function is known only at a discrete set of spatial grid points. In general an individual line of sight passes the medium without intersecting one or more grid points. Hence the resulting integration and interpolation problem must be solved by a sophisticated method.

In the case of the two-dimensional binary geometry it is expedient to reduce the corresponding interpolation problem. The source function can be derived for the crossing points between the space-grid surfaces and the line of sight by means of a one-dimensional interpolation procedure. Now the source function can be represented by a polynomial spline along the line of sight with high accuracy. The ray integral (23) may be solved by any appropriate quadrature algorithm. Numerical experiments demonstrate the superiority of a self-correcting procedure with a preselected increment of the optical depth.

A first series of test calculations are performed in order to study the accuracy of Hempe's method (i.e. Sobolev's source function approximation with correct ray-tracing) concerning the binary problem. For the absorption coefficients I chose a representation of the form

$$k(x,r') = \frac{k_o}{\left(\frac{r'}{R_B}\right)^2 \left(1 - \frac{R_*}{r'}\right)^\beta} \; u_\infty \; \Phi(x) \, , \qquad (24)$$

where r' denotes the distance from the giant and u_∞ the dimensionless terminal velocity. The above representation implies that the opacity parameter k_o is proportional to $\dot{M}/v_\infty^2$. The velocity parameter β used in the presented computations is set to be 2.5. Using typical values of the binary separation and stellar radii I have calculated an extensive sample of line profiles with different opacities and velocities. In Fig. 2 I have plotted theoretical profiles at four different position angles ($\alpha = 0°$ stands for eclipse). All profiles are computed with the assumption of pure scattering (i.e. $\varepsilon = 0$). At eclipse we can see only scattered light from the extended scattering shell around the B star. The emission profile is asymmetric and slightly redshifted, since the emitting region is mainly moving away from the observer. The second position angle ($\alpha = 36°$) results in a degenerated P Cygni profile. Due to the long line of sight through the envelope we have

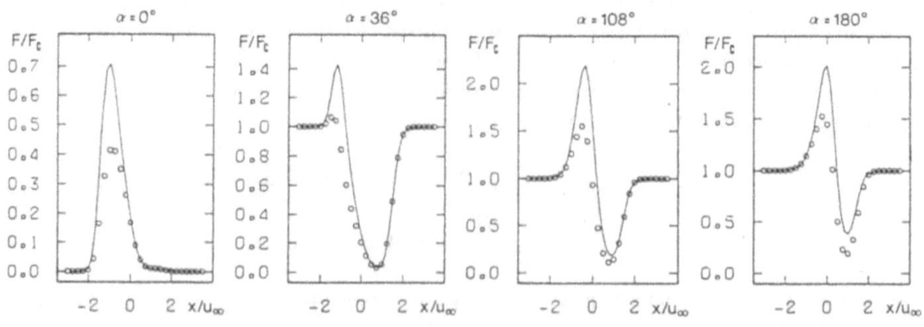

Fig. 2. Theoretical flux profiles for different position angles of a ζ Aur model system. Comparison of integral-operator results (solid curves) and approximate solutions calculated with Hempe's code (circles). Wind parameters: k_o = 1000 and u_∞ = 2

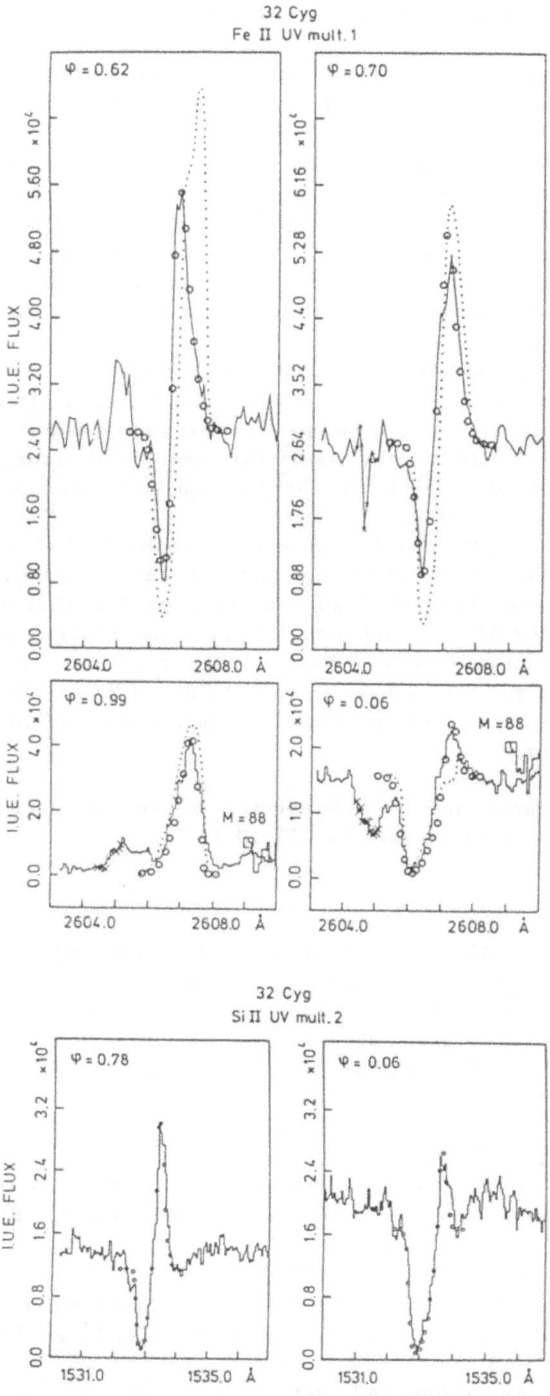

Fig. 3. Comparison of observed and calculated wind line profiles for Fe II and Si II (circles, this work; dotted lines, Che et al., 1983)

a strong absorption component. Finally, the configurations with $\alpha = 108^\circ$ and $\alpha = 180^\circ$ generate normal P Cygni profiles.

A comparison of Hempe's method with the exact treatment of the radiation transfer (Fig. 2) shows that the Sobolev approximation is inadequate for the quantitative analysis of ζ Aur winds. Only in the case of high velocity amplitudes ($u_\infty \gg 10$) or small opacities ($k_0 \leq 100$) the escape probability approach yields satisfactory results. Mainly the emission components are drastically underestimated by applying Hempe's code to resonance lines of the predominant ions . Che et al. (1983) have postulated a deviation from pure scattering to give a consistent match of the observed profiles in ζ Aur systems. Considering the small electron densities this hypothesis must be refused a priori.

The new transfer code based on the integral-operator method is applied to the system 32 Cyg. In Fig. 3 I present a selection of profile fits for wind lines at various phases. A detailed comparison between theoretical and observed profiles of resonance lines yields the final wind model: $v_\infty = 90$ km s^{-1}, $v_{sto} = 30$ km s^{-1}, $\dot{M} = 1.5\ 10^{-8}\ M_\odot$ yr^{-1} (mean value, assuming solar abundances).

Obviously, the assumption of pure scattering appears to be sufficient to explain the strengths of the observed emission components at all phases. Remaining discrepancies between theoretical and observed profile shapes have its origin in the inadequacy of the wind model. Especially, stochastic processes or local density fluctuations may have dominant effects. UV pumping processes and ionization effects have to be considered by an adequate multi-level / multi-ion analysis. An appropriate generalization of the computer code is in preparation.

Acknowledgements. This research has been supported by the Deutsche Forschungsgemeinschaft under grant number Re 353/20-1.

References

Avrett, E.H., Loeser, R.: 1984, *Methods in radiative transfer,* ed. W. Kalkofen, Cambridge University Press, p. 341
Baade, R.: 1986, *Astron. Astrophys.* **154**, 145
Baade, R.: 1988, *Ph. D. Thesis*, Univ. Hamburg
Che, A., Hempe, K., Reimers, D.: 1983, *Astron. Astrophys.* **126**, 225
Hempe, K.: 1982, *Astron. Astrophys.* **115**, 133
Hempe, K.: 1984, *Astron. Astrophys. Suppl. Ser.* **56**, 115
Jones, H.P.: 1973, *Astrophys. J.* **185**, 183
Jones, H.P., Skumanich, A.: 1970, *Spectrum Formation in Stars With Steady-State Extended Atmospheres*, eds. H. Groth, P. Wellmann, NBS Spec. Publ. No. 332, Washington, p. 138
Jones, H.P., Skumanich, A.: 1973, *Astrophys. J.* **185**, 167
Mihalas, D.: 1978, *Stellar Atmospheres*, 2d ed., W.H. Freeman, San Francisco
Reimers, D.: 1987, *Circumstellar Matter*, IAU Symposium No. 122, D. Reidel Publishing Company, p. 307
Schröder, K.-P.: 1985, *Astron. Astrophys.* **147**, 103
Schröder, K.-P.: 1986, *Astron. Astrophys.* **170**, 70

Accretion Disks in Close Binaries

W.J. Duschl

Institut für Theoretische Astrophysik, Universität Heidelberg,
Im Neuenheimer Feld 561, D-6900 Heidelberg, Fed. Rep. of Germany

Abstract. *We discuss models for accretion disks in close binary systems in which the turbulent viscosity and convective energy transport are fully coupled in the convectively structured zones. Thus, at least in these regions, it is no longer necessary to introduce an ad hoc viscosity parameter removing one free parameter of previously described models. The implications and predictions of the models are described. Finally the assumption of isotropy of the turbulence is discussed.*

1. Introduction

Accretion disks are among the oldest concepts in astrophysics. In 1755, Immanuel Kant proposed a model for the origin of the planetary system that, today, we would call an accretion disk scenario. First theoretical descriptions of the physical processes in accretion disks, i.e. of the inward transport of matter with an outward transport of angular momentum, date back to von Weizsäcker (1943) and, especially, Lüst (1952). Another two decades later, Shakura and Sunyaev (1973) and Novikov and Thorne (1973) introduced what one nowadays calls the *standard accretion disk model*. Osaki (1974) predicted that geometrically thin accretion disks are unstable and that this is the reason for dwarf nova outbursts and related phenomena. In the following years progress was made in searching for such an instability (e.g., Hōshi, 1979); in 1981 Meyer and Meyer-Hofmeister described a full limit cycle that corresponds to Osaki's instability. In many respects this standard model proved to be very successful in explaining observed features in a wide variety of astrophysical systems; ranging from close binaries (dwarf novae, symbiotic stars, low mass X-ray binaries, etc.) to the powerhouses of active galaxies and quasars.

Nonetheless, to some degree, the success of this model is surprising. The surprise does not come from the fact that the model seems to be applicable to so many so different systems. That it is useful in so many respects only means that the underlying physical process is of very general validity, and this can – and indeed does – mean that it is important in many different environments and situations. The surprise comes much more from the fact that the model works as well as it does, although one basic material function is virtually unknown, namely viscosity.

One can easily show that the standard sources of viscosity (e.g., molecular viscosity) are too small by many orders of magnitude. This led to the concept of turbulent viscosity as the agents in accretion disks. While Shakura and Sunyaev's parameterization of turbulent viscosity (see Eq. (1)) made the theory easy to handle, the lack of knowledge of the underlying physics was, at best, hidden.

With the beginning of time dependent modelling of accretion disk outbursts, due to the above mentioned instability, it soon became clear that the viscosity parameter was not a simple numerical constant. Furthermore, it showed that convection plays an important rôle in the structure of accretion disks. So one is left with the situation that, in principle, two entirely unrelated turbulent motions should be present at the same moment at the same place. Although this model proved to be successful (surprisingly!), it left a somewhat uneasy feeling.

Remedies of some kind or other were discussed (e.g., magnetic fields). In the following we will restrict ourselves to the hydrodynamic case* and discuss new results obtained within the standard model framework; the one, but important, exception is that in these models the two turbulences are coupled, so that they are regarded as only one turbulence that manifests itself in two ways: energy transport (predominantly in vertical direction; called *convection*) and mass/angular momentum transport (mainly in radial direction; called *turbulent viscosity*). This directional preference of the two manifestations is only a result from the physical structure of geometrically thin accretion disks: the radial temperature and pressure gradient is virtually negligible compared to the vertical one (⇒ convection works mainly in vertical direction), while the gradient of the macroscopic velocity field is negligible in vertical direction but plays the dominant rôle in radial direction (⇒ viscosity causes predominantly radial motion of mass and angular momentum).

2. A Self-Consistent Viscosity in Convective Regions

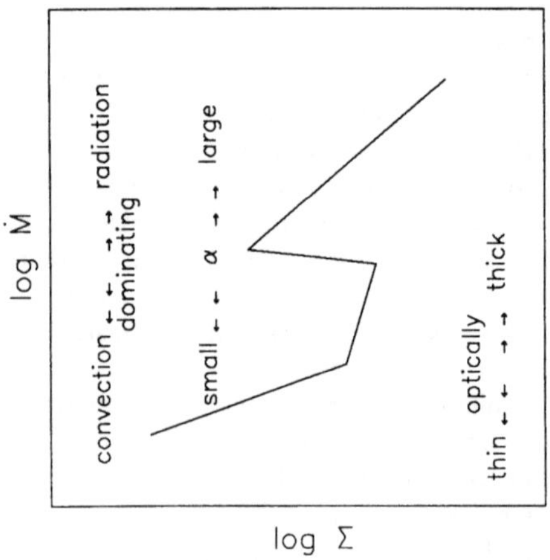

Figure 1: *A typical mass flow rate* ($\dot{M}$) *– surface density* (Σ) *– relation for close binary accretion disks calculated using the assumptions of the standard model. The domains of some of the relevant phenomena and processes are marked in the figure.*

In close binary systems (dwarf novae, symbiotic stars, etc.), accretion disks are geometrically thin, i.e. their extent perpendicular to the plane of rotation is small compared to their radial distance from the object that dominates the gravitational potential. In such disks, only the s-ϕ-component of the stress tensor, $\mathcal{T}$, is important; in the standard accretion disk models, this tensor element is described by

$$\mathcal{T}_{s\phi} = \alpha \cdot P. \tag{1}$$

Here, P is the pressure and α a dimensionless proportionality constant. In principle, the underlying physics should allow one to determine α (and thus make an ad hoc ansatz as in Eq. (1) no longer necessary). But due to the lack of knowledge concerning turbulent processes in fluids, one regards α as a *free parameter*, allowing one to fit observed time scales. Outburst models for accretion disks have shown that for hot accretion disks (i.e., with temperatures of more than about 8,000 K) one needs $\alpha \approx \mathcal{O}(1)$, while for lower temperatures $\alpha \approx \mathcal{O}(0.01 \ldots 0.1)$ is necessary for reasonably good fits to observations. On the other hand, one found when modelling stationary accretion disks that, for the

* We will not give an introduction to the basic physics of the standard model of accretion disks here. For details, see the book by Frank, King, and Raine (1985) and the reviews by Pringle (1981), Meyer (1985), and Osaki (1989).

lower temperature range, convection can play an important rôle and even dominate the vertical energy transport. Thus, in the standard models, one is left with two turbulences at the same time at the same place. These different turbulences are not necessarily of the same order of magnitude. Figure 1 summarizes the relevant phenomena and processes in a typical close binary accretion disk calculated according to the standard model. The optically thin parts of the accretion disks were modelled by isothermal disks ($10^{3.6}$ K) following results by Williams (1980) and Tylenda (1981).

Here we propose, instead of introducing ad hoc values for α, to stay as self-consistent as possible. This can be done by assuming that the same turbulence that transports energy in the vertical direction is also responsible for transporting mass and angular momentum in radial direction (or vice versa): both are equivalent; it is not the case that one is the cause for the other. First, we take an isotropic turbulence and write in convective regions:

$$\alpha = v_{\rm conv}/c_{\rm s} \tag{2}$$

where $v_{\rm conv}$ is the velocity of convective motion (which we calculate according to the mixing length description) and $c_{\rm s}$ is the sound velocity. We want to emphasize that this does not mean that convection drives viscosity. It can also be read the other way round, i.e., Eq. (2) means a full coupling between the two processes. In a later section we will discuss how anisotropy can influence the results.

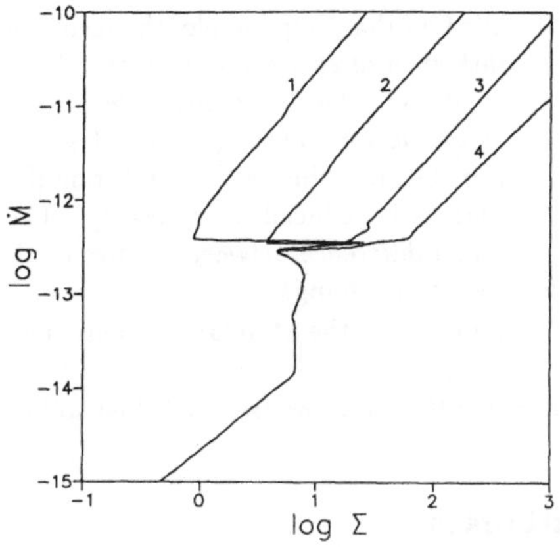

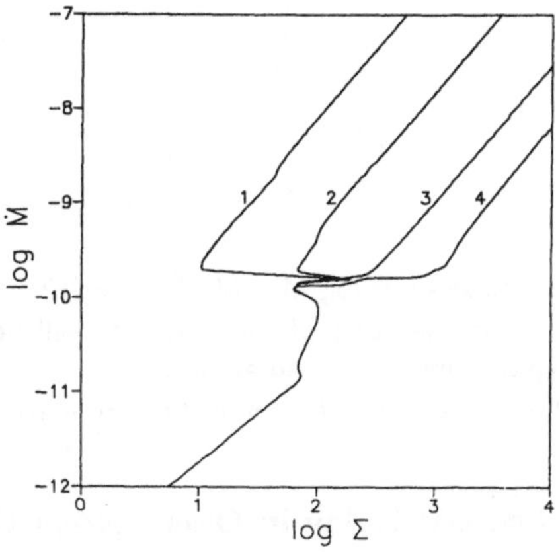

Figures 2 *(left)* **and 3** *(right)*: *The constant mass flow rate,* $\dot{M}$, *as a function of the surface density,* Σ, *for two radial distances (Fig. 2: log* s*/cm = 9.0; Fig. 3: 10.0) from a 1* $M_\odot$ *accretor.* $\alpha_{\rm rad}$ *= 1.0 (label* **1** *in the figures), 0.1* (**2**), *0.01* (**3**), *and 0.001* (**4**). *Here and in the following figures,* $\dot{M}$ *is given in* $M_\odot$*/yr, and* Σ *in g/cm*^{2}.

While Eq. (2) gives a prescription for α in convective regions, it does not say anything about radiative parts of accretion disks. There we have a different situation: the inconsistency of the standard model comes from the fact that the only turbulence present, the viscosity, is not taken into account when calculating the energy transport. Duschl (1983) has shown that in this context, self-consistency only marginally is violated. While, in principle, the viscosity turbulence, in a radiatively structured region would act as a *negative convection*, this poses no problem, as the amount of energy to be transported as compensation is always very small. This picture is not in conflict with *Eddington's rule* (Eddington, 1930; Hazlehurst, 1989) as we are not dealing with a self-sustaining convective flux. So in the following we still have to introduce an ad hoc α in radiative regions, $\alpha_{\rm rad}$.

We calculated the disk structure for typical accretion disks in close binaries (for details, see Duschl (1989)), and take as an example an accretion disk around a white dwarf of 1 $M_\odot$. Figures 2 and 3 show the resulting surface density, Σ, for different constant mass flow rates, $\dot{M}$, at two different radii (log s/cm = 9.0 (Fig. 2), and 10.0 (Fig. 3)), for four different α_{rad} (1.0, 0.1, 0..01, and 0.001).

In Fig. 4, we compare the results for log s/cm = 9.0, $\alpha_{rad} = 1.0$ with a standard model for the same parameters but $\alpha = 850 \cdot (z/s)^{3/2}$. The latter is the description by Meyer and Meyer-Hofmeister (1983): it has the required features for the hot and the cool temperature domains and is based on dimensional analysis of the action of small scale magnetic fields; it is gauged such that the upper branch corresponds to $\alpha \approx 1$.

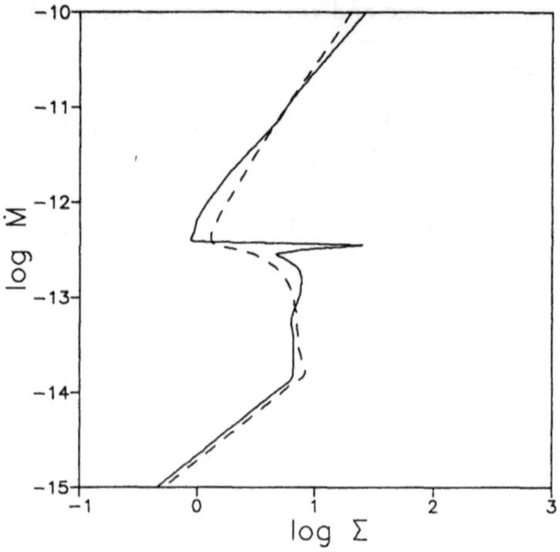

Figure 4: *Comparison of a standard accretion disk model (dashed line; $\alpha = 850 \cdot (z/s)^{3/2}$) with a model calculated as described in the text (full line) with $\alpha_{rad} = 1.0$ for a 1 $M_\odot$ accretor at log s/cm = 9.0.*

We find that, in principle, the resulting disk structures are very similar in both solutions. This is remarkable insofar as the new models require one less parameter, i.e., the freedom in fitting the models is reduced considerably. The major difference between the two models is the spike at log $\dot{M}$/ ($M_\odot$/yr) $\approx$ -12.4. This means two things:

- the basic features of the outbursts will be the same as in the standard models, i.e., typical time scales and amplitudes.
- due to the spike, the detailed rise to maximum will differ from models published so far.

3. The UV-Delay in Dwarf Nova Outbursts

One major problem of standard accretion disk models was that considerable changes in some of the details of the structure were necessary to explain the so-called UV-delay. This feature appears in a number of dwarf novae and manifests itself in the following way: while the outburst in visual lightbands already proceeds and reaches its maximum, the outburst in the UV-band begins only a day later (see the review by Verbunt (1987), and references therein). To explain this, one has to separate the rise to characteristic values of the visual bands (6,000 – 7,000 K) from the rise to higher temperatures characteristic of UV-emission ($\geq$ 10,000 K) (Pringle et al., 1986).

There have been several attempts to remedy this within the standard description (Meyer-Hofmeister, 1987; Mineshige, 1988; Meyer and Meyer-Hofmeister, 1989): while the last of these succeeds best in reproducing the UV-delay, it requires major modifications of the underlying outburst cycle.

A final statement can be made only after finishing time dependent model calculations. But estimates of the time scales indicate that the spikes found in the self-consistent models have just the right properties to explain the UV-delay: locally a typical outburst will follow the cycle *lower branch* → *middle branch (= spike)* → *upper branch* → *lower branch* (Fig. 5). While the jumps from one branch to the other proceed on much faster time scales (Meyer, 1984), the evolution on the branches proceeds on viscous timescales. One finds (Duschl, 1989) that a typical time scale on the spike is of the order of a few 10^5 sec. So one can expect that a typical outburst starts with a jump from the lower to the middle branch causing the *optical outburst*, this evolves (locally) by viscous dissipation with a time scale of the order of a day. Only afterwards the *UV-outburst* takes place.

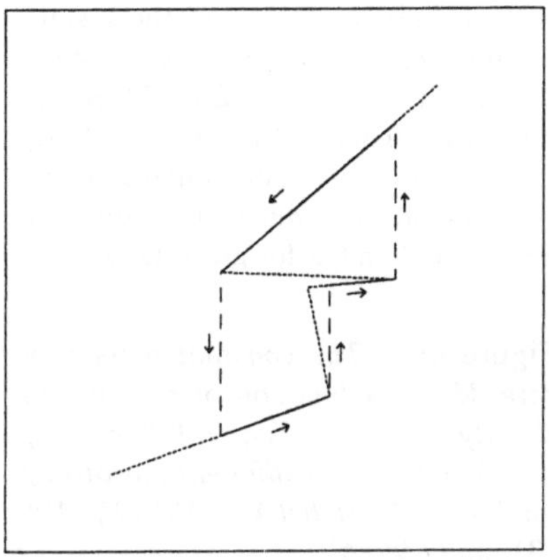

Figure 5: *A typical local limit cycle: the dashed lines (long dashes) indicate evolution on the (comparatively) fast front moving time scale, while evolution along full lines proceeds on the (slower) viscous time scale. The short dashes give the* $\dot{M} - \Sigma$ *– relation over a larger range.*

Here we argued only in terms of the local limit cycle. This is justified only as long as one is interested in the principle features; local instabilities have to proceed into the neighbouring environment; only this produces the "real" outburst. Furthermore, the details of the limit cycle depend strongly on the choice of α_{rad}. As one can see from Figs. 2 and 3, too low an α_{rad} means that an outburst encompassing the full range from very low to very high rates is not to be expected. While for α_{rad} near the limiting value, the radial coupling can overcome this, a limit for α_{rad} will exist below which no outbursts are possible that ressemble those observed in dwarf novae. From the models presented here it seems that this limiting value lies between 0.1 and 0.01. We emphasize that this limiting value is not a "natural constant" but depends on chemical composition, radial extent of the disk, etc.. The possible consequences of different α_{rad} for the limit cycles (locally and globally) are discussed elsewhere (Duschl, 1989).

4. Anisotropic Turbulence

In Eq. (2) the coupling between convection and viscosity is described. There we assumed that the turbulence is isotropic. There are arguments saying that due to strong rotational effects in accretion disks, this overestimates the viscosity. In the framework of mixing length theory, anisotropy would mean that the "eddy velocity" is not the same in radial and vertical direction. To describe this, we rewrite Eq. (2) in the following way:

$$\alpha = \frac{v_{\mathrm{conv}}}{c_s} \cdot \mathrm{tg}\ \gamma. \tag{3}$$

Here we introduced an *anisotropy angle* γ that describes the "preferred" direction of turbulent motion. $\gamma = 45°$ corresponds to the isotropic case; fully vertical motion would mean $\gamma = 0°$, while the relative maximum effect of radial matter/angular momentum transport would be reached for $\gamma = 90°$, i.e., $\gamma \in \{0° \ldots 90°\}$. Heuristically, one can understand tg γ as the ratio between the characteristic turbulent velocities in s- and in z-direction:

tg $\gamma = v_{\text{turb},s}/v_{\text{turb},z}$. For $\gamma = 45°$, Eq. (3) becomes identical with Eq. (2): $v_{\text{turb},s} = v_{\text{turb},z} = v_{\text{conv}}$. We emphasize that γ is a tool for a numerical experiment to check how sensitive the results are to a possible anisotropy of the turbulence. γ is simply a free parameter introduced for this experiment. If one wants to calculate a "real" model including anisotropy one needs a physical theory describing the processes leading to anisotropy. Physically realistic anisotropy will not be characterized by a single anisotropy angle; we expect that – in our terms – γ will be a function of the physical variables describing the disk. As here we want to check only the principle sensitivity of the results on anisotropy, introducing a constant γ is a suitable means. Eq. (3) means not only a scaling of α by tg γ. As α, c_s, and v_{conv} are fully coupled, these three quantities adjust in a way that Eq. (3) is fulfilled for given tg γ.

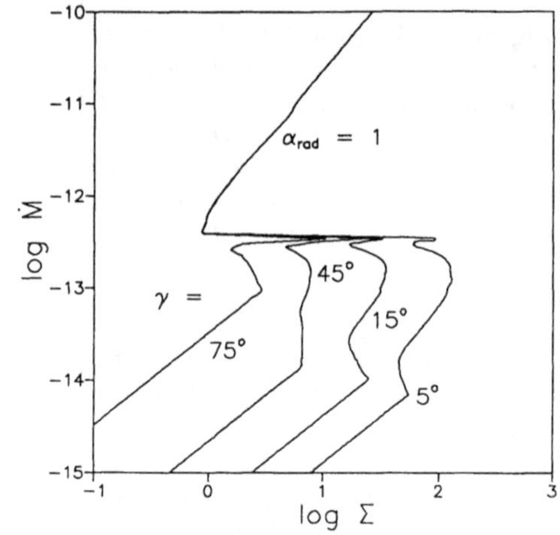

Figure 6: *The constant mass flow rate, Ṁ, as a function of the surface density, Σ, at log s/cm = 9.0 for* α_{rad} *= 1.0 and for four different anisotropy angles, γ: 5° (label* **1***), 15° (***2***), 45° (***3***), and 75° (***4***).*

In Fig. 6, the resulting disk structures for four angles γ are shown ($\gamma = 5°$, 15°, 45°, and 75°) for log s/cm = 9.0, and $\alpha_{\text{rad}} = 1.0$.

$\gamma \in \{45° \ldots 90°\}$ means, in terms of the viscosity, an increase of α compared to the isotropic case, while $\gamma \in \{0° \ldots 45°\}$ corresponds to a decrease. The amount by which α in-/decreases is determined self-consistenly by the vertical disk structure. When tg $\gamma > 1$, in principle $\alpha > 1$ is possible as well. This corresponds to supersonic turbulence: as it is highly unlikely that supersonic turbulence could be maintained under the circumstances discussed here, we limit the viscosity parameter to $\alpha \leq 1$.

The results in Fig. 6 show that:

- The general features of the $\dot{M} - \Sigma$ – relations are, to a large extent, independent of the anisotropy of the turbulence: the spike, for example, is present in all relations, and the overall width of the instability region (for fixed α_{rad}) is strongly changed only for extreme values of γ. For instance, for $\gamma = 15°$, Δ log Σ changes only from 1.4 (for isotropic models) to 1.6, while for $\gamma = 15°$, Δ log $\Sigma \approx 1.1$ (in all three cases: $\alpha_{\text{rad}} = 1.0$).
- Details of the relations vary characteristicly with a variation of γ:
  > The relative width of the spike (compared with the overall width of the instability region) becomes smaller with smaller angles γ. As long as the spike is still extended enough to make an intermediate viscous evolution possible, it clearly separates the optical and the UV outburst because the time scale is determined by the width of the spike. If the spike is no longer large enough to enforce an intermediate viscous phase, its existence still influences the outburst behaviour, but with a less clear separation between the two phases.
  > The smaller γ is, the more mass is stored locally in the disk. This is reflected by the higher values of the surface density Σ. Additionally this constrains the domain of optically thin disks to smaller and smaller $\dot{M}$. The existence of the UV delay plus the

observation of – at least partially – optically thin disks during minimum state gives a handle to estimate the anisotropy of the turbulence. Both facts favour γ-values that are not too small ($\gamma \gtrsim 20°$).

> For larger γ, we find that the total width ($\Delta \log \Sigma$) of the limit cycle becomes smaller for fixed α_{rad}. As is already known from the standard model, too small a width of the limit cycle makes modelling of the outburst time scales difficult. This gives another handle on the range of "allowed" γ and rules out extremly large γ.

> As was discussed by Duschl (1989), the structure of the $\dot{M} - \Sigma$ – relation puts a lower limit on α_{rad}: if α_{rad} is smaller than some limiting value, there is no longer a limit cycle encompassing the entire unstable structure; instead we get two seperate limit cycles (cf. Fig. 2 for $\alpha_{rad} = 0.01$). Although the radial coupling will soften this effect, this only means a somewhat lower limit. So one finds that smaller γ means a less stringent lower limit on α_{rad}. On the other hand, large γ makes the possible range for α_{rad} very small. Numerical models in the framework of the standard theory point towards $\alpha_{rad} \in \{ 0.1 \ldots 1 \}$, but not necessarily exactly 1. This indicates that the anisotropy of the turbulence does not favour the domain where $v_{turb,s} \gg v_{turb,z}$, i.e., we expect an upper limit of γ not much larger than the isotropic value.

Detailed outburst characteristics can be clarified by forthcoming numerical models of the evolution of such disks.

5. Conclusions

One can summarize the results of the model calculations presented in this paper in the following way:

1. It is possible to construct models of accretion disks with the properties required by a close binary environment (dwarf novae, symbiotic stars, etc.) that do not need an ad hoc turbulence for viscosity in convective regions. Instead it is assumed that the same turbulence is the driving mechanism of convection and viscosity. Thus a major inconsistency of the standard accretion disk theory can be solved to a large degree.
2. The UV-delay observed in some dwarf nova systems can be understood in the framework of these models. The details of the form of the $\dot{M} - \Sigma$ - relation allow for a stable intermediate branch with *a)* temperatures characteristic for sources radiating mainly in the optical but not the UV, and *b)* a viscous evolution time scale of the order of a day. This branch is due to the transition from convectively to radiatively structured accretion disks.
3. Anisotropy of the underlying turbulence may give rise to certain features in the structure of the disks that influence details in the evolution of the disks, but that do not change the overall basic features of these evolutions.
4. Comparison with observations point in the direction that the anisotropy of the turbulence is not very strong. Otherwise one would expect consequences that are well within what one can observe.
5. In the present state, the limiting values can only be estimated. Numerical models of the evolution of these disks are under way; they will allow to give much firmer limits.

Acknowledgments

I thank Prof. Werner M. Tscharnuter and Dr. Davina Innes for many very helpful comments on this paper.

References

Duschl, W.J., 1983: *Astron. Astrophys.*, **121**,153
Duschl, W.J., 1989: *Astron. Astrophys.*, in press
Eddington, A.S., 1930: *The Internal Constitution of the Stars*, Cambridge University Press, Cambridge, U.K.
Frank, J., King, A.R., Raine, D.J., 1985: *Accretion Power in Astrophysics*, Cambridge University Press, Cambridge, U.K.
Hazlehurst, J., 1989: *Observatory*, **109**,91
Hōshi, R., 1979: *Progr. Theor. Phys.*, **61**,1307
Kant, I., 1755: *Allgemeine Naturgeschichte und Theorie des Himmels*, Königsberg
Lüst, R., 1952: *Zeitschr. Naturforschung*, **7a**,87
Meyer, F., 1984: *Astron. Astrophys.*, **132**,143
Meyer, F., 1985: in: *Recent Results on Cataclysmic Variables*, ESA SP-**236**,83
Meyer, F., Meyer-Hofmeister, E., 1981: *Astron. Astrophys.*, **104**,L10
Meyer, F., Meyer-Hofmeister, E., 1989: *Astron. Astrophys.*, **221**,36
Meyer-Hofmeister E., 1987: *Astron. Astrophys.*, **175**,113
Mineshige, S., 1988: *Astron. Astrophys.*, **190**,72
Novikov, I.D., Thorne K.S., 1973: in: *Black Holes*, eds: C. DeWitt, B.S. DeWitt; Gordon and Breach Science Publishers, New York, USA; pp.343ff.
Osaki, Y., 1974: *Publ. Astron. Soc. Japan*, **26**,429
Osaki, Y., 1989: in: *Theory of Accretion Disks*, eds: F. Meyer, W.J. Duschl, J. Frank, E. Meyer-Hofmeister; Kluwer Academic Publishers, Dordrecht, The Netherlands; pp.183ff
Pringle, J.E., 1981: *Ann. Rev. Astron. Astrophys.*, **19**,137
Pringle, J.E., Verbunt, F., Wade, R., 1986: *Monthly Notices Royal Astron. Soc.*, **221**,169
Shakura, N.I., Sunyaev, R.A., 1973: *Astron. Astrophys.*, **24**,337
Tylenda, R., 1981: *Acta Astronomica*, **31**,127
Verbunt, F., 1987: *Astron. Astrophys. Suppl.*, **71**,339
von Weizsäcker, C.F., 1943: *Zeitschr. Astrophys.*, **22**,319
Williams. R.E., 1980: *Astrophys. J.*, **235**,939

Index of Contributors

MIX
Papier aus verantwortungsvollen Quellen
Paper from responsible sources
FSC® C105338

If you have any concerns about our products,
you can contact us on
ProductSafety@springernature.com

In case Publisher is established outside the EU,
the EU authorized representative is:
Springer Nature Customer Service Center GmbH
Europaplatz 3, 69115 Heidelberg, Germany

Printed by Libri Plureos GmbH
in Hamburg, Germany